COASTAL ECOSYSTEMS
Hazards, Management and Rehabilitation

COASTAL ECOSYSTEMS
Hazards, Management and Rehabilitation

Editor
Rattan K. Datta

CENTRE FOR SCIENCE & TECHNOLOGY OF THE
NON-ALIGNED AND OTHER DEVELOPING COUNTRIES
(NAM S&T CENTRE)

2021

Daya Publishing House®
A Division of
Astral International Pvt. Ltd.
New Delhi – 110 002

Reprinted, 2021

ISBN: 9788170359869 (INT. EDITION)

Centre for Science and Technology of the Non-Aligned and Other Developing Countries (NAM S&T Centre)
Core-6A, 2nd Floor, India Habitat Centre, Lodhi Road,
New Delhi-110 003 (India)
Phone: +91-11-24644974, 24645134, Fax: +91-11-24644973
E-mail: namstct@gmail.com
Website: www.namstct.org

Published by : **Daya Publishing House®**
A Division of
Astral International Pvt. Ltd.
– ISO 9001:2015 Certified Company –
4736/23, Ansari Road, Darya Ganj
New Delhi-110 002
Ph. 011-43549197, 23278134
e-mail: info@astralint.com
Website: www.astralint.com

Foreword

The Centre for Science and Technology of the Non-Aligned and Other Developing Countries (NAM S&T Centre) is actively involved with the promotion of South-South collaboration in science and technology through conducting seminars, workshops and training on emerging technologies, and subjects of interest to developing countries.

In line with these efforts, the Centre had organized an international workshop-cum-training programme on 'Coastal Ecosystems: Hazards, Management and Rehabilitation' at Purwokerto, Indonesia during August 2006 jointly with the Centre for Tropical Marine Ecology, Bremen, Germany (ZMT–Bremen) and Jenderal Soedirman University (UNSOED) of Indonesia. The location of the workshop was not far from the source of Tsunami of 26th December 2004, and representatives from most of the countries which were worst affected participated and presented their reports in this workshop. This gave excellent opportunity to the scientists and managers to share their experiences and learn useful lessons to face any future eventuality.

As a follow up on the above event, Dr. Rattan K. Datta, an eminent meteorologist in India has edited the material presented during the workshop, which has been compiled to form this publication entitled 'Coastal Ecosystems: Hazards, Management and Rehabilitation'. The publication has been divided in four parts, namely, (*i*) General Topics on Physical Oceanography; (*ii*) Impact Studies Related to Tsunami, Tropical Cyclones and Toxic Contamination; (*iii*) General Coastal Management Issues including Use of GIS and Remote Sensing Technology; and (*iv*) Coastal Ecosystem: Country Reports and Case Studies. The publication has a good mix of tutorials from the experts including cause/effects of tsunami, general atmospheric and oceanographical modeling, country wise reports and interesting impact studies, Prof. Dr. Venu Ittekkot, Director, ZMT-Bremen has given a preface for the publication.

In this context, it gives me great pleasure to share that Government of India, through Ministry of Earth Sciences (MoES), has been implementing an R&D programme on Integrated Coastal and Marine Area Management (ICMAM). The significant achievements made under ICMAM programme are: Implemented model projects successfully on (*i*) Application of GIS and Remote Sensing for management of critical habitats such as mangroves, coral reefs, and areas of high biodiversity; (*ii*) Determination of Waste Assimilation Capacity of coastal waters; (*iii*) Developed guidelines for conducting Environmental Impact Assessment studies for ports and harbours, waste disposal and tourism; and (*iv*) Developed ICMAM Plans for Chennai, Goa and Gulf of Kachchh, and demonstrated the application of scientific tools and techniques in the management of coastal areas. Besides, the Ministry has developed (*i*) Shoreline management plans for Ennore, Munambum to Kayamkulam (Kerala), and (*ii*) Tidal inlet management plans for Nethravathi estuary (Karnataka) and Vellar estuary (Tamil Nadu). These Plans contain solutions to solve the erosion/accretion problems at these sites. In addition, the concept of application of ecosystem in a limited area of Kochi backwaters (Kerala) and Chilka Lake (Orissa). Also, the advantages of increasing tidal amplitude to enhance productivity demonstrated in Vedaranyam (Tamil Nadu) lagoon and many more coastal area management activities are progressing with active participation of coastal State Governments for implementation. After the incident of tsunami occurred on 26th December 2004, the MoES has set up a 'Tsunami Warning System' at Indian National Centre for Ocean Information Services (INCOIS), Hyderabad, an-autonomous Centre towards coastal disaster management

This publication, I am sure will provide wealth of information at one place for the earth scientists as well as managers of coastal ecosystems. I am pleased to recommend this publication.

P.S. Goel
Secretary
Ministry of Earth Sciences,
Government of India

Preamble

The predicted increase in the number of people living in coastal regions of the world implies that by the year 2050, almost two-thirds of the world's population will reside within 150 km of a coastline. Their lives will be sustained mostly by the goods, services and the income provided by coastal ecosystems, as an abundant source of food, as living and recreational area, and as dumping grounds for waste. If current trends in the use of coastal resources continue, most of them will be living in severely polluted, ecologically degraded coastal mega-cities of more than 10 million inhabitants. For most countries, this will have disastrous consequences for the sustainability of their ecological, social and economic systems. Furthermore, coastal ecosystems provide protection to shores from erosion and function as physical buffers against ocean-related disasters. Maritime needs of the coastal states reflect the necessity of finding solutions to these problems.

Added to these are international pressures arising, for example, from: (1) the diversification of ocean uses from extended maritime jurisdiction, (2) the necessity to handle and manage large quantities of data from multi- and interdisciplinary ocean surveys as well as, (3) the need to formulate national marine policies and to participate in international arrangements.

The skills, expertise and the infrastructure needed for dealing with these maritime issues are still rudimentary in many countries reducing their ability to make the best use of their coastal seas and their resources.

In general, building and maintaining capacity and infrastructure is an expensive affair, despite the good intentions of many nations, the resources available and set aside for such activities are still inadequate. International cooperation here could play a constructive role although this should in no way can and should replace national efforts. There have been in place several schemes and programs directed towards capacity building conducted by several international organizations. Experience gained during the last few years has shown that international cooperation in capacity building is most effective and at its best when it is seen to bring benefits each of the cooperating partners. Very often "nationally funded" projects that encourage and allow, in our case, excellent German scientists to work on scientific issues in developing countries have also mobilized resources, and contributed towards mobilizing resources and exceptionally good scientists in those countries. While enabling us to work together and learn from each other as equal and contributing partners, it has helped us to move to a new stage in capacity building in marine sciences, from the assistance to partnerships, where partners can without reservations articulate their needs and priorities.

An example of such a project is the bilateral German-Indonesian research project SPICE (Science for the Protection of Indonesian Coastal Marine Ecosystems) which is coordinated by the ZMT and funded by the Ministry of Education and Research in Germany (BMBF) and the Ministries of Marine Affairs and Fisheries (DKP) and of Research and Technology (RISTEK) in Indonesia. The project partners are Research Institutions and Universities in Germany and Indonesia, and among them the Jenderal Soedirman University at Purwokerto (UNSOED), which together with the NAM S&T Centre and the ZMT, co-organized the workshop. The association with the NAM S&T Centre and its network partners has enabled both the ZMT and the UNSOED to compare and contrast scientific issues related to coastal zone management from environments that differ widely in their structure and functions. The contacts we have made at the workshop will facilitate these exchanges also in future.

I would like to thank Prof. Arun Kulshreshtha, Director, NAM S&T Centre for introducing us to this new circle of colleagues and for agreeing to join us in contributing to building capacity also in the field of marine sciences. I hope our joint activities will encourage and enable more and more young women and men from developing countries to get involved in the study of the seas.

Vengopalan Ittekkot

V. Ittekkot

Director,

Centre for Tropical Marine Ecology,

Bremen

Preface

The present Marine Research Program on the land-sea interface is part of a new environmental research strategy in the frame work of which research is carried out to develop a concept for sustainable development. In the next few years, we will have to start by narrowing the current gaps in our basic understanding of the marine ecosystem and the interdependencies between ocean and coastal zones.

In view of the scope and complexity of the existing challenges, there is a need for global coordination of research. Global problem can only be effectively addressed within the frame work of an international effort, which is also part of our strategy.

Human activities that currently affect the coastal zones include fishing, aquaculture, coastal agriculture, forestry, engineering and coastal construction, mining, ship building, oil extraction, transfer and transportation, electric power plant, navigation sea port and naval operation, tourism and recreation. These influence climate change and affect many of regulatory and socio-economic as well as the function of coastal zones. The coastal zones are also affected by both gradual and recurrent process such as accretion and erosion and from calamities such as floods, hurricane, typhoon, tsunamis, earth-quakes and severe storms. These and the increase of economic activity within the coastal zones result in substantial damages to property, society, economy and environment.

The international workshop-cum-training course on Coastal Ecosystem: Hazards, Management and Rehabilitation shall provide some alternative means to solve the ecological as well as socio-economical problems of the coastal zone. I believe that to address the existing problems in coastal ecosystem mentioned above shall be best achieved by promoting and strengthening mutual partnership between government and non-government organizations involved in undertaking research on sustainable coastal zone management. Participation of other stakeholders in such effort should also be encouraged.

Presentations of important papers followed by fruitful discussion during the workshop facilitated the exchange of information on the status of developments practiced in participant countries in the field of integrated coastal zones management with particular emphasise on Hazard Management and Rehabilitation.

Rubijyanto Misman and Edy Yuwono
Jenderal Soedirman University,
Purwokerto, Indonesia

Editor's Note

Earth scientists in the Indian subcontinent were aware of the phenomenon of tsunami more as an academic issue. The tropical cyclones, storm surges and their associated destruction potential were well known and were considered as important phenomena for detailed studies both for prediction and disaster preparedness as well as management.

Tsunami on 26th December, 2004 and associated destruction and loss of human lives changed the whole situation. I was aware of the international workshop-cum-training program on "Coastal Ecosystems, Hazard Management and Rehabilitation" at Purwokerto, Indonesia through newsletters of the NAM S&T Centre. When I received an invitation to attend this event, I was quite happy because of the importance of the topic not only for India but also for the humanity as a whole. While at Purwokerto, I was asked to take up the responsibility of editing the proceedings of the workshop.

Producing a publication out of the papers presented in a seminar-cum-training workshop is generally an onerous task because authors after going back to their work places get busy with other priority assignments. But I was lucky that the staff of NAM S&T Centre followed it up and initial material was handed over to me for editing.

An initial browsing through the papers led me to distribute them into four parts. In case of some papers there could be some doubt in selection of the part, but in general these seem to fit in well in four distinct parts. These are:

Part I: General Topics on Physical Oceanography

Part II: Impact Studies related to Tsunami, Tropical Cyclones and Toxic Contamination

Part III: General Coastal Management Issues including Use of GIS and Remote Sensing Technologies

Part IV: Coastal Ecosystem: Country Reports, Case Studies

Part I has nine papers, which are of general nature, more in the form of training material. The first paper by Peter Damm is a tutorial on Basics of Physical Oceanography starting from the basic governing equations describing concepts of flow, tides and Tsunami waves. Similarly, the second paper is specific on concepts of Tsunami including institutional framework on detection and warning system with lead-time limitations. The third paper is a description of fundamentals of mathematical modeling for weather prediction, which is backbone of the National Centre for Medium Range Weather Forecasting in India and other world centres. Papers at 4 and 5 are general reviews on continuous changes being brought either by nature or through the human activities. The authors discuss these especially with reference to mangrove ecosystem and hazard management. Kankara in his paper gives special emphasis on the application of GIS for analysis and developing strategies for hazard management. The next paper is also a general paper describing conceptual frame work for risk management in the Indian coastal ecosystem. The paper by Supermoko provides frame work for evaluation of coastal ecosystem in terms of mangrove, coral reef, sea grass and wetland. The valuation takes into account the direct as well as indirect cost and also at the national and international level. The last paper in this part describes significance of management of mangrove ecosystem for protection against natural hazards.

Part II deals with impact studies related to natural phenomena like tsunami and tropical cyclones on one side and through human activities on the other hand. Both lead to serious damage to the coastal ecosystems. Out of the seven papers in this part, the first two papers are related to the recent tsunami. Sri Lanka was one of the worst affected countries, Rupasingha and Pierera have described the impact of tsunami on the coastal ecosystem, especially the mangrove. Sri Lanka is not that vulnerable to the tropical cyclones as the coastal regions of India. The next paper by Muchlisin also describes the impact of tsunami on aquaculture ponds in Aceh, Indonesia. He compares the observations with the analysis based on Spot-5 multi spectral images. The next two papers by Ali of Pakistan and Amargos *et al.*, of Cuba describe the impact of typical tropical cyclone/hurricane on coastal ecosystem. The last three papers highlight the damage to the ecosystem through human activities like affluence from the industry, over exploitation for tourism and oil spill.

Part III consists of five papers, the first two of which relate to coastal management for the coasts of Columbia and Kenya. The third paper, also from Kenya, describes the applications of GIS and remote sensing for the analyses of coastal ecosystem, prediction of hazards as well as delineation of regions of high vulnerability, especially related to mangrove management. The last two papers are case studies from Cuba and India for sustained rehabilitation of coastal ecosystem under the pressure of human activities.

The last part contains five papers mainly dealing with respective country reports from Kuwait, Mauritius, Myanmar, Syria and Vietnam.

As the publication is now ready to go for printing, I look back and feel happy that I agreed to the suggestion by Prof Arun Kulshreshtha to edit this publication. My wife, Meena, was also a great motivator.

During the process of review I was very actively supported by the NAM S&T Centre. I would especially like to place on record my special appreciation for the valuable support by Mr. Bandyopadhyay, Ms. Prinyanka Ahuja and Ms. Ruchi Singh under the over all guidance of Prof. Kulshreshtha.

I also thank all the authors for their contribution to this common cause.

I am also grateful to Dr. P. S. Goyal, Secretary, Ministry of Earth Sciences, Government of India for providing a Foreword, to Prof. Ittekkot, Director, Centre for Tropical Marine Ecology, Bremen for writing the Preamble, to Prof. Misman and Dr. Yuwono of Purwokwerto University for writing the Preface, and to Prof. Kulshreshtha for writing the introduction.

Noting the distribution of papers, I am quite confident that this publication will be equally useful for the students of oceanography, mathematical modelers, researchers in the area, as well as for the planners. In one volume we have reports from almost 14 countries, besides applications of recent technologies.

It may be useful to emphasis that India has attained good skill for atmospheric and oceanographic modeling as well as techniques for hazard prediction and management. As such India can take a lead role to help train the scientific community of the NAM and other developing countries.

Rattan K. Datta
Founder Director,
National Centre for Medium Range Weather Forecasting (NCMRWF),
Government of India

Introduction

The Earth is unique among the known celestial bodies as it has water that covers three-fourths of the surface of this planet and constitutes 60-70 wt per cent of the living world. Water bodies are particularly vital for human life. The living organisms in the seas produce a third of the oxygen that we breathe, offer a valuable source of protein and moderate the global climatic change. The oceans provide abundant valuable natural resources of immense economic significance for the human civilization. It is therefore logical that over the ages human settlements are being established on the riverbanks and vast coastal belts.

Throughout the history, humans have been directly or indirectly influenced by the coastal ecosystem. Coastal zones contain many of the Earth's most complex and diverse ecological systems, productive in both the biological and economic senses. Coastal region includes inshore waters, inter-tidal areas and extensive tracts of contiguous land. It is well recognized that terrestrial and marine components of the coastal region are often closely connected through hydrologic linkages between upland catchments and coastal wetlands. A good example of these inter-linkages is seen where estuaries convert to lagoons with reduced water flow and/or increased silt load. Thus the coastal region could be considered to include an area of land or water where ecological processes of both land and marine environments are inter-related with a further linkage with human activities in the region.

From large estuarine systems to salt marshes the coastal ecosystems are known for their key ecological function and societal value. They provide habitat and nursery grounds for commercially and recreationally important finfish and shellfish. Reefs, mangroves, wetlands and tidelands act as the nursery and feeding areas for many marine species, help in the recycling of the nutrients, and provide buffer for storm protection and controlling erosion. Disturbances caused due to natural calamities and the adverse impact of human settlements threaten these important ecosystems. The existence of coastal marsh is being severely endangered due to increasing development, land subsidence, erosion and sea level rise. In some areas, invasive species have displaced native species, threatening commercially important biological resources by altering the habitat and productivity of the marsh. Increase in suspended solids and growth of algae affect water clarity, inhibiting the growth of submerged aquatic vegetation.

The need for ocean resources is steadily increasing and a significant part of the world's new economic potential is available in the coastal zone. Coastal zones are not only economically valuable for fisheries, commerce, navigation and recreation, but also include some of the most ecologically important environments. Estuaries and wetlands are highly productive, providing food, refuge and nursery functions to a large variety of biota including finfish, decapods, shellfish and birds. In addition, shoreline and coastal features such as dunes, berms, reefs and barrier islands protect the inland from storm events and high-energy waves. Coastal and marine ecosystems play an important role in supporting the economies of many nations and in providing livelihoods for the world's poor. Regrettably, these economic, social, and environmental benefits are at risk as the coastal zones are affected by both gradual and recurrent processes such as accretion and erosion and from extreme natural events such as floods, hurricanes, typhoons, tsunamis, earthquakes and severe storms. This is further aggravated due to increased economic activity within the coastal zones resulting in costly damages to property, society, economy and environment. This combination of natural and human forces and the uncertainties involved in their origins and impacts presents major challenges to coastal managers.

In order to protect against continued degradation and loss of coastal ecosystem services, all concerned must have a better understanding of the structure and functioning of coastal systems in their complexity and interactions, their response to natural and anthropogenic pressures and devising ways to restore their functionality to provide products and services. In the short term however, it will be necessary to create the necessary institutional framework and infrastructure to combat the adverse impacts of natural hazards on coastal populations and provide them adequate relief and rehabilitation facilities in the immediate post hazard period. The integration of environmental and livelihoods concerns should be a major objective that can be best achieved by promoting and strengthening partnerships between governmental and non-governmental organisations involved in implementing sustainable coastal zone management measures.

With the aforementioned in view, the Centre for Science and Technology of the Non-Aligned and other Developing Countries (NAM S&T Centre) organised an International workshop–cum–training course on 'Coastal Ecosystems: Hazards Management and Rehabilitation' at Purwokerto in Indonesia during 8–17 August, 2006 in association with Jenderal Soedirman University (UNSOED), Purwokerto and the Centre for Tropical Marine Ecology (ZMT), Bremen, Germany. 43 scientists, experts and professionals from 14 countries, viz. Colombia, Cuba, Egypt, Germany, India, Indonesia, Kuwait, Malaysia, Mauritius, Myanmar, Pakistan, Sri Lanka, Syria and Vietnam participated in this event, which was conducted in eight technical sessions and five training sessions.

On behalf of the NAM S&T Centre, I express sincere gratitude to Prof. Dr. Venugopalan Ittekkot, Director, ZMT, Bremen for taking this unique initiative and having been the real force behind the successful management of the workshop. I would like to have special words of appreciation to Prof. Rubiyanto Misman and Dr. Edy Yuwono of UNSOED for hosting and making splendid arrangements for holding the event in Indonesia. I would also like to thank Dr. Rattan K. Datta, Former Director of the National Centre for Medium Range Weather Forecasting (NCMRWF), Government of India for editing this volume. My grateful thanks are also due to the Ministry of Earth Sciences (MoES), Government of India for providing generous financial support to bring out this publication. Last, but not the least, I would like to express my happiness with the efforts put in by Ms. Ruchi Singh in asseduously checking the final manuscript and greatly appreciate the valuable services provided by the entire team of the NAM S&T Centre under the guidance of Mr. M. Bandyopadhyay, particularly by Ms. Priyanka Ahuja, and Mr. Pankaj Buttan, who coordinated the work on the organisation of the workshop and subsequently compiled the material for bringing out the publication.

I am confident that the material presented in this volume will be highly useful and serve as a valuable reference material for scientists, professionals and people working in this area.

Arun P. Kulshreshtha

Director, NAM S&T Centre

Contents

Part III: GENERAL COASTAL MANAGEMENT ISSUES INCLUDING USE OF GIS AND REMOTE SENSING TECHNOLOGIES

Part IV: COASTAL ECOSYSTEM: COUNTRY REPORTS AND CASE STUDIES

Part I

General Topics on Physical Oceanography

Chapter 1

An Introduction to Physical Oceanography

Peter E. Damm

Institute of Oceanography, University of Hamburg, Germany

E-mail: damm@ifm.uni-hamburg.de

Governing Equations

Equation of State for Sea Water

In general the density of matter is defined as mass per unit volume [g/cm³]. Distilled water at a normal atmospheric pressure (1015 mbar) and a temperature of 4°C has a density of 1 [g/cm³]. Its density decrease for lower or higher temperature than 4°C; its freezing temperature point is 0°C.

The development of the density of sea water is slightly different to distilled water. It is given by an empirical function:

$$\text{Density Rho Seawater} = f\,[T, S, p]$$

T: Temperature

S: Salinity

p: Pressure

The density of sea water depends on:

Temperature: Increase of temperature → decreasing water density

Salinity: Increase of salinity → increasing water density

Pressure: Increase of pressure → increasing water density

For instance, the temperature of a sample is T = 20°C and its salinity S = 35 [g salt per 1Kg water], the density of sea water at atmospheric Pressure (*i.e.* at the sea surface) is then 1.02478 [g/cm³].

The relatively high salt content of sea water is importance in the oceanography. In the open ocean the salinity varies, within limits, of about 34 [g salt per 1Kg water] to 38. The average is close to 35. Landlocked adjacent seas usually have a much lower salinity, especially if they are located in a humid zone with strong river run-off from land. Examples are the inner parts of the Baltic Sea and some fjords where the salinity may be as small as 0.5 or 1. Water with a very low salinity is called brackish water. Landlocked adjacent seas in arid regions, where evaporation exceeds precipitation and river run-off, have a much higher salinity, up to 43 or 45, as in the Red Sea, or even higher in some of the lagoons off the coast in the Gulf of Mexico.

It has been found that, regardless of the absolute concentration, the relative proportions of the different major constituents (NaCl, $MgCl_2$, $MgSO_4$, $CaSO_4$ etc.) in sea water are constant to within the second decimal place of the salinity expressed in g salt per 1Kg water, except in the region of high dilutions or brackish water near shore, where deviations may occur. Differences at the third decimal place have been found in the different oceans.

The nomenclature of sea water density *rho* is as follow:

$\rho_{s,t,p}$ Consider T, S, p

$\rho_{t,s}$ Consider T, S but no pressure

For achieving a more feasible value of the sea water density, a transformation from *rho* (physics) to *sigma* (oceanography) is done:

A (numerical) example of this transformation is:

$$\rho_t = 1.02478 \quad \mapsto \quad \sigma_t = 24.78$$

The development of the density of sea water as function of temperature and salinity for normal atmospheric pressure conditions is shown in a TS-diagram (Figure 1.1). There, the salinity ranges from freshwater to extreme oceanic salinities on the axis of the abscissa; on the axis of the ordinate the temperature range is shown. The curved lines in the diagram are reflecting the sea water density (isopycnals) in values of sigma (0...30).

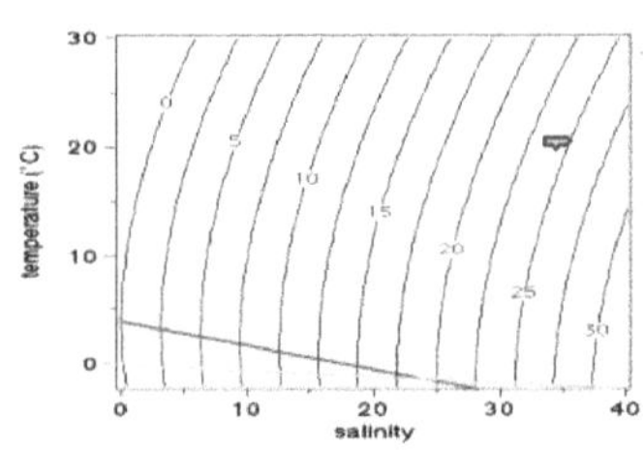

Figure 1.1: TS-diagram

The green dot at the right upper corner in the shown TS-diagram represents the position of the density demonstrated in the example above (sigma: 24.78, temperature: 20°C, salinity: 35). The density maximum as function of T and S is shown in the colour scheme as the red line, the development of the freezing point is given by the light blue line (start position left: 0;4 and 0;0 respectively). The intersection of the red and the light blue line is located at the temperature value of –1.33°C and at the salinity value of 24.695 ~= 24.7.

Notice that the density maximum is above the freezing point for salinities less than 24.7 but below the freezing point for salinities larger 24.7. This affects the thermal convection:

If S < 24.7: The water cools until it reaches maximum density; then, when the surface water becomes lighter (*i.e.* after the density maximum has been passed) cooling is restricted to the wind-mixed layer, which could freeze. The deep basins are filled with water of maximum density.

If S > 24.7: Convection always reaches the entire water body (column). Cooling is slowed down because a large amount of heat is stored in the water body. This is because the water reaches freezing the point before the maximum density is attained.

Law of Conservation of Mass (Continuity Equation)

An important law for calculations in the field of oceanography is the conservation of water mass.

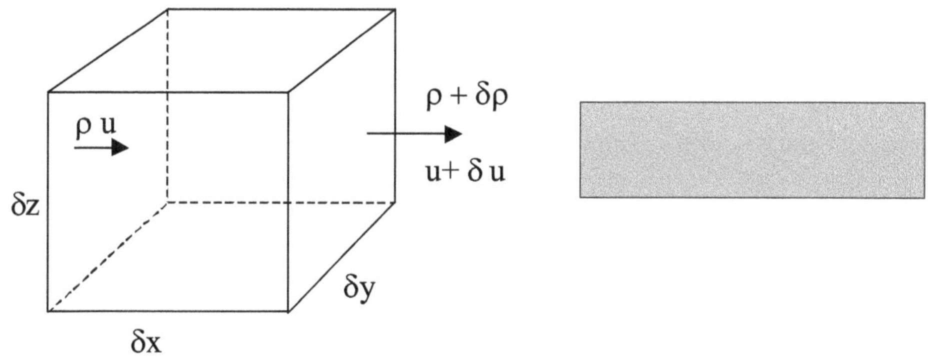

Figure 1.2: Visualization of the Equation of Continuity and its Mathematical Formulation for a Homogenous Density Distribution in Space

A very useful simplification of the equation in Figure 1.2 can be obtained by considering those fluid motions for which the density is constant in time and space. This assumption is not exactly correct for water, but for a number of applications it is a satisfactory approximation. If the density is constant all terms vanish except the partial derivates motion, so that the equation takes the simple form:

$$\frac{\partial u}{\partial x} + \frac{\partial v}{\partial y} + \frac{\partial w}{\partial z} = 0$$

Momentum Equation

Newton's second law that force equals mass multiplied by acceleration is expressed in vector form and used to obtain the equations of hydrodynamics. This law is applied to a continuous volume of fluid, as differentiated from the application of the law to a solid object. Consequently, the equations must account for the fact that different parts of a fluid are moving in different directions at different times and places and that two fluid elements can not occupy the same part of space at the same time. This requirement is met by adjoining another equation that provides for the conservation of mass.

Coming back to Newton's second law, one may write:

$$\mathbf{F} = m\,\mathbf{a}$$

forces (**F**) = mass (m) multiplied by acceleration (**a**); bold characters or characters with arrows on the top indicate vectors, italic letters indicate scalars

or

In fluid dynamics this equation is expressed in terms of forces per unit mass

$$\mathbf{F}' = \mathbf{F}/m$$

$$\mathbf{F}' = \mathbf{a} = \mathrm{d}\mathbf{v}/\mathrm{dt}$$

where,

v = (u,v,w) is the velocity in component notation along the axes x,y,z with x pointing east, y north and z downward (**a** = d**v**/dt is the (total) acceleration)

If there is more than one force, Newton's Second Law applies to the sum of all forces involved. Taking all forces acting on the fluid in account, the equation takes the form (characters with an arrow on top are vectors):

$$\frac{d\vec{v}}{dt} = -\frac{1}{\rho}\,grad(p) + \vec{K}_G + \vec{K}_T + \vec{K}_{friction}\,; \quad grad\,(p) = \left(\frac{\partial}{\partial x}, \frac{\partial}{\partial y}, \frac{\partial}{\partial z}\right) p$$

K_G, K_T and $K_{friction}$ represent gravitational, tidal and frictional forces.

This equation is called the "NAVIER-STOKES Equation", its terms from the left to the right in detail: (total) acceleration, pressure gradient, gravity, tidal force, friction, *i.e.* the acceleration is balanced by the pressure gradient, gravity, tidal force, and the friction.

For investigations of the hydrodynamics of the ocean a virtual force–*the Coriolis Force*–has to be taken into consideration, because the co-ordination system of the basic NAVIER-STOKES-Equation is fixed relative to the earth. But the earth is rotating about its polar axis at a constant angular velocity. Therefore an adequate additional term is necessary to consider the rotation of the earth. The "Equation of Motion" include the Coriolis force and is given by:

$$\frac{d\vec{v}}{dt} = -\frac{1}{\rho}\,grad(p) - 2\,\vec{\Omega}\times\vec{v} + \vec{g} + \vec{K}_T + \vec{K}_{friction}.$$

with its special components.

Some remarks concerning the forces acting on the ocean.

The pressure gradient is negative because the acceleration produced by a pressure gradient is directed opposite to the gradient, so the associated water movement is "down the gradient". As on the right shown in Figure 1.3.

$$\vec{v} = \vec{v}\,(u, v, w) =: velocity\ vector$$

$$\rho =: density\,; \quad p =: pressure$$

$$\vec{\Omega} = \frac{2\,\pi}{86164\,s} =: terra\ rotation\ [fixed\ star]$$

The Coriolis force–a virtual force–vanish at the equator and increases from the equator to the poles. It is proportional in magnitude to the flow speed and directed perpendicular to the direction of the flow. It acts to the right of the flow in the northern hemisphere and to the left in the southern hemisphere.

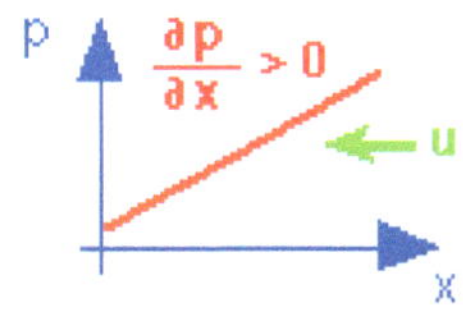

Figure 1.3: Acceleration Produced by a Pressure Gradient

The gravity does not exert a horizontal force and thus cannot produce horizontal accelerations; it is important for movements, which involve vertical motions (convection, waves).

The tidal force must only be considered for tidal problems; it can be ignored in a discussion of the general oceanic circulation.

The frictional force acts in the water body as molecular, turbulence and also acts on the surface of the ocean (wind stress) and on the bottom of the sea.

The Equation of Motion in Components Notation

The equation of motion which is the NAVIER-STOKES-Equation including the Coriolis forces can be written in its components notation. There are three components, the first for the west-east-direction, the second for the south-north-direction and the third for the earth centre-surface-direction:

West-East-Direction

$$\frac{\partial u}{\partial t}+u\frac{\partial u}{\partial x}+v\frac{\partial u}{\partial y}+w\frac{\partial u}{\partial z}=-\frac{1}{\rho}\cdot\frac{\partial p}{\partial x}+2\,\Omega\,v\,\sin\varphi+Fx$$

South-North-Direction

$$\frac{\partial v}{\partial t}+u\frac{\partial v}{\partial x}+v\frac{\partial v}{\partial y}+w\frac{\partial v}{\partial z}=-\frac{1}{\rho}\cdot\frac{\partial p}{\partial y}-2\,\Omega\,u\,\sin\varphi+Fy$$

Earth Centre-Surface-Direction

$$\frac{\partial w}{\partial t}+u\frac{\partial w}{\partial x}+v\frac{\partial w}{\partial y}+w\frac{\partial w}{\partial z}=-\frac{1}{\rho}\cdot\frac{\partial p}{\partial z}-g+Fz$$

On the left hand side of these three (component) equations, the total acceleration is split into the local time derivation term and into the advection (horizontal component equations) and convection (vertical component equation) terms respectively. The gravity term occurs in the third equation only, the terms F_x, F_y and F_z include the tide and the friction force.

Principles for Hydrodynamic-Numerical Calculations

For a hydrodynamic-numerical calculation of the ocean a set of equations is necessary which describe the physical processes of the water body. In general, there is no analytic solution available for this set of equations. Therefore, a numerical iteration scheme is used to solve the set of equation set approximately.

As an example for a 2-dimensional (x-, y-direction and vertical integrated) hydrodynamic-numerical investigation a set of 5 equations is given. These equations are formulated in a Cartesian *x, y, z* coordinate system (*x* positive eastwards, *y* positive northwards, *z* positive upwards):

1. The momentum equations in x- and y-direction,
2. The hydrostatic approximation,
3. The equation of continuity,
4. The equation of state for the sea water

$$\frac{\partial U}{\partial t}+u\frac{\partial U}{\partial x}+v\frac{\partial U}{\partial y}+w\frac{\partial U}{\partial z}-f\cdot V=$$
$$-\frac{h}{\rho}\cdot\frac{\partial p}{\partial x}+\frac{\partial}{\partial x}\left(A_{Ih}\frac{\partial U}{\partial x}\right)+\frac{\partial}{\partial y}\left(A_{Ih}\frac{\partial U}{\partial y}\right)+\frac{\partial}{\partial z}\left(A_{Iv}\frac{\partial U}{\partial z}\right),$$

$$\frac{\partial V}{\partial t}+u\frac{\partial V}{\partial x}+v\frac{\partial V}{\partial y}+w\frac{\partial V}{\partial z}+f\cdot U=$$
$$-\frac{h}{\rho}\cdot\frac{\partial p}{\partial y}+\frac{\partial}{\partial x}\left(A_{Ih}\frac{\partial V}{\partial x}\right)+\frac{\partial}{\partial y}\left(A_{Ih}\frac{\partial V}{\partial y}\right)+\frac{\partial}{\partial z}\left(A_{Iv}\frac{\partial V}{\partial z}\right),$$

$$\frac{\partial p}{\partial z}=-g\cdot\rho,$$

$$\frac{\partial\zeta}{\partial t}=-\left(\frac{\partial U}{\partial x}+\frac{\partial V}{\partial y}\right),$$

$$\rho=f(T,S,p)$$

With u,v,w: velocity components; U,V: water mass transports; h: thickness of the considered water layer; A_{Ih}, A_{Iv}: horizontal, vertical kinematic viscosity, respectively.

To get a numerical solution of the above given set of equations a Taylor Series Expansion of each equation and each term have to be carried out. In general, various series expansions (as example: central, forward, backward differences) are possible.

A "forward" Taylor Series expansion at point x is given by differences between points B and C as shown in Figure 1.4. It follows:

$$N_{x+\Delta x} = N_x + \Delta x \frac{\partial N_x}{\partial x} + \frac{1}{2}\Delta x^2 \frac{\partial^2 N_x}{\partial x^2} + \frac{1}{6}\Delta x^3 \frac{\partial^3 N_x}{\partial x^3} + \frac{1}{24}\Delta x^4 \frac{\partial^4 N_x}{\partial x^4} + \cdots$$

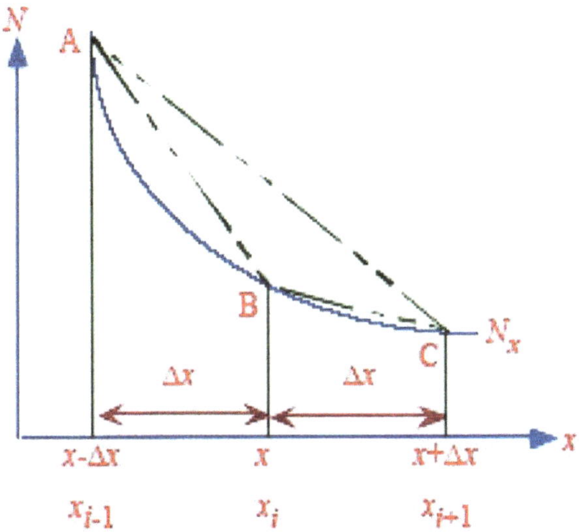

Figure 1.4: Illustration of Taylor Series Expansion

To get an approximate solution for the first derivative, only the first two terms of the Taylor Series are used:

$$\frac{\partial N_x}{\partial x} \approx \frac{N_{x+\Delta x} - N_x}{\Delta x}$$

This is a first-order forward difference approximation of a first derivative.

The same procedure can be applied for a "backward" Taylor Series expansion.

$$N_{x-\Delta x} = N_x - \Delta x \frac{\partial N_x}{\partial x} + \frac{1}{2}\Delta x^2 \frac{\partial^2 N_x}{\partial x^2} - \frac{1}{6}\Delta x^3 \frac{\partial^3 N_x}{\partial x^3} + \frac{1}{24}\Delta x^4 \frac{\partial^4 N_x}{\partial x^4} + \cdots$$

This is a first-order backward difference approximation of a first derivative.

$$\frac{\partial N_x}{\partial x} \approx \frac{N_x - N_{x-\Delta x}}{\Delta x}$$

These steps of Taylor Series expansions are the fundamental basis to design a numerical program-code for solving partial differential equations.

Flow Phenomena

Inertial Flow

If a water parcel is given some momentum (a push) and if it is then left to itself, the only force acting on it, is the Coriolis force. Newton's Second Law states that the parcel will accelerate at a constant rate. Since the acceleration is produced by the Coriolis force it is directed perpendicular to the parcel's path; in other words, the acceleration takes the form of a constant change of direction. The result is that the parcel moves in a circle. Inertial motion is very common in the ocean, where it is usually found superimposed on other movements (such as geostrophic or wind-driven flow discussed below).

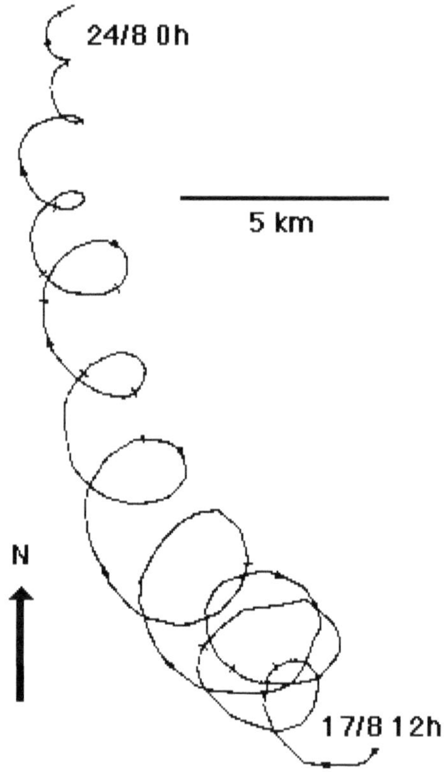

Figure 1.5: Inertial Motion Observed in the Baltic Sea

Figure 1.5 shows the trajectory of a water parcel driven by an inertial motion in the Baltic Sea, superimposed on a slow north-westward drift. Cross marks along the particle path indicate noon and midnight. The water received a "push" (probably produced by a wind burst of several hours) some time after August 17. It is then seen to move in circles. Friction reduces the inertial circles, until the inertial motion has nearly died away on August 24.

Geostrophic Flow

A circulation which is driven by the balance between the pressure gradient force and the Coriolis force is known as geostrophic flow. In geostrophic flows particles move along isobars (contours of constant pressure), with high pressure on their right in the northern hemisphere and on their left in the southern hemisphere. Since the pressure at any depth is determined by the weight of the water above, high and low pressure are equivalent to high sea level and low sea level. Below about 100 m depth and about 100 km away from coastlines all currents are geostrophic; closer to the surface and to boundaries they are modified by additional forces.

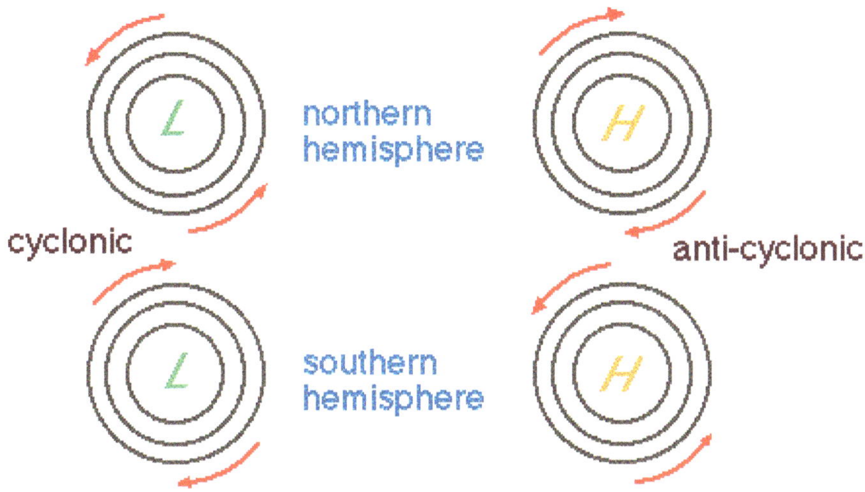

Figure 1.6: Geostrophic Flow. Around High and Low Pressures Centres for Instance

Ekman Flow, Upwelling

Currents in the approximately upper 150 m of the ocean are directly affected by the wind, *i.e.* by transfer of momentum from the atmosphere to the ocean. The balance of forces therefore involves frictional forces, which cause a departure from geostrophic flow; water moves across isobars from areas of high pressure to areas of low pressure. The layer in which the flow is non-geostrophic is known as the Ekman layer. The direction of water movement in the Ekman layer varies with depth. The details are complicated, but when only the steady state is considered, an important result is that the net (*i.e.* vertically averaged) transport in the Ekman layer is directed perpendicular to the wind direction, to the right in the northern and to the left in the southern hemisphere.

The variation of the current in the Ekman layer with depth is shown in Figure 1.7. The current there is shown from the surface (z = 0) to the bottom of the Ekman layer (z = D) in depth increments of 1/10 of the Ekman layer thickness. D is the thickness of the Ekman layer.

In the northern hemisphere the current is 45° to the right of the wind at the surface and rotates clockwise as it decreases with depth. The total wind-driven component of water transport is directed perpendicular to the mean wind stress to the right in the Northern Hemisphere. The net water mass transport (vertically average over the Ekman layer) may cause coastal upwelling phenomena: In the northern hemisphere, coastal upwelling can generate by winds from the north, blowing along the west coast of a continent. Water masses moving offshore are replaced by cold, deep, nutrient-laden water masses. A high biological productivity in the area is the consequence of the upwelling phenomenon.

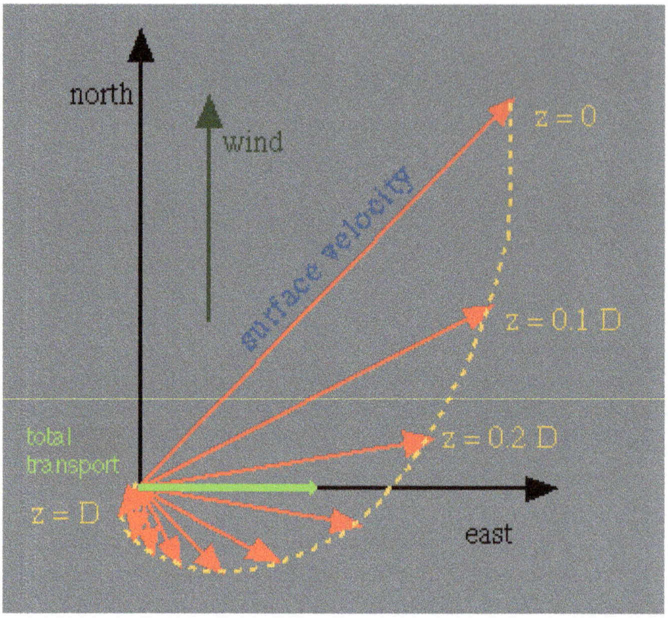

Figure 1.7: Current Distribution in the Ekman Layer (Ekman Helix)

Waves and Tides

Classification of Waves

Ocean waves can be classified in various ways. One classification uses the forces, which generate the waves. In ascending order of wave lengths we have:

1. Meteorological forcing (wind, air pressure); sea and swell belong to this category.
2. Earthquakes; they generate tsunamis, which are shallow water or long waves.
3. Tides (astronomical forcing); they are always shallow water or long waves.

"Shallow water waves" or "long waves" are in a physical view the same, an explanation is given with the dispersion of waves later.

Another classification is based on the frequency spectrum of all oceanic waves and distinguishes between capillary waves, gravity waves, long periodic waves, tides and trans-tidal waves. A third classification is based on the restoring forces responsible for returning the water particles to their average position in the water column (illustrated in Figure 1.8).

As shown in Figure 1.8, the wave-generating forces divide the whole oceanic wave spectrum into two parts. Wind generated waves cover the higher frequency scale from 10^{-2} to 10^{2} Hz (corresponds to periods of 5 minutes–less 0.1 seconds). The

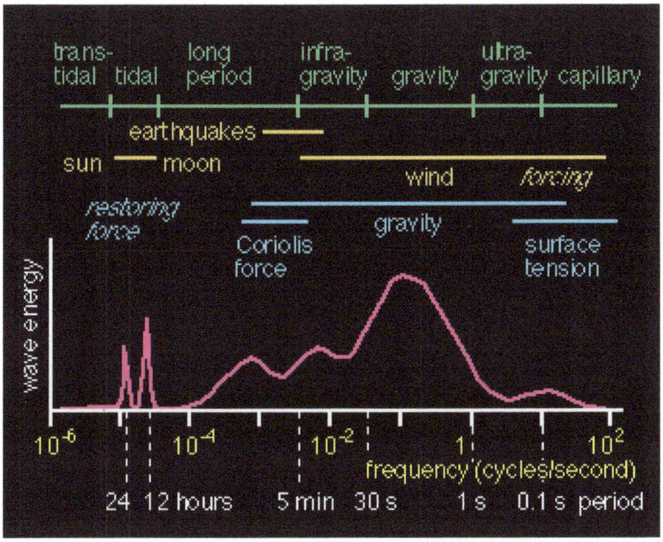

Figure 1.8: Classification of Waves (explanation see text)

lower frequency scale from 10^{-6} to 10^{-2} Hz (10 days–1 minute) is covered by tidal and earthquake waves.

The relative amount of wave energy has a characteristic distribution along the frequency spectrum (red line in Figure 1.8). Most of the energy is generated by wind stress with an ocean wave frequency in the range of 10^{-1} Hz (period: 10 seconds). Two small local energy maxima occur at the tidal periods of 24 and 12 hours.

The oceanic wave classification which is defined by periods, divides the spectrum in seven parts: capillary, ultra gravity, gravity, infra gravity, long period, tidal, and trans-tidal waves (see Figure 1.8 top line).

The third classification–based on the restoring force–covers only the higher spectrum of oceanic waves: form surface tension forces (10^{2}–1 Hz), gravity forces (1–10^{-2} Hz), to the Coriolis force (10^{-2} to 10^{-3} Hz).

In generally, oceanic waves are periodic deformations of an interface. The most known oceanic waves are surface waves, *i.e.* deformations at the atmosphere-ocean interface.

The deformations propagate with the wave speed, while the particles describe orbital or oscillatory motions at particle speed and remain at the same position on average:

1. In shallow water, the particle paths flatten to ellipses. The particle paths remains constant essentially over the entire water depth (Figure 1.9).
2. In deep water, particle paths are circles. Particle movement decreases rapidly (exponentially) with depth (Figure 1.10).

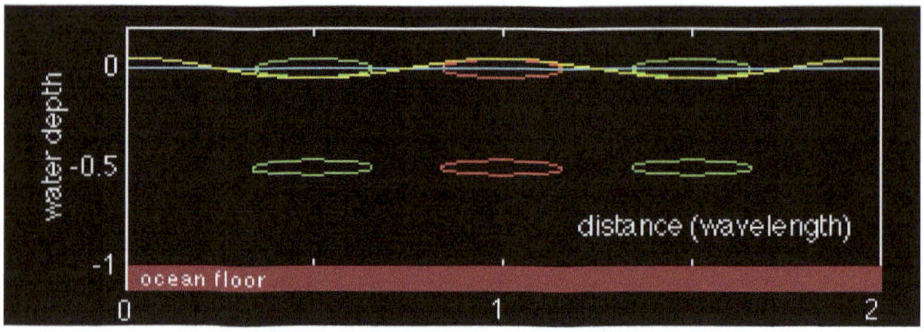

Figure 1.9: Shallow Water Waves

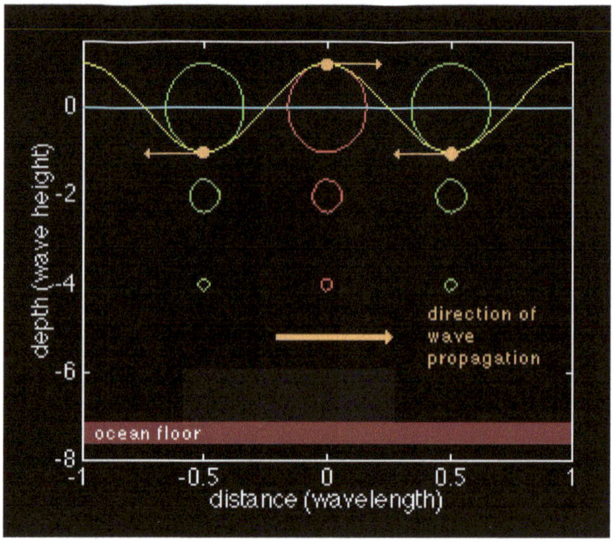

Figure 1.10: Deep Water Waves

In both cases, particles under wave crests move in the direction of wave propagation; particles under wave troughs move against the direction of wave propagation.

The change from deep to shallow water waves occurs when the wavelength L becomes larger than twice the water depth *h*. It is therefore useful to distinguish between deep water waves (short waves) and shallow water waves (long waves).

General Description of Waves

The simplest way to characterise oceanic waves is by regarding waves as harmonic oscillations without any deformations, *i.e.* without linear or non linear superposition. A harmonic wave can be described by a number of parameters. In particular oceanic waves are described by (see Figure 1.11):

1. Period t
2. Frequency w = 2 p/t

3. Wavelength L
4. Wave speed c = L/t
5. Wave height $H = 2A$ (A = amplitude)
6. Wave steepness d = H/L

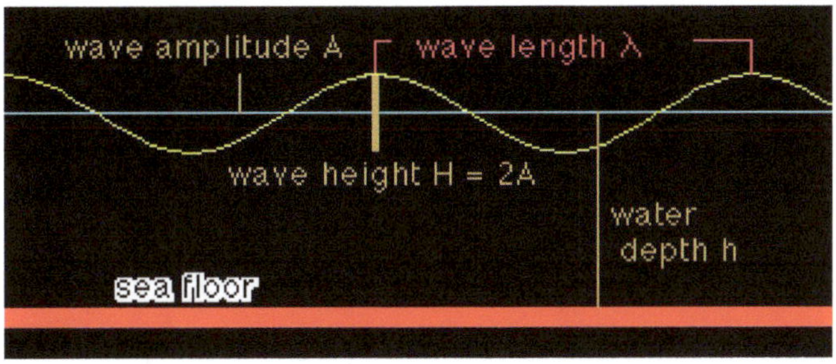

Figure 1.11: Harmonic Wave and its Description Parameter in Oceanography

Superposition of two waves with nearly equal frequencies w_1 and w_2 produces wave groups or packets. Individual wave crests travel with phase velocity c (identical to the wave speed); wave packets travel with group velocity c_g.

$$c_g = \frac{d\omega}{dk} = c - \lambda \frac{dc}{d\lambda}$$

The waves after a superposition (yellow lines in Figure 1.12) travel at wave speed c (phase velocity), the wave groups or packets (red lines in Figure 1.12) travel at group velocity cg.

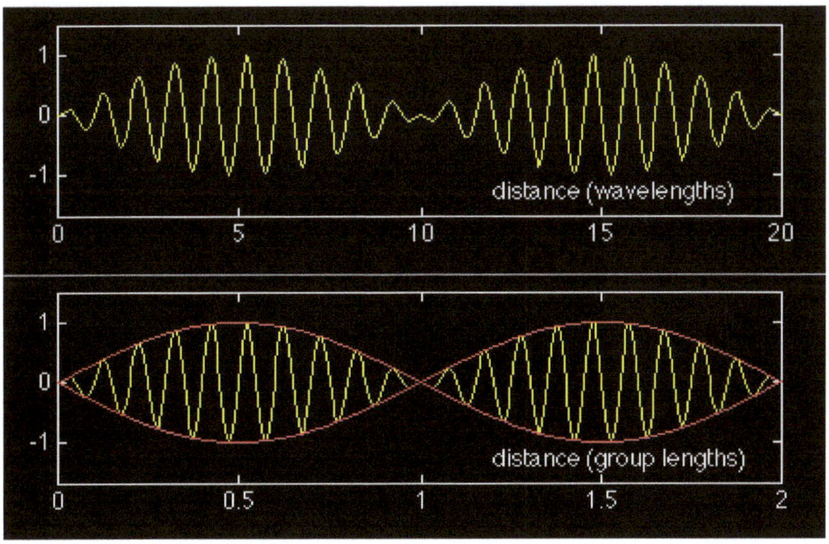

Figure 1.12: Superposition of Two Waves, Distance of Wave Length (Upper Panel) and of Wave Group Length (Lower Panel)

The packets transport the energy contained in the wave field, which means that the energy also travels at group velocity cg.

The phase velocity *c* of waves is dependent on the wave length. This dependence is called dispersion. In most oceanic situations the wave steepness is very small and the phase velocity is then given by the following dispersion relation:

$$c = \sqrt{\left(\frac{g\lambda}{2\pi}\tanh\left(\frac{2\pi h}{\lambda}\right)\right)}$$; h is the water depth.

The phase velocity *c* of deep water waves (identical to short waves, because the water depth is large compared to the wave length) is approximately given by:

$$h \geq \frac{1}{2}\lambda \Leftrightarrow \frac{h}{\lambda} \geq \frac{1}{2} \quad \Rightarrow \quad \tanh\left(\frac{2\pi h}{\lambda}\right) \to 1$$; therefore

$$c = \sqrt{\left(\frac{g\lambda}{2\pi}\tanh\left(\frac{2\pi h}{\lambda}\right)\right)} \quad \to \quad c = \sqrt{\frac{g\lambda}{2\pi}}; \quad c \approx \sqrt{\lambda}$$

The consequence of this estimation is that the phase velocity of deep water waves (short waves) is proportional to its wave length.

The phase velocity *c* of shallow water waves (identical to long waves, because the water depth is small compared to the wave length) is approximately given by:

$$h \leq \frac{1}{20}\lambda \Leftrightarrow \frac{h}{\lambda} \leq \frac{1}{20} \quad \Rightarrow \quad \tanh\left(\frac{2\pi h}{\lambda}\right) \to \frac{2\pi h}{\lambda}$$; therefore

$$c = \sqrt{\left(\frac{g\lambda}{2\pi}\tanh\left(\frac{2\pi h}{\lambda}\right)\right)} \quad \to \quad c = \sqrt{g\,h}; \quad c \approx \sqrt{h}$$

The consequence of this estimation is that the phase velocity of shallow water waves (long waves) is proportional to the water depth.

As an example, consider waves produced by a distant storm. In the open ocean such waves travel as deep water waves; their wave speed therefore depends on their wavelength L. Thus long waves travel fastest and arrive at distant locations first. They are experienced as swell. When swell approaches a beach the water depth decreases and there comes a point when the waves change from deep water waves to shallow water waves. As a consequence the wave speed *c* decreases as the depth decreases and the waves bend inwards; the wave front becomes progressively more parallel to the beach (Figure 1.13).

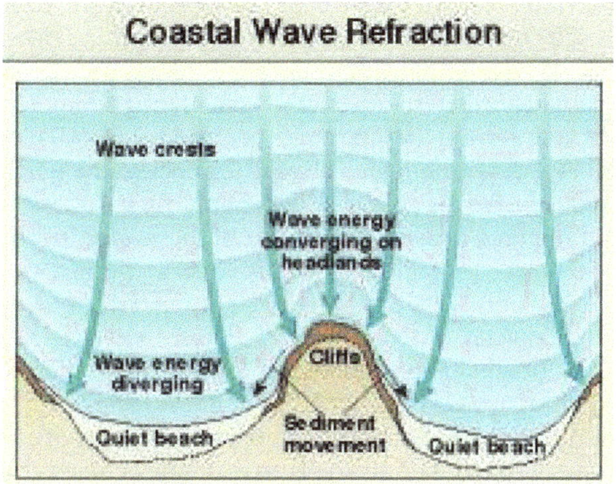

Figure 1.13: The Wave Front Becomes Progressively More Parallel to the Beach

The general dependence of the phase velocity *c* of oceanic waves to the wave length can be put into relation to the wave group velocity *cg*, which transports the energy of the wave field. There are three dispersion relations established, *i.e.* the relation between the phase velocity *c* and the group velocity cg.

Normal Dispersion (Deep Water Waves)

In normal dispersion, c increases with the wave length, *i.e.* the crests of long waves travel faster than the crest of short waves. As a consequence $cg < c$, *i.e.* the energy travels slower than the wave crests. This occurs with gravity waves such as swell.

Non-Dispersive Waves

Here, $cg = c$, *i.e.* all wave crests travel at the same speed, and energy propagates at the same speed.

Anomalous Dispersion

This is found when $cg > c$. Capillary waves are an example. Energy propagates faster than wave crests, and short waves travel faster than long waves.

Tsunami

Tsunamis are long waves generated by submarine earthquakes. Tsunami is Japanese for "harbour wave". They are often called tidal waves; but this is a misnomer, since tsunamis have nothing to do with tides.

A strong tsunami in known history was produced by the eruption of the Krakatau of the Sunda Island group in 1883. It reached a wave height of 35 m and claimed 36,830 lives. Four tsunamis with heights in excess of 30 m have been documented in the Pacific Ocean since 684 A.D. Another strong tsunami in the Atlantic Ocean was observed in 1955 after an earthquake near Lisbon (Portugal); destroyed houses and boats are shown in Figure 1.14. The recent strong tsunami in December 24, 2004 was generated offshore west of North Sumatra and along the Nicobar and Andaman Islands (approximately 1000 km of a tsunami generating zone). This tsunami affected a lot of adjacent countries of the Andaman Sea and accessorily India, Sri Lanka, the Maldives Islands and with a lower impact it affected the north-easterly coast of Africa. In the most destroyed area, in North Sumatra, Indonesia, more than 230,000 people died or remained missing and hundred thousands lost their homes (Figure 1.15).

Internal Waves

It was said before that waves are periodic movements of interfaces. If the water column consists of an upper layer and a denser lower layer, the interface between the layers can undergo wave motion. This motion, which does not affect the surface and is mostly not observable at the surface, is an example of an internal wave.

The restoring force for waves is proportional to the product of gravity and the density difference between the two layers (the relative buoyancy). At internal interfaces this difference is much smaller than the density difference between air and water (by several orders of magnitude). As a consequence, internal waves can attain much larger amplitudes than surface waves. It also takes longer for the restoring force to return particles to their average position, and internal waves have periods much

Figure 1.14: Destroyed Houses and Boats Near Lisbon, Portugal, After Tsunami in 1955

Figure 1.15: Both Photos Show a Cargo Ship which was Pushed by the Tsunami Wave 2004 in North Sumatra from a 2.5 km afar River Over Inhabited Land to this "Last" Place

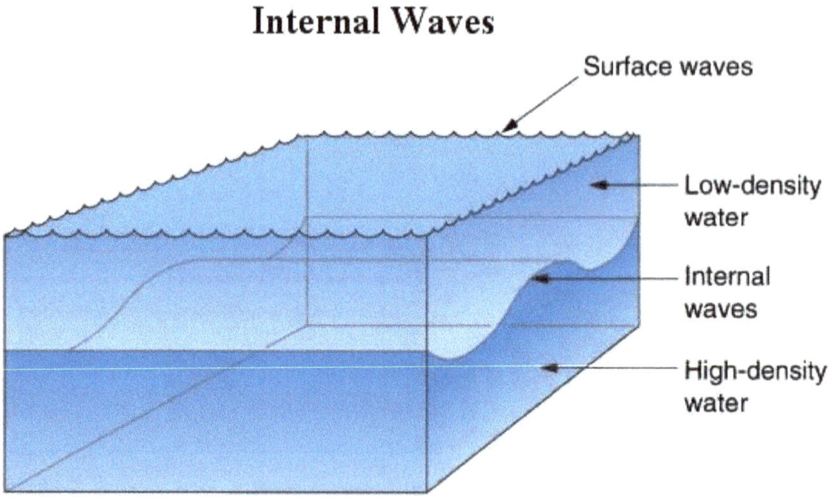

Figure 1.16: Visualisation of Internal Waves

longer than surface gravity waves (from 10–20 minutes to several hours, compared to several seconds or minutes for surface gravity waves).

Tides

Tides are generated by the gravitational potential of the moon and the sun. Their propagation and amplitude are influenced by friction, the rotation of the earth (Coriolis force), and resonances determined by the shapes and depths of the ocean basins and marginal seas.

Tides are long waves, either progressive or standing. The dominant period usually is 12 hours 25 minutes, which is 1/2 of a lunar day. The most obvious expression of tides is the rise and fall in sea level. Equally important is a regular change in current speed and direction; tidal currents are among the strongest in the world ocean. The generation of tidal forces illustrate in Figure 1.17. As the earth revolves around the gravitational centre of the sun/earth system (or moon/earth system), the orientation of the axis of the earth in space remains the same. This is called revolution without rotation.

The tide generating force is the sum of gravitational and centrifugal forces. In revolution without rotation the centrifugal force is the same for every point on the surface of the earth, but the gravitational force varies. It follows that the tide generating

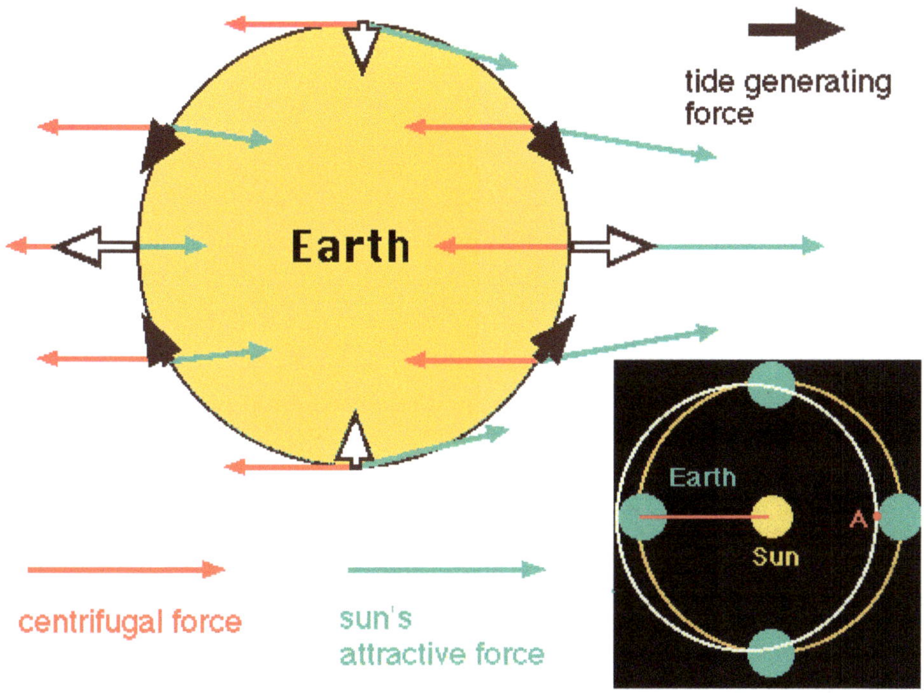

Figure 1.17: Generation of Tidal Forces. The small image in the lower right corner illustrates the revolution without rotation of the sun/earth system. The tide generating force (brown arrows) is the sum of gravitational (green arrows) and centrifugal (red arrow) forces.

force varies in intensity and direction over the earth's surface. Its vertical component is negligibly small against gravity; its effect on the ocean can be disregarded. Its horizontal component produces the tidal currents, which result in sea level variations.

The gravitational force exerted by a celestial body (moon, sun or star) is proportional to its mass but inversely proportional to the square of the distance. The larger distance between the sun and earth, compared to the distance between the moon and earth, means that the sun's gravitational force (and hence its tide-generating force) is only about 46 per cent of that from the moon. Other celestial bodies do not exert a significant tidal force.

The tides can be represented as the sum of harmonic oscillations, plus harmonic oscillations of all its superposition (such as inequalities). Each oscillation, known as a tidal constituent, has its amplitude, period and phase, which can be extracted from observations by harmonic analysis. Hundreds of such oscillations have been identified, but in most situations and for predictions over a year it is sufficient to include only M_2, S_2, K_1 and O_1. Practical predictions produced on computers for official tide tables use significantly more terms than these four. The source and period of these four main partial tides are:

1. Tides produced by the moon
   (*a*) M_2 (semidiurnal lunar) 1/2 day = 12h 25min
   (*b*) O_1 (diurnal lunar) 1 day = 25h 50min
2. Tides produced by the sun
   (*a*) S_2 (semidiurnal solar) 1/2 day = 12h
3. Tides produced by the moon and sun
   (*a*) K_1 (diurnal lunar-solar) 1 day = 23h 56min

The amount of these for tides is different at different locations on the earth. To classify the development of the tide (at a special location) over a period of approximately one month, a form factor F is used:

$$F = \frac{K_1 + O_1}{M_2 + S_2}$$

Four categories are distinguished:

Value of F	*Category*
0–0.25	semidiurnal
0.25–1.5	mixed, mainly semidiurnal
1.5–3	mixed, mainly diurnal
> 3	diurnal

An example of tidal classification is given in Figure 1.18.

Only the largest water bodies can accommodate directly forced tides. On a non-rotating earth the tides would be standing waves; they would have the form of seiches, *i.e.* a back and forth movement of water across lines of no vertical movement (nodes). On a rotating earth the tidal wave is transformed into a movement around points of no vertical movement known as amphidromic points (Figure 1.19).

1. At amphidromic points the tidal range is zero.
2. Co-range lines (lines of constant tidal range) run around amphidromic points in quasi-circular fashion.
3. Co-phase lines (lines of constant phase, or lines which connect all places where high water occurs at the same time) emanate from amphidromic points like spokes of a wheel.

Tides in marginal seas and bays cannot be directly forced; they are co-oscillation tides generated by tidal movement at the connection with the ocean basins. Depending on the size of the sea or bay they take the shape of a seiche or rotate around one or more amphidromic points.

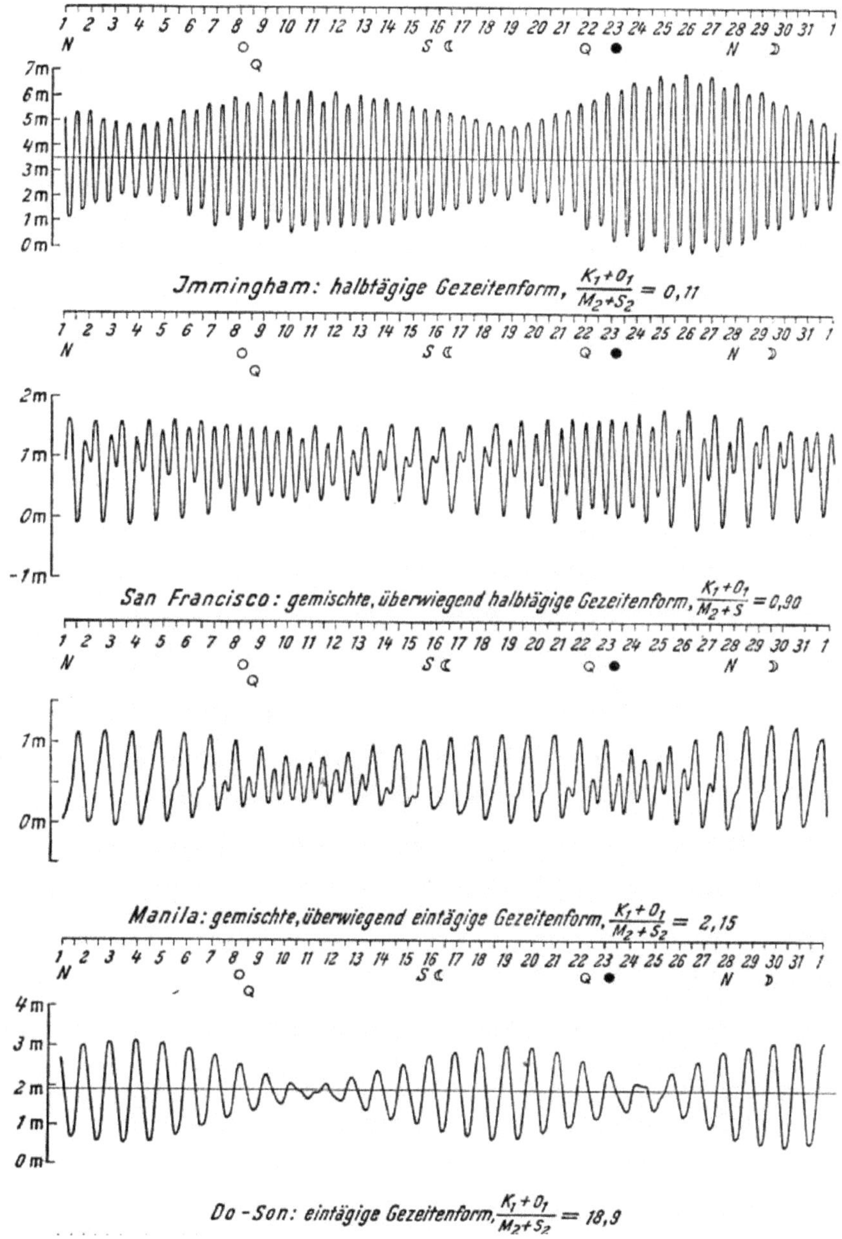

Figure 1.18: Examples of Tidal Classification

Top Panel: Value of F = 0.11 ; semidiurnal tide; Immingham, Great Britain
Upper Panel: Value of F = 0.90 ; mixed, mainly semidiurnal tide; San Francisco, USA
Lower Panel: Value of F = 2.15 ; mixed, mainly diurnal tide; Manila, Philippines
Lowest Panel: Value of F = 18.9 ; diurnal tide; Do-Son, Gulf of Tonkin, Vietnam

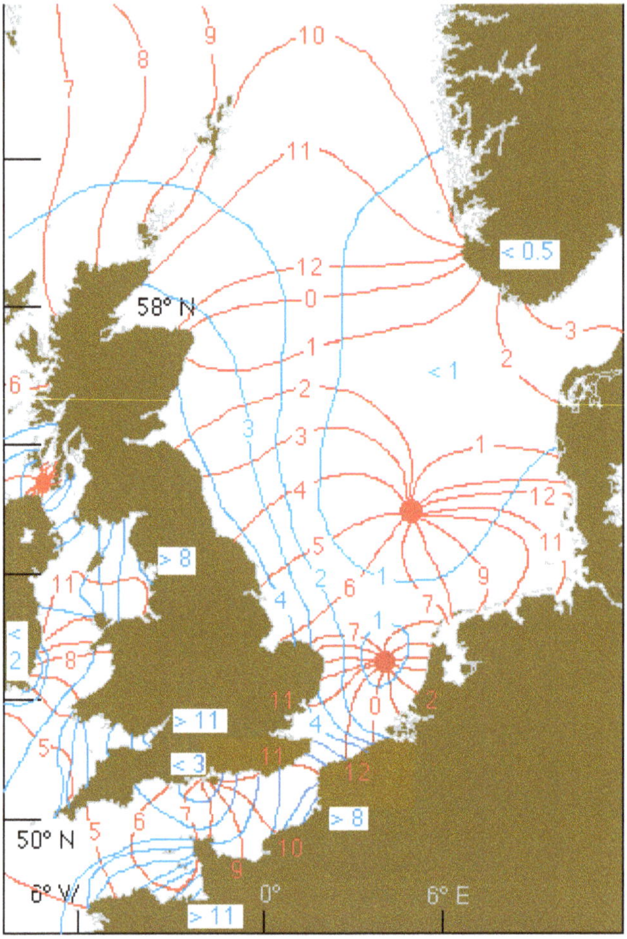

Figure 1.19: Tides in the North Sea Red Lines are Co-phase Lines of the M_2 Tide, Labelled in Hours After the Moon's Transit Through the Meridian of Greenwich. Blue lines give the mean tidal range at spring tide (co-range lines of the sum of M_2 and S_2).

If the tidal forcing is in resonance with a seiche period for the sea or bay, the tidal range is amplified and can be enormous. This produces the largest tidal ranges in the world ocean (14 m in the Bay of Fundy on the Canadian east coast; 10 m at St. Malo in France, 8 m on the North West Shelf of Australia and at the extreme north of the Gulf of California in Mexico; all are mainly semidiurnal tides). Tidal range is then largest at the inner end of the bay, in accordance with the dynamics of seiches in open basins. Modest amplification is experienced in Spencer Gulf of South Australia where the tidal range at spring tide is 3 m at the inner end, while it is less than 1 m at the entrance to the Gulf.

A special tidal propagation–a bore tide is shown in Figure 1.20. A bore tide is caused by topographic conditions around the river mouth area. It is running up the river with a "water wall" as tidal front.

Figure 1.20: The High Water Tide Running Up the River Tsien-tang, China, as a Bore Tide. The height of the "water wall" is up to 8m (spring tide), the "wall speed" is about 20 km per hour.

Some Definitions Concerning the Tide

High water	a water level maximum ("high tide")
Low water	a water level minimum ("low tide")
Mean Tidal Level	the mean water level, relative to a reference point (the "datum") when averaged over a long time
Tidal range	the difference between high and low tide
Daily inequality	the difference between two successive low or high tides
Spring tide	the tide following full and new moon
Neap tide	the tide following the first and last quarter of the moon phases

Acknowledgment

I like to thank my colleagues Dagmar Hainbucher and Thomas Pohlmann for kind assistance and discussions and Prof. Ittekkot from the ZMT, Bremen, for inviting me to the Jenderal Soedirman University, Purwokerto, Indonesia, to participate on the International Workshop–cum–Training Course on "Coastal Ecosystems: Hazards Management and Rehabilitation".

Chapter 2

Tsunami and Coastal Disaster Risk Management in Indonesia

Eberhard Krain, Dewi Yanurita and Anne-Katrin Link
Centre for Tropical Marine Ecology (ZMT), Bremen, Germany
E-mail: eberhard.krain@zmt-bremen.de

Introduction

Natural extreme events, such as floods, earthquakes or storms, can cause much damage to people, property, and nature. During the 26th December 2004 Tsunami off the coast of northern Sumatra, Indonesia, many coastal communities along the shores of the Indian Ocean were affected. In Indonesia more than 230,000 people died or remained missing and hundred thousands lost their homes. This study explains and summarize the physical characteristics of tsunamis and reflect the current state of tsunami risk management, particularly with respect to Indonesia where the authors gathered some insights, including risk perception of the communities and the legal framework. This study is likely to equip students and practitioners in coastal management initiatives with an in-depth understanding of tsunamis and provide them with some hands-on guidance for preparing coastal disaster risk management plans or to incorporate these into broader coastal zone management plans and initiatives.

Causes of Tsunamis

Tsunamis are caused by earthquakes, landslides (above or underwater), volcanic eruptions, or nuclear explosions. Even impacts from outer space (meteorites, asteroids, and comets) or portions of icebergs breaking into the water (so-called "fjord tsunamis") in a fjord, a narrow ocean inlet surrounded by cliffs, can generate tsunamis. But the most frequent cause for tsunami generation is an earthquake. In the following text and diagrams the scenarios of the plate tectonics before, during and after an earthquake in Indonesia, resulting in a tsunami wave, will be described similar to the December 26th, 2004, tsunami in the Indian Ocean off Aceh, Indonesia.

At the start of the earthquake, the Indian-Australian plate moved against and underneath the Eurasian plate (Figure 2.1). Then the Indian-Australian plate flexed down, by which it was warped spring-like due to the built up tension, resulting in a rise of the coastal zone (Figure 2.1b). When the tension was too high, the plate "snapped" at the plate boundary. Hereby, the nose of the overriding plate pushed out and moved the water body lying above it virtually triggering the tsunami. After the tension was released, the coastal area was lowered again (Figure 2.1c).

This was the "birth" of the tsunami: the water body above the plate was lifted up by the snapped plate (Figure 2.1d). After about ten minutes waves radiated and progressed outwards from the centre of the earthquake. Firstly; a depression occurred at the earthquake centre (Figure 2.1e). After a short while, the propagating movement

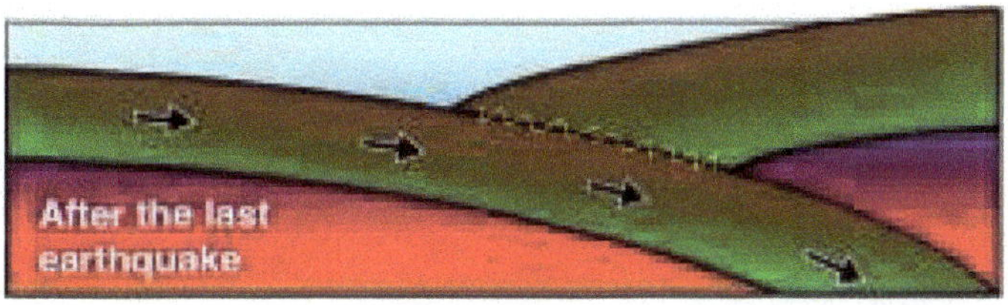

Figure 2.1a

Figure 2.1b

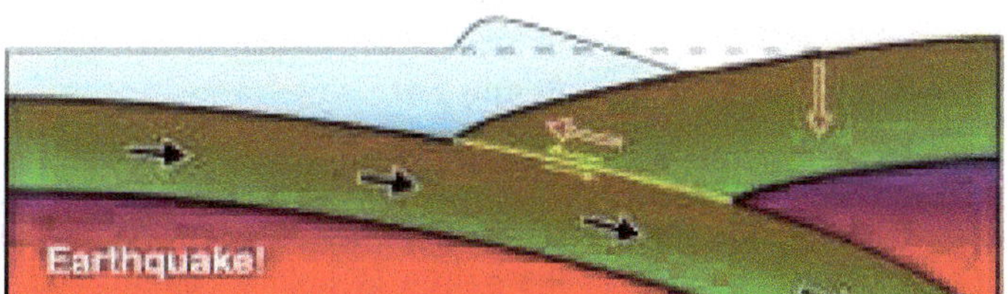

Figure 2.1c

Figure 2.1a–c: Tsunami Generation. (2.1a) The Australian plate moves against and underneath the Eurasian plate. (2.1b) The Indian-Australian plate flexes down, and is warped spring-like, resulting in a rise of the coastal zone. (2.1c) The nose of the overriding plate pushes out and moves the water body lying above it, and the coastal area is lowered again. *Source*: SOEST (2006).

of the wave uplifted the water body in the centre again, and the radiating waves propagated towards the coastline (Figure 2.1f).

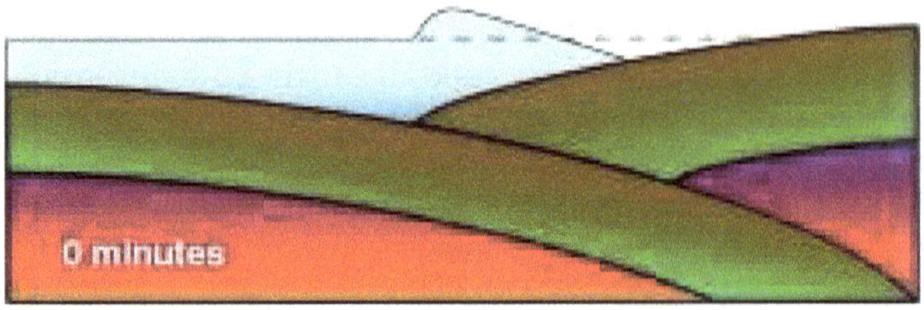

Figure 2.1d

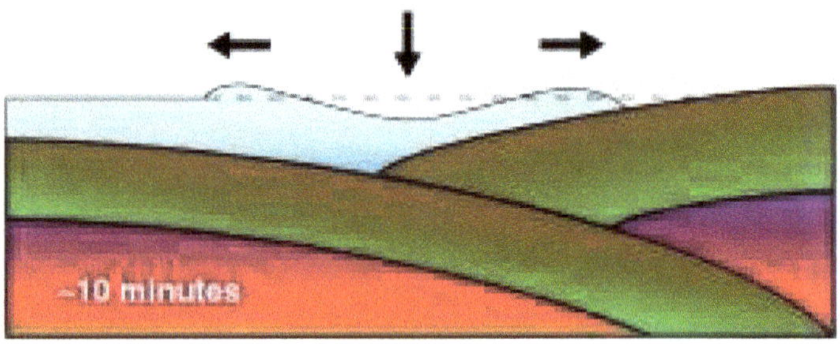

Figure 2.1e

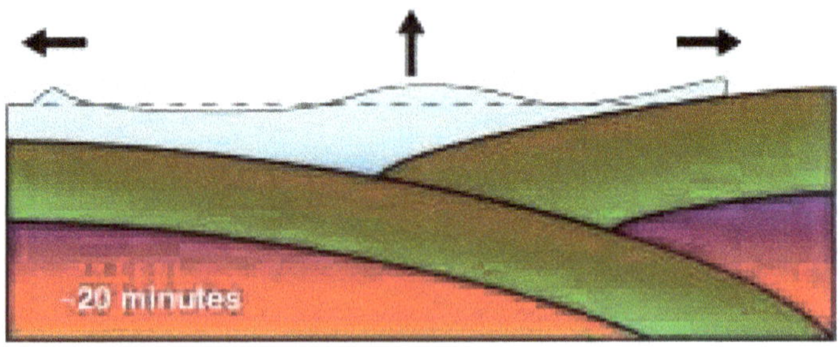

Figure 2.1f

Figure 2.1d–f: Tsunami Generation. (2.1d) The water body above the plate is lifted up by the snapped plate. (2.1e) A depression occurs at the earthquake centre; (2.1f) The water body in the centre is uplifted again, and the waves propagate towards the coastline. *Source*: SOEST (2006).

Physical Properties of Tsunami Waves

What is a Wave?

In general we can say that a wave is any periodic, vertical displacement of a surface. Waves are produced by a generating force and a restoring force; it then propagates away from the source of disturbance.

Figure 2.2b shows a vertical profile of an idealized ocean wave.

Regular, symmetrical waves can be described by their parameters, which are

1. Wave height, which is the distance from trough to crest
2. Wave length, which is the distance from crest to crest or trough to trough and the
3. Wave period. The wave period is the time span which is needed for one complete wavelength to pass through a fixed point.

Waves can be classified according to their wave period (Figure 2.2a and 2.2b). A tsunami is a series of ocean waves of extremely long wave length and long period

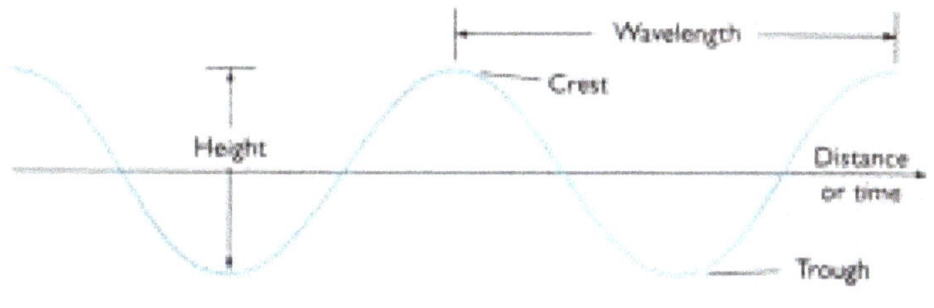

Figure 2.2a: Wave Properties: Wave Parameters

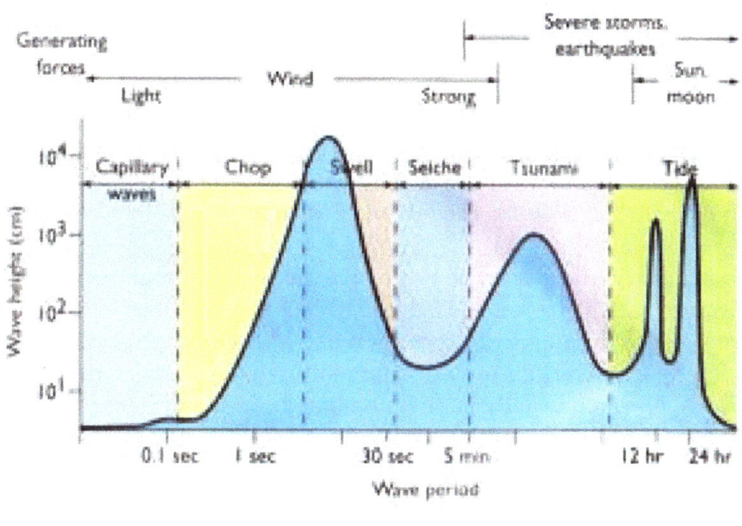

Figure 2.2b: Idealized Wave Spectra

Figure 2.2a–b: Wave Properties. (2.2a) Wave parameters: regular, symmetrical waves are described by their height, wavelength and period (the time one wavelength takes to pass a fixed point). (2.2b) Idealised wave spectrum: waves are classified according to their wave period. The scales of wave height and wave period are not linear, but logarithmic, being based on the power of 10. *Source*: Pinet (1998: 229).

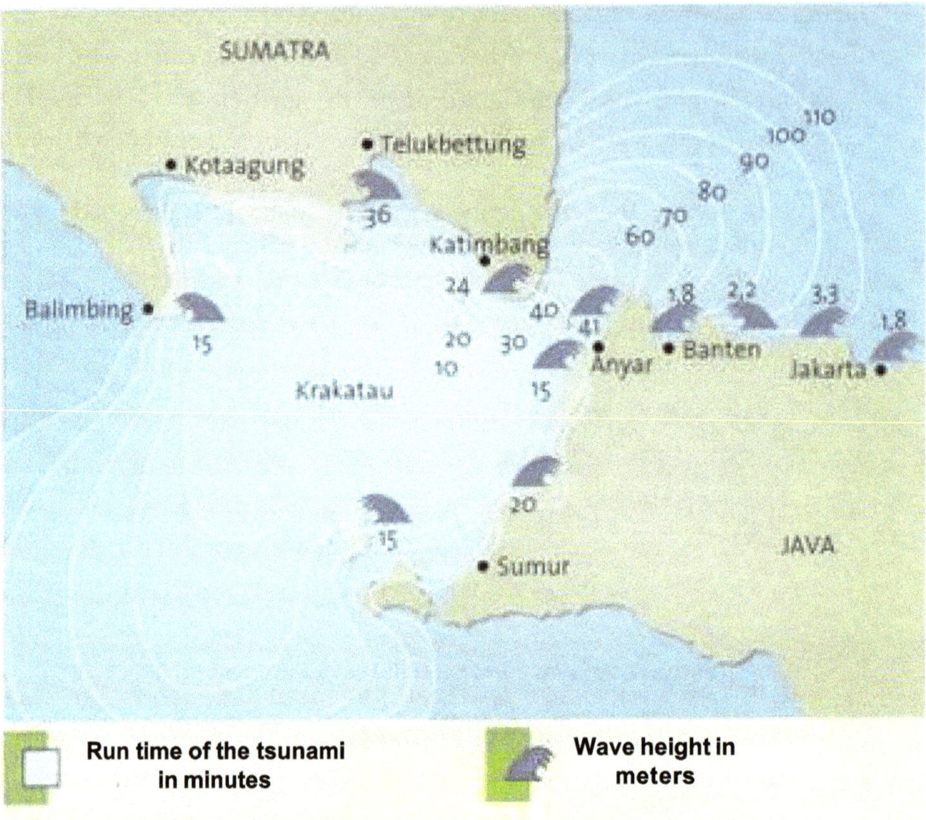

Figure 2.3: The Famous Krakatau Eruption in 1883: Wave Height in Meters and Run Time of the Tsunami in Minutes
***Source*: Niedek and Frater (2004:22).**

generated in a body of water by an impulsive disturbance that displaces the water. Tsunami waves usually have a wave speed of 500–600 mph (800 -1000 km per hour), a wave period of 10 minutes to 2 hours apart and a wave length of 60–300 miles apart (100–500 km apart). The anatomy of a tsunami wave is similar to that of a wind-generated wave, which usually has a wave speed of 5-60 mph (8 -100 km per hour), a wave period of 5-20 seconds apart and a wave length of 300-600 feet apart (100–200 meters apart) (http://library.thinkquest.org/04oct/01797/dynamics/birth.html). The velocity of a wave depends on the depth of the ocean. The deeper the ocean is, the faster the wave travels! Figure 2.9 shows the propagation of a tsunami wave during the famous eruption of the Krakatau in 1883 in the Sunda Strait (the strait between Java and Sumatra).

Tsunami Waves vs. Wind-Generated Waves

Most of the waves present on the ocean's surface are wind-generated waves. Size and type of wind-generated waves are controlled by:

1. Wind velocity,
2. Wind duration,
3. Fetch, and
4. The original state of the sea surface.

As wind velocity increases, wave length, period and height increase (but only with sufficient wind duration and fetch!). Where waves are generated, *e.g.* within a storm or very windy area, waves with many periods and heights and lengths are present. Due to the wave interference the seas are irregular within the fetch area (Figure 2.4). Outside of the fetch area ocean swells develop as they sort themselves. With distance from the storm the length of the wave crest increases, and swells lose their energy very slowly, and can even travel the entire length of the Pacific, from the Antarctic to Alaska! The significant wave height is the average wave height of the highest 1/3 of the waves present and is a good indicator of potential for wave damage.

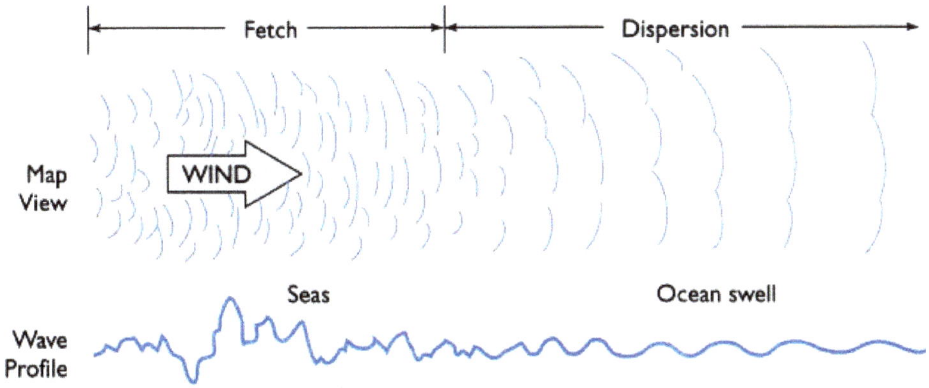

Figure 2.4: Map View and Wave Profile of Wind-generated Waves
***Source*: Pinet (1998:239)**

As a matter of fact, two wave trains (frequency of wave) can merge (Figure 2.5a)of slightly different wave lengths (but the same amplitudes), to form wave groups (Figure 2.5b)

Tsunami waves are characterized by

1. Long wave lengths (> 150 km),
2. Small wave heights (0.5-1m), and
3. High velocities (700-900 km/hour).

They may cross an entire ocean within 24 hours. In contrast to that, wind-generated waves have shorter wave lengths.

Energy Transformation

Waves move through the open ocean. As the waves encounter water depths smaller than 1/2 wavelength, friction removes energy, wave length decreases and

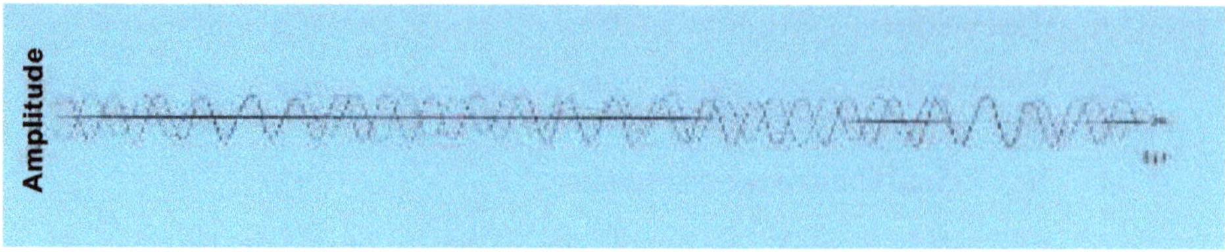

Figure 2.5a: Wave Trains: Merging of 2 Wave Trains
***Source*: Schwarzer (2006)**

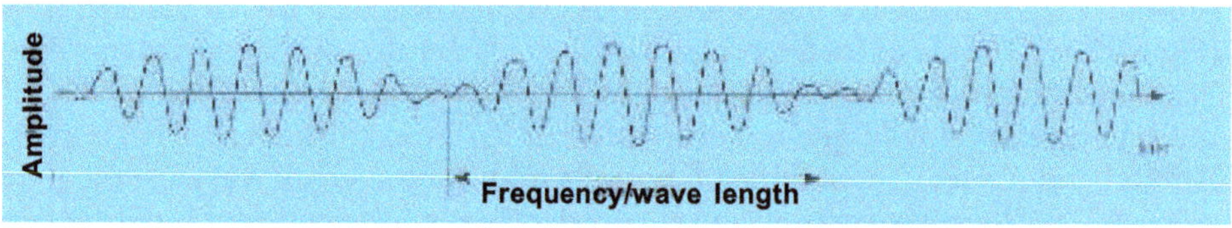

Figure 2.5b: Wave Group Formation of Wave Groups

wave height increases. The wave period stays constant (Figure 2.6). The waves decrease in height relative to width and become longer on the top than on the bottom. This results in greater net motion of the water in the direction of the wave. The period of the wave stays the same, but length and speed decrease. The wave becomes steeper/ higher (since steepness is height/length) and when steepness is >1/8 it breaks. Breaking releases the wave's energy and leads to a transport of water and sediment. If the water is very shallow compared to the wavelength (D/L<1/20), then the speed of a wave is strongly determined by the water depth.

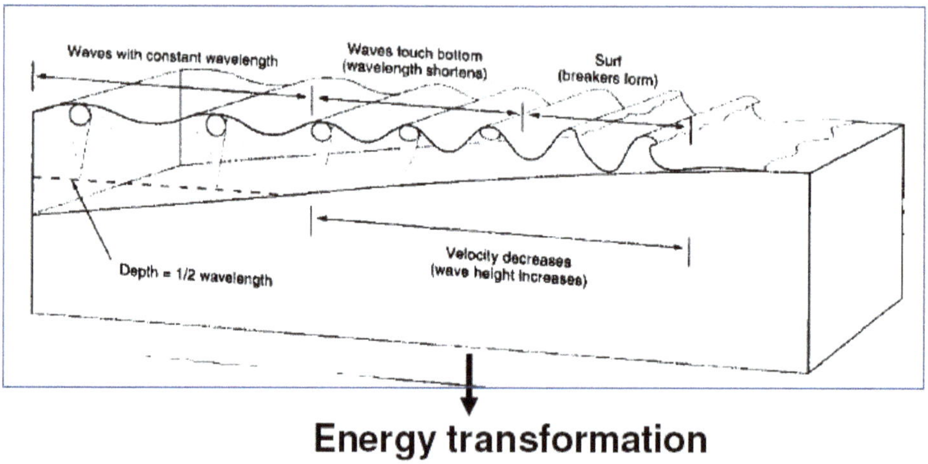

Figure 2.6: Wave Energy Transformation at the Shoreline
***Source*: Schwarzer (2006).**

Deep-water vs. Shallow-water Waves

Deep-water waves are defined as waves moving through water that is deeper than its own wave base, which is one half its wave length. Intermediate-water waves travels through water depths that are between one half and one twentieth the wave length, whereas shallow water waves occur in water less than one 20th of the wave length.

In deep-water waves, wave motion is not detectable at depths deeper than half a wave length. This border is called "wave base". In Figure 2.7a, the Blue arrows symbolise the motion of water particles beneath the waves. As one wave length passes, a water particle at the surface moves through a circular orbit with a diameter equal to the height of the wave. Particles below the surface also move through orbits with the passage of a wave. The orbital diameter decreases with depth (Figure 2.7b). The orbits beneath the waves are slightly open, which results in a net displacement of water (mass transport), which also allows the drifting of water particles (Figure 2.7c). This shows that the wave energy, not the water particles, travel across and through the water!

Figure 2.8a shows that the particle orbits in deep-water waves are in circles and decrease in diameters with depth. In opposite to that, shallow-water wave orbits are compressed vertically into elongated ellipses and progressively flatten their orbits towards the sea-bed (Figure 2.8b). When a wave enters shallow water, it is slowed down by friction. This causes the orbits, especially in the lower part of the wave, to become elliptical. BUT: the speed of these elliptical movements does not decrease with water depth as in deep water waves.

Definition of Shore Face Border

Johnson defined the border of the *shore face border* as the point where the wave base first meets the seafloor, and where the energy transformation starts (Figure 2.9). For tsunami disaster management it is important to consider the local geomorphology, as it affects the behavior of the wave onshore.

Wave Refraction

Wave behavior changes when they approach shallow water. They start to bend when they propagate towards the coastline (Figure 2.10a). Waves travel more slowly in shallow water (shallower than the wave base) than in deeper water. This causes the wave front to bend so it is more parallel to shore (Figure 2.10b).

Refraction causes waves to approach the shoreline at low angles. In bays energy is spread out by refraction (Figure 2.11a). At headlands the wave energy is concentrated by refraction (Figure 2.11b).

However, what happens when (like in the case of a tsunami) the wave travels on shore? On gently sloping coasts a giant tsunami can develop, and on sudden steep coasts a tsunami can be blocked towards the coast. In addition to that, the morphology of a coastline affects the height of the approaching wave in such a way, that in bays the waves get higher due to the funnel-like shape of the coast, while on headland the waves are diverted and get smaller.

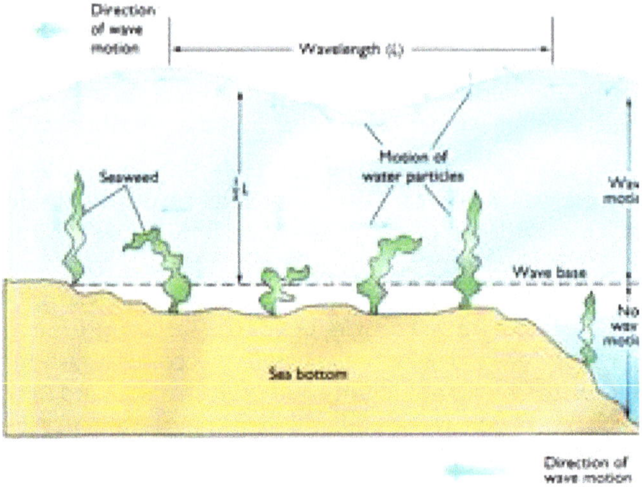

Figure 2.7a: Wave Motion with Depth

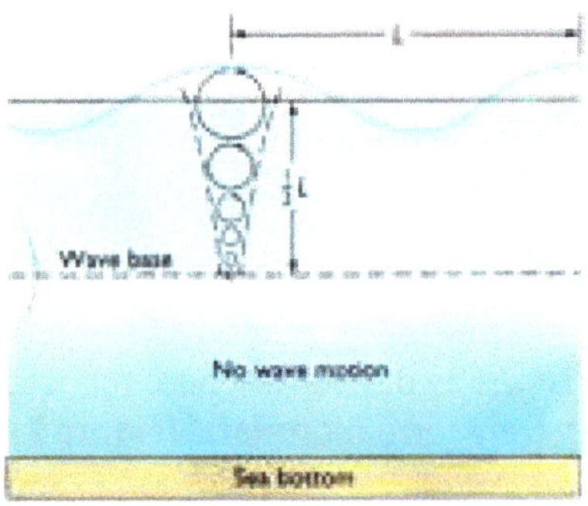

Figure 2.7b: Orbits of Water Particles

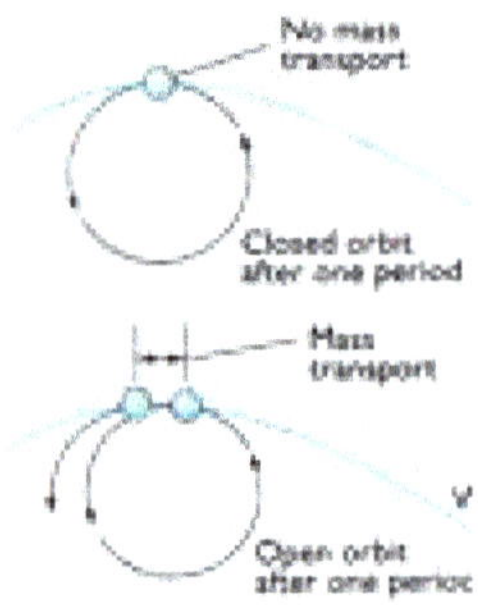

Figure 2.7c: Stokes Drift

Figure 2.7a–c: The Motion of Water Particles Beneath Waves. (2.7a) Wave motion with depth: the arrows at the sea surface denote the motion of water particles beneath waves. With the passage of one wavelength, a water particle at the sea surface describes a circular orbit with a diameter equal to the wave's height. Water particles below the sea surface also describe orbits with the passage of each wave, as indicated by the back-and-forth motion of seaweed attached to the ocean floor.(2.7b) Orbits of water particles: the orbital diameters described by water particles beneath waves decrease rapidly with distance below the surface. Wave-induced water motion essentially ceases at depth (wave base) that is equal to one half of the wavelength. (2.7c) Stokes drift: the orbits described by water particles beneath waves are not closed, but slightly open. This results in a net displacement of water, called mass transport (Stokes drift), in the direction of the wave advance. *Source*: Pinet (1998:233).

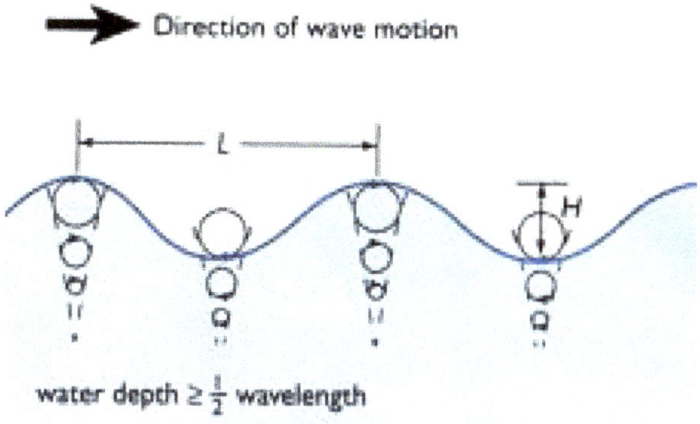

Figure 2.8a: Deep-water Wave

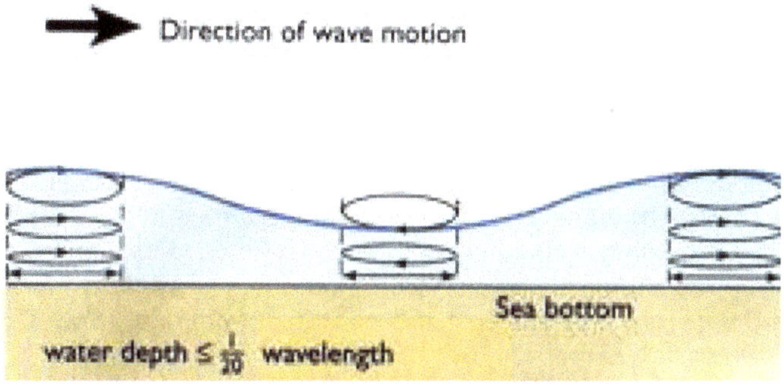

Figure 2.8b: Shallow-water Wave

Figure 2.8a–b: The Distortion of Water-particle Orbits in Shallow Water. (2.8a) Deep-water wave: for deep-water waves, the orbits described by water particles beneath waves are circular. (2.8b) Shallow-water wave: for shallow-water waves, the orbits of water particles are greatly compressed vertically into elongated ellipses. *Source*: Pinet (1998: 235).

Characteristics and Dimensions of Tsunamis

Tsunami Trough and Crest

In many reports on tsunamis it is described that a wave trough was visible first. During the 26th December 2004 tsunami the trough arrived at the coast of Sumatra first; but contrary to that, on the Lankan/Indian coastline the crest arrived first. In a simulation (Figure 2.12) it can become visible how the wave crest (indicated in red) might have propagated towards India, whereas the wave trough (indicated in blue) moved towards the Indonesian coastline. It depends on the plate tectonic how a tsunami is generated and whether a wave crest or wave trough will arrive first at the coast. Off Sumatra usually first a trough arrives and it takes about 20-30 minutes

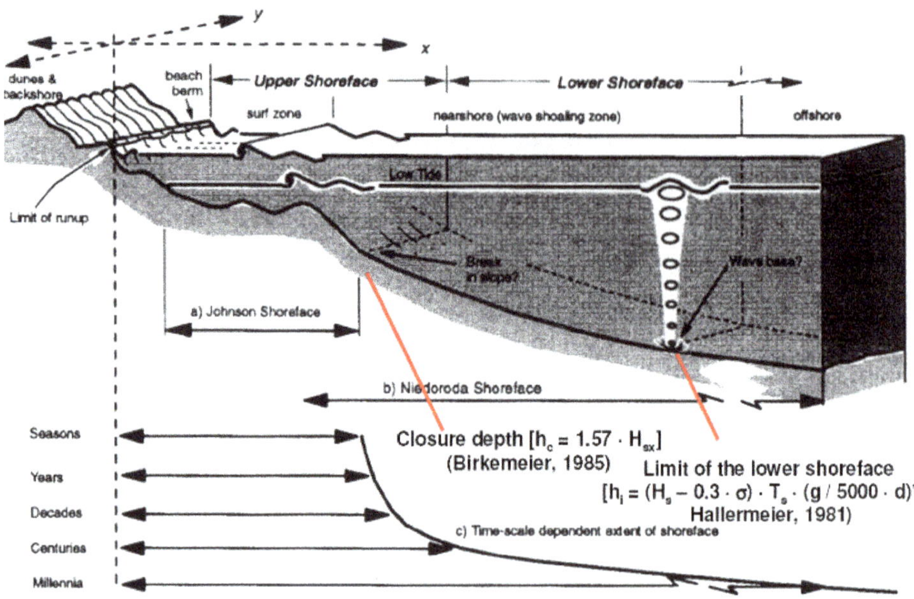

Figure 2.9: The Shore Face Definition. The shore face as defined (a) explicitly by Johnson (1919). (b) loosely by Niederoda and Swift (1991) and (c) by Cowell *et al.* (1999) by extending Johnson's definition in relation to the natural geodynamic continuum. The curve at the bottom schematically depicts the relationship between time scale and scale of morphological change. (from Cowell *et al.*, 1999). *Source*: Schwarzer (2006).

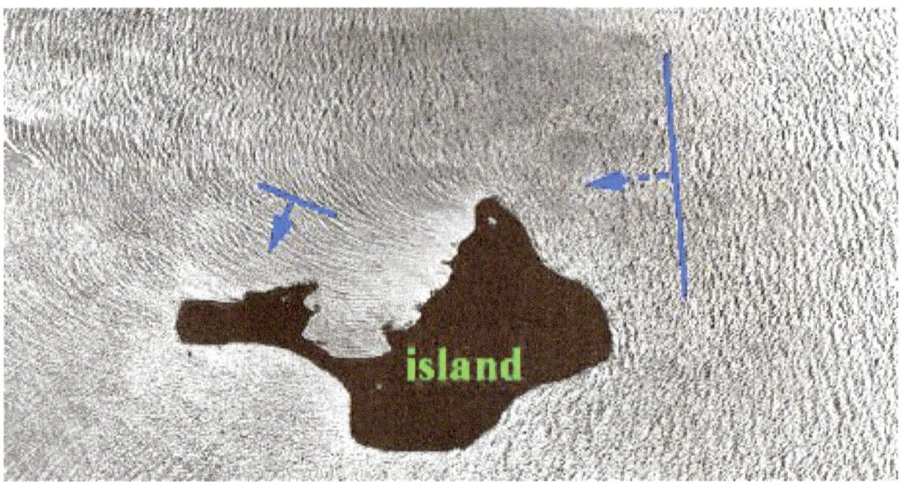

Figure 2.10a: Wave Refraction in Shallow Water.
Change in wave direction in shallow water. *Source*: Schwarzer (2006)

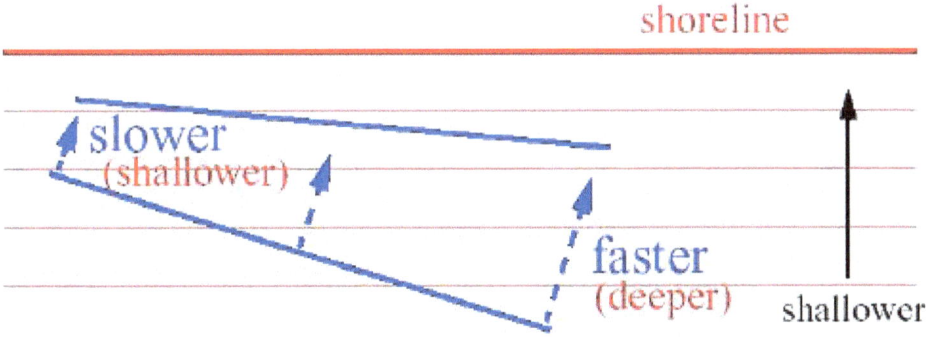

Figure 2.10b: Change in Velocity in Shallow Water

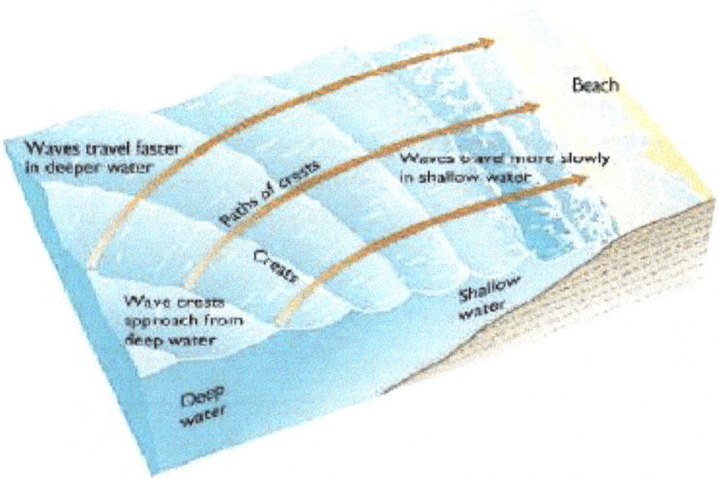

Figure 2.11a: Wave Refraction at the Shoreline.
Wave refraction in bays. *Source*: Schwarzer (2006)

until the wave crest reaches. This gives people about 10 to 20 minutes to run to higher grounds after noticing a wave trough.

How Many Tsunami Waves Are Common?

It is a common belief that the first tsunami wave is the worst and strongest. This belief can be fatal. There can be several waves, and according to records in Hawaii there were up to 12 of them. During the 26th December 2004 Tsunami wave no. 4 was the most destructive, and in one of the tsunamis striking Hawaii it was wave no. 8.

Size of Tsunamis

The size of a tsunami depends on the size of the trigger (most commonly: earthquakes), the morphology (bathymetry) and the substrate and vegetation (coral

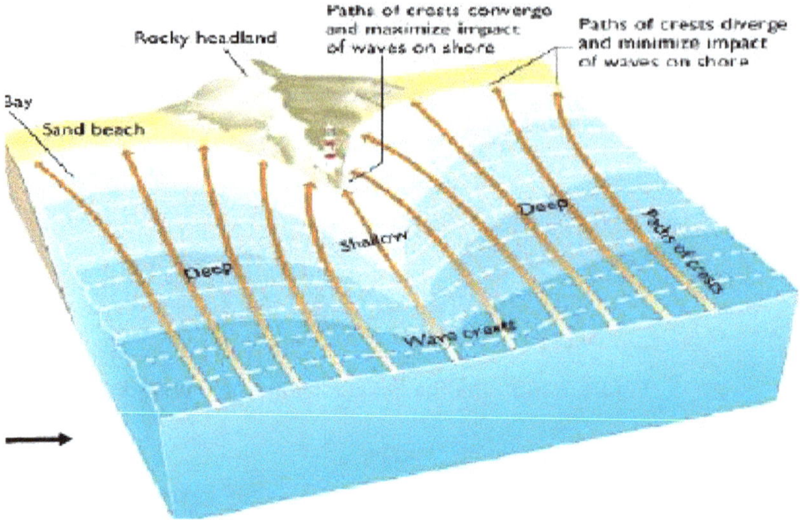

Figure 2.11b: Wave Refraction at Headlands

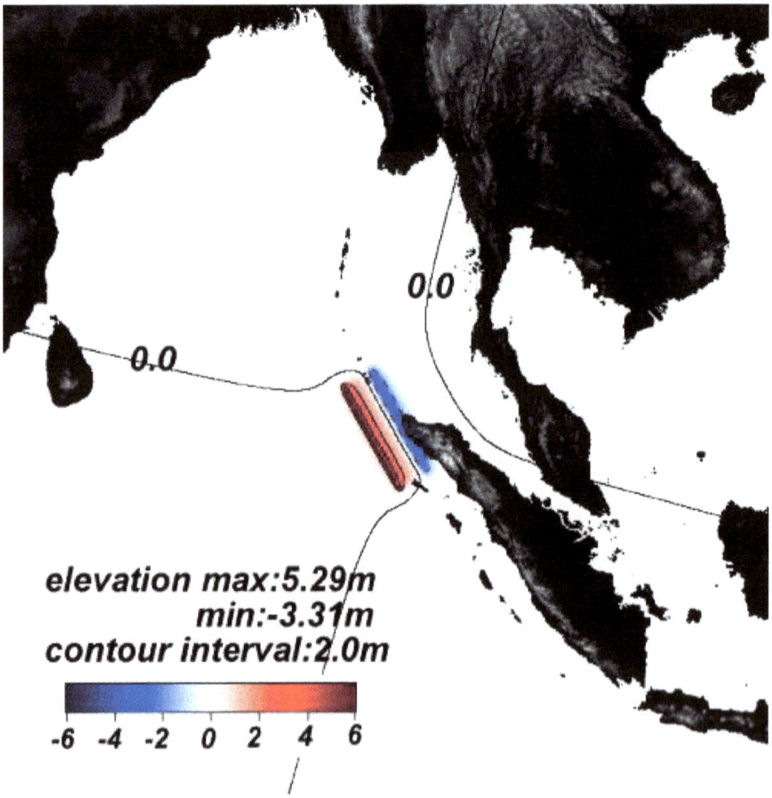

Figure 2.12: Model of Wave Crest (Red) and Trough (Blue) of the Indian Ocean Tsunami. *Source*: Research Centre for Disaster Reduction Systems (2005)

reefs/mangroves present?) of the shoreline. On the sea it is very hard to "see" a tsunami wave, unless a tsunami early warning system detects it. Commonly the wave crest is only about 0.5 to 1 m high in the open ocean and does not differ much to the other wind-generated waves around it. Approaching the land, water walls ("bore") can emerge which can be up to 30–40 m high. Tsunamis can be very destructive due to water walls/flooding, but also due to the back-flow of the flood pulling back into the sea, which might be loaded with items from the coastline (cars, trees, etc.).

Where do Tsunamis Occur?

Potentially threatened coastlines worldwide are displayed in Figure 2.13. According to records, 80 per cent of all tsunamis strike in the Pacific Ocean, and have affected countries along the western coastline of the American continent, South-East Asia and the Pacific island groups. Due to the high frequency of earthquakes, the Indian Ocean is also at high risk. There are regions of the Mediterranean Sea and New Zealand, both with regions of high volcanic activity, who also have a higher tsunami risk. Predictions were made for "Cumbre Vieja", an active volcano of the Spanish Canary Islands, from which 500 km^3 of rock might slide into the sea at any time in the future. This is indeed a potentially very dangerous tsunami generator in the Atlantic Ocean, and waves could even reach distant areas such as the USA!

The "Tsunamis hazard map of the world" shows the potential tsunami risk for different regions in the world (Figure 2.14). Indonesia has the greatest potential risk

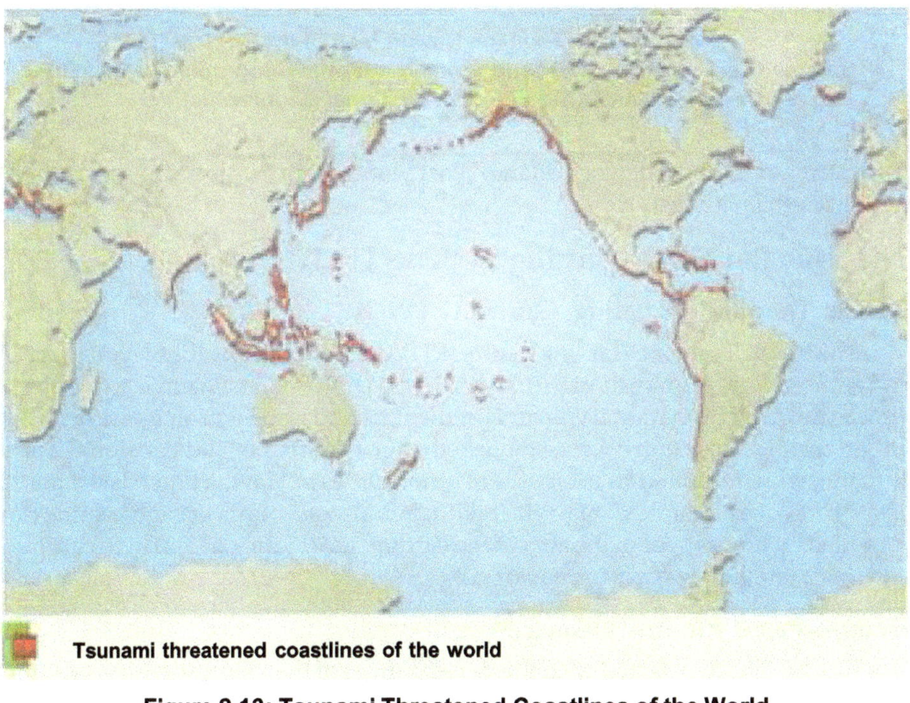

Figure 2.13: Tsunami Threatened Coastlines of the World
***Source*: Niedek and Frater (2004:20).**

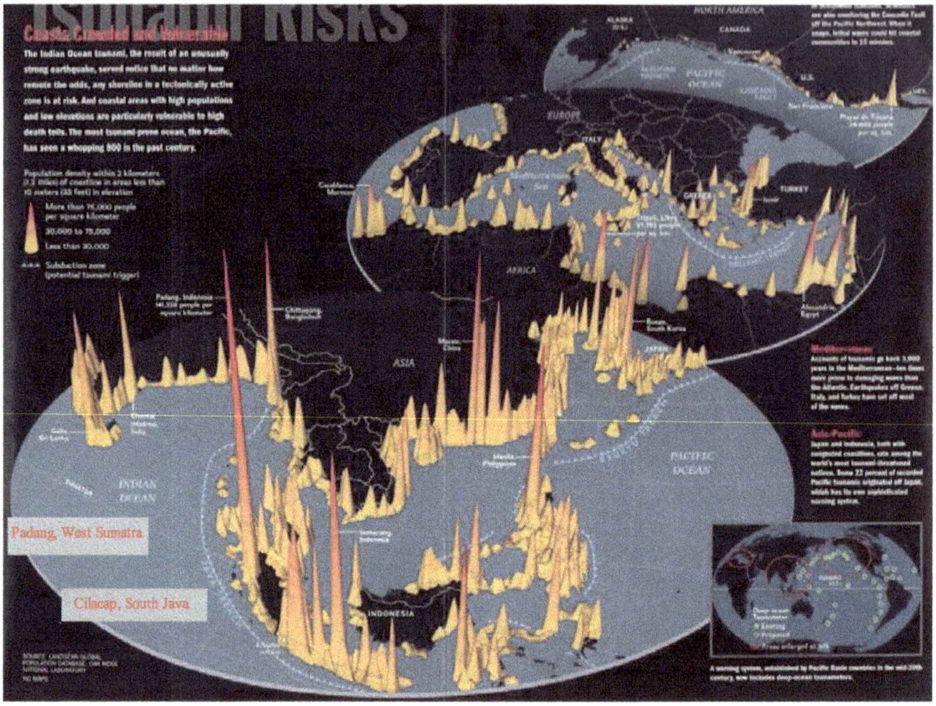

Figure 2.14: Tsunami Hazard Map of the World
***Source*: Oak Ridge National Laboratory in National Geographic (April 2005)**
[Indonesian version of the National Geographic]

from all regions, in particular Padang, West Sumatra. But vulnerability is also high for the South Java coast.

The Tsunami Early Warning Systems (TEWS)

Pacific Tsunami Warning Centre (PTWC)

The Pacific Tsunami Warning Centre (PTWC) is operated by the National Oceanic and Atmospheric Administration (NOAA) in the USA. It is a tsunami warning system which can give international tsunami predictions and issue warnings for the Pacific Ocean region. This centre was established after Aleutian Island earthquake and a tsunami, which resulted in 165 losses of human lives in Hawaii and Alaska in 1946. The PTWC uses seismic data to start with, but also accounts oceanographic data, provided by tide gauges in the area of the earthquake (Figure 2.15). These can assure whether a tsunami wave is formed or not.

Deep-ocean Tsunami Detection

In 1995, NOAA started with the development of the "Deep-ocean Assessment and Reporting of Tsunamis" (DART) system. The DART system consists of seafloor bottom pressure recorder (BPR), and a surface buoy. Data from the BPR on the seafloor

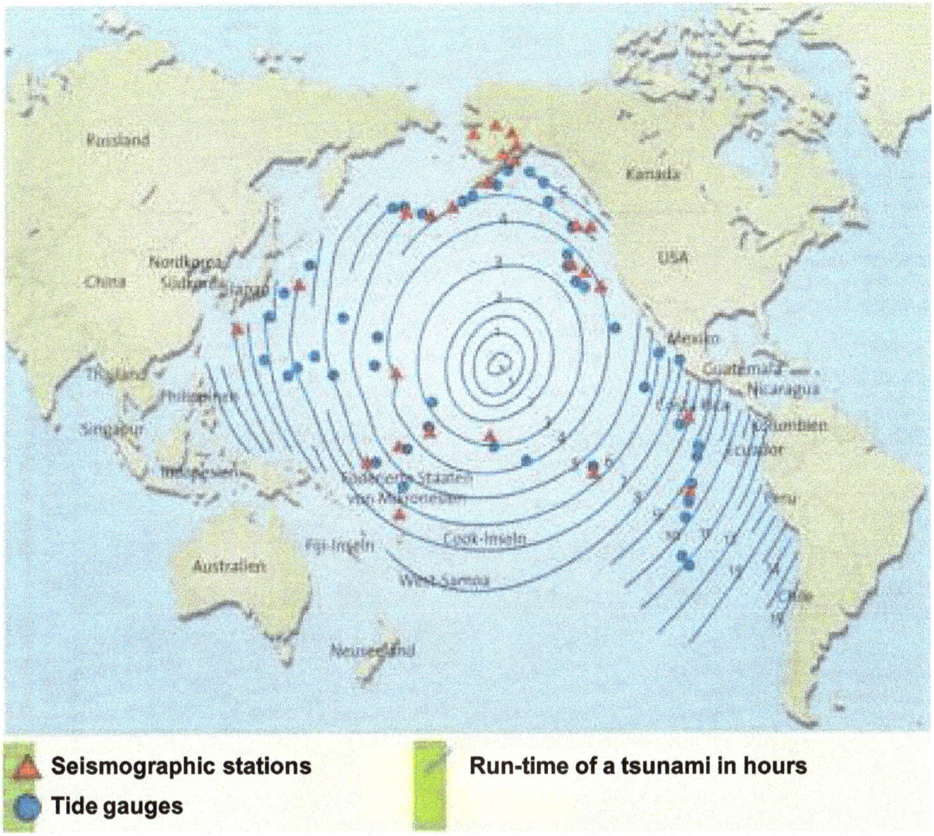

Figure 2.15: The Pacific Tsunami Warning Centre (PTWC)
***Source*: Niedek and Frater (2004:25)**

is transmitted to the surface buoy by an acoustic signal. Then the data is transmitted to ground stations via satellite, where the data then can be analyzed and possible threats detected (Figure 2.16). Up to 2001 only six stations had been deployed. The advantage of this system is that buoys can give information on the tsunami wave while it is still far out in the ocean, providing more time for evacuation and other measures. After the Indian Ocean Tsunami, plans were made to extend the number of DART buoys by 32 until 2007.

The German-Indian Ocean Tsunami Early Warning System (GITEWS)

The GITEWS (German-Indonesian Tsunami Early Warning System) project, funded by the BMBF (Bundesministerium für Bildung und Forschung), aims to establish an effective Tsunami Early Warning System for Indonesia and the Indian Ocean, and to enhance capacity building in Indonesia. The project is coordinated by the GFZ-Potsdam (Geo-Research Centre Potsdam), which facilitates studies on tsunami detection measures, including ground-based or space-borne radar sensors. In addition to that it conducts research by modeling, investigating for example

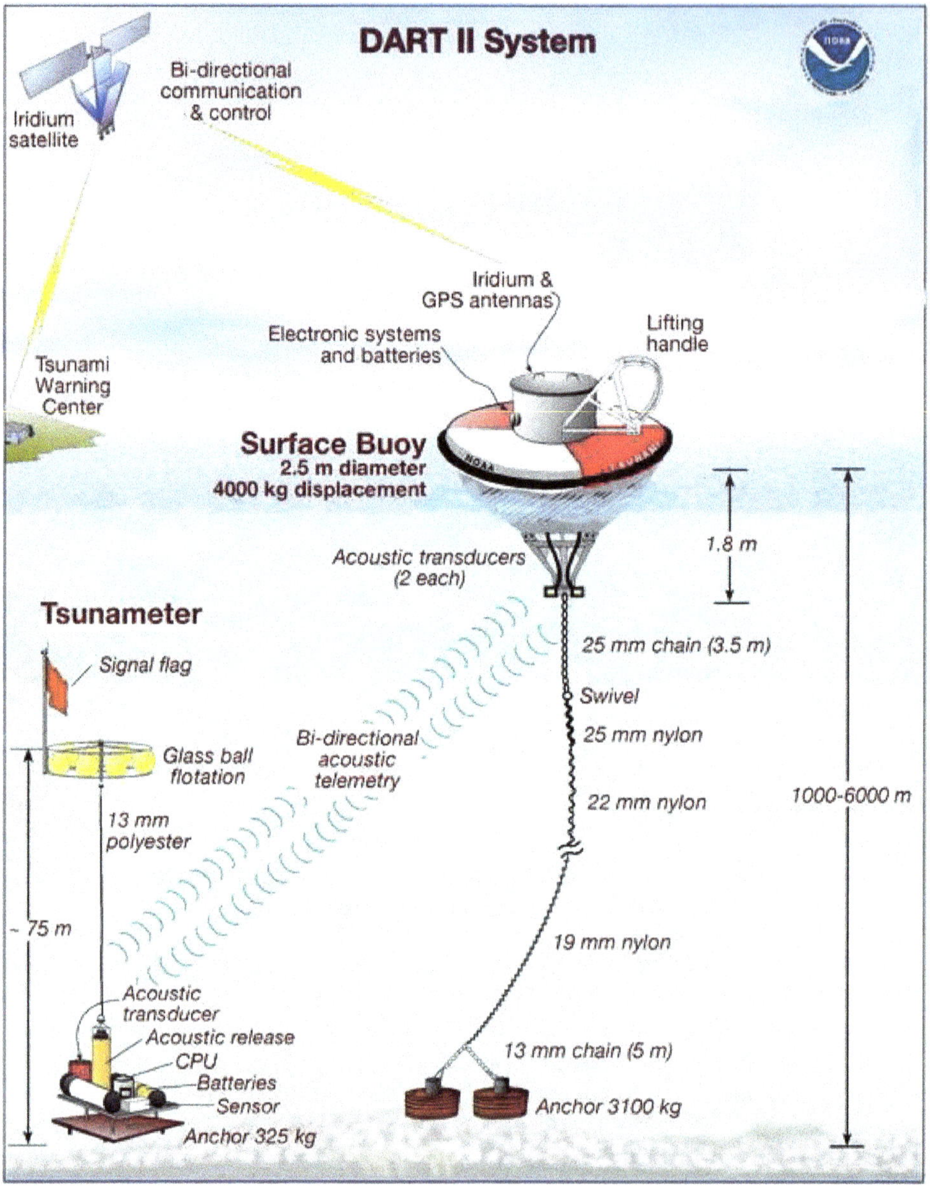

Figure 2.16: The Deep-ocean Assessment and Reporting of Tsunamis (DART) System. *Source*: NOAA (n.p.)

earthquakes as tsunami triggers, wave propagation in deep and shallow waters, and wave propagation on coast. Figure 2.17 shows the planned setup of the GITEWS.

The project faces tremendous technical and management challenges. Many places along Sumatra or Java only have about 5–15 minutes for early warning (while there

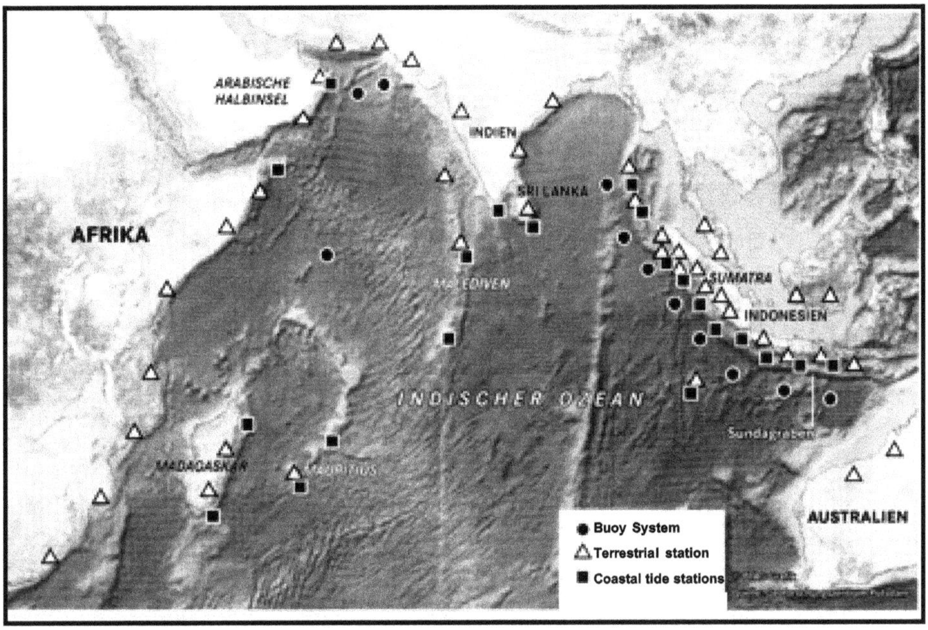

Figure 2.17: The German Indian Ocean Tsunami Early Warning System (GITEWS) as Planned by the GFZ, Showing the Location of the Planned Buoy Systems, Terrestrial and Coastal Tide Stations.
***Source*: Geoforschungszentrum Potsdam, Weser Kurier (2006).**

are about 2-3 hours for India and Sri Lanka and half a day for countries along the East African shores). The safety of the buoys cannot be taken for granted. There might be pirates and ordinary fishermen taking the equipment. The maintenance cost are also high (about 1 million Euro per year). Indian Ocean nations could not agree (yet) on establishing a central warning agency comparable to the Pacific Tsunami Warning System in Hawaii. However, national warning centres around the Indian Ocean shall be established and coordinated by UNESCO. In 2008 the German-Indonesian Tsunami Early Warning System shall be operational technically.

What to do if a Tsunami is Approaching

NASA has underlined various Do's and Don'ts "If you are in a dangerous area, immediately turn off the water, gas, and electricity and quickly move to higher ground. Remember, once a Tsunami Warning is issued, it could be a matter of minutes, or even seconds, before the wave hits.

If a Tsunami Warning is issued, NEVER go down to the beach to watch the waves come in; you won't live to tell the story. Remember also that a tsunami is a series of waves. Often the first wave may be the least dangerous. The waves may get progressively worse as time goes by. Listen to a portable radio to learn when it is safe to return home.

After a tsunami has hit, all food and water should be tested for contamination before it is eaten. All buildings should be checked for gas leaks and electrical shorts before anyone enters. Administer first aid only if you know what to do.

Stay tuned to local radio and TV stations, especially NOAA Weather Radio, for evacuation orders if a Tsunami Warning has been issued. Do NOT return to low-lying areas until the tsunami threat has passed and the "all clear" is announced."

Natural Buffers for Coastlines

Coral Reefs

Coral reefs are the first buffer in front of the coastline, and are said to provide a significant protection by breaking the force of a tsunami in- and under-water before the waves hit the shoreline. It is assumed that the buffer is functioning more effectively the healthier the reefs are; and pre-damaged reefs appear to provide much less shore protection. For instance, mined corals reefs in Sri Lanka showed that the tsunami found gaps through the reefs and hit the coast much more forceful; where there were no corridors the Tsunami only went up to 50 meters on the coast; in one instance with a corridor (from coral mining) the Tsunami went 1.5 km inland and hit a passenger train killing 1,700 people!

Mangroves

Mangrove forests give the best natural coastal protection due to the special anatomy of their roots (stilt, prop-roots and pneumatophores), which break the waves and can take out a high amount of the tsunami wave's energy. According to Mazda *et al.* (1997) a mangrove forest may reduce a tsunami wave by 20 cm every 100 m. Moreover, it may save people's lives. When the wave recedes it might prevent them from getting swept into the sea.

Coastal Terrestrial Vegetation

Trees, tree crops, and plantations (oil palms and coconuts) may also provide a significant buffer to some extent. Less effective are annual or seasonal crops, and rather ineffective is bare land or shrimp ponds.

Disaster Management in Indonesia

On average, 200-300 disasters occur in Indonesia each and every year, by which 300-400 people lose their lives, thousands of people get homeless, and millions of dollars are lost due to losses to the sea, and damage to properties. Indonesia is at high risk of natural disasters as it is located in an area where three tectonic plates border. With 190 volcanoes Indonesia is the country with the highest amount of volcanoes world-wide (Figure 2.18), and earthquakes and natural disasters happen quite frequently.

With assistance of the German-Indonesian GeoRisk project various types of disasters were systematically investigated and analyzed with respect to the causes, the vulnerability and interlinkages.

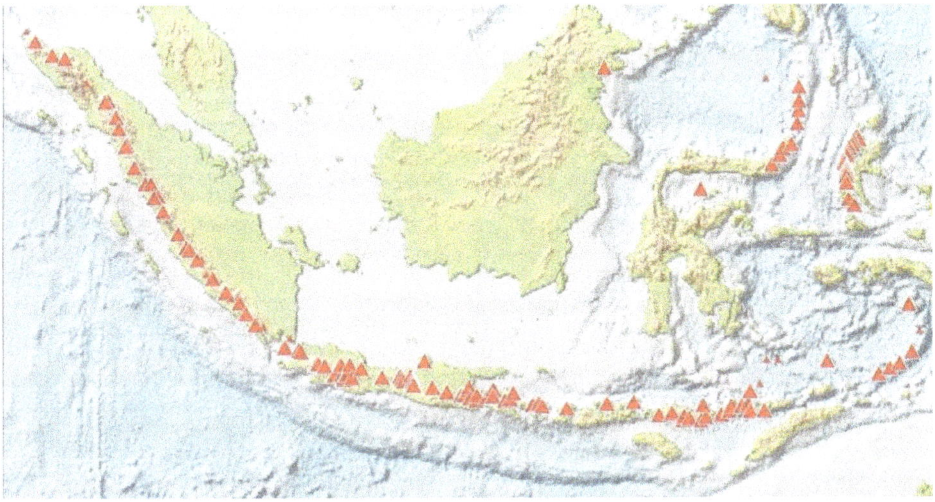

Figure 2.18: Active and Potentially Active Volcanoes in Indonesia
***Source*: Effendi *et al.* (2004:7)**

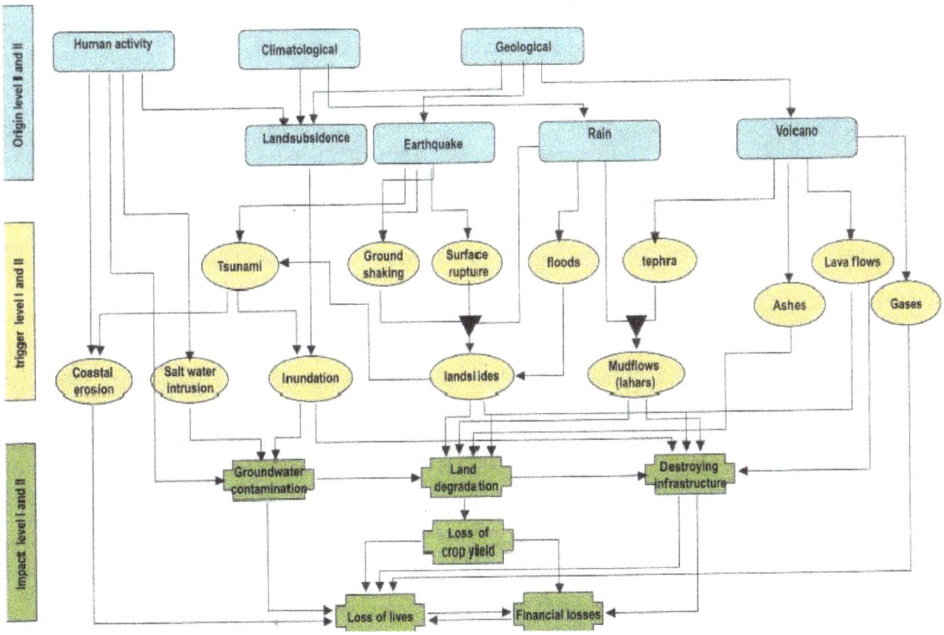

Figure 2.19: Conceptual View of the Geohazard System
***Source*: Effendi *et al.* (2004:9)**

Institutional Framework for Natural Disaster Management in Indonesia

Decentralization Laws and Handling of Disasters

With promulgation of the laws on decentralization (No. 22/1999 and 25/2001 on Regional Autonomy), disaster management became the task of local governments. The law intends an increasing public participation at all levels of political decision making. Disasters that cross the border of a district Kabupaten are then following under the jurisdiction of the province. If a natural disaster happens, several provinces central government will become responsible.

Disaster Risk Management on National Level

The "National Coordination Board for Natural Disaster Management" (BAKORNAS PBP) was created by decentralization law and Presidential Decrease 3 and 11 of 2001. It was first a non-structural body and since December 2005 a structural body (having their own staff and budget) to work out, coordinate, guide, monitor, and standardize the national approach for disaster management. It is also responsible for Internally Displaced Persons (IDP) and refugees, and guides and monitors actions directed against man-made hazards. It is placed under the Vice-President of Indonesia; and a secretariat of the board has been established under the Ministry of Social Welfare. The board integrates key ministries like Finance, Health, Home Affairs, Social Welfare, Public Works, the provincial governments and representatives of the military and the police.

Disaster Risk Management on Regional and Local Level

SATKORLAK PBP, the Provincial Unit for Coordination of Disaster Risk Management, is responsible on provincial level and is chaired by the provincial governor. On district or municipal level, SATLAK PBP District Unit for Implementation of Disaster Risk Management, chaired by the head of the district ("Bupati"), becomes responsible. SATGAS, the Task Force for Disaster Risk Management works on local, sub-district level.

Problems in Disaster Risk Management

A big problem that has become evident is that a conducive and coherent legal framework for natural disaster management in Indonesia is missing so far. Laws and presidential decrees stipulating functions of bodies and ministries are often contradicting and/or overlapping, and there is a heavily fragmented line of responsibilities which makes a coherent handling of emergencies difficult. Most local governments are neither technically, administratively, nor financially capable to comply with the tasks of disaster management, plus there is a general lack of specific disaster mitigation knowledge, skills and information. Up to today, a *reactive* rather than a *preventive* disaster approach is taken. Budgets in annual planning documents are usually missing for disaster management.

The Relationship Between Coastal Disaster Risk Management and Integrated Coastal Zone Management

It is usually said that coastal disaster risk management should be made part of integrated coastal zone management. However, in Indonesia (and possibly in many other countries, too) there is a divide between these two functions and there is also a divide on how the management of disasters and regional development should be approached.

With respect to the organizational divide: Integrated coastal zone management is part of the functions of the national or regional development planning agencies (in Indonesia: BAPPENAS and BAPPEDA), while disaster risk management is executed by the Vice President and the Minister for Social Welfare on national level and Governor/Bupati and an agency directly subordinated to him on provincial or district level, respectively. Thus for an effective disaster risk management these different organizations have to work closely together (and often they don't).

With respect to the management approach divide: Integrated coastal zone management should usually work in a participatory manner involving all stakeholders. This is usually a lengthy matter. However, once a disaster has struck, fast action and a command-like approach is needed. Thus these two functions (integrated coastal zone management and disaster risk management) appear to need be tackled in fundamentally different ways!

Though these two functions appear difficult to be reconciled, this is not so drastic when one attempts to differentiate between the different phases of good disaster risk management, as can be seen in the following compilation:

Table 2.1: Different Management Styles by Disaster Phases

Sl.No.	*Phases of a Disaster*	*Management Styles/Mechanisms*
1.	Before a disaster: disaster preparedness	Usually enough time available to analyze problems and to come to agreement between all stakeholders, government and non-government agencies. A participatory management style can be well employed.
2.	At a disaster: organizing relief measures	Usually very little time is available to save human lives and prevent further damages to infrastructure. A fast and commando-like style is necessary.
3.	After a disaster: Rehabilitation and reconstruction	Here usually there is again time to involve stakeholders properly and to coordinate between the various government and non-government agencies. Here a participatory management style can be employed again.

People's Attitude

Attempts were made to find out more about the attitude towards natural disasters of people living in areas that are at risk of a tsunami. So far there is only scattered anecdotal evidence.

Makassar, South Sulawesi, Provincial Government

While organizationally regional development planning and management appeared strong (BAPPEDA), disaster risk management appeared rather weak. In Makassar, at the South Sulawesi Provincial Government no strategy, vision or plan was in place for disaster risk management, while there were such strategies and plans prepared for regional development planning.

Makassar, Barrang Lompo Island (near Makassar, South Sulawesi)

Little or no preparedness for disasters could be noticed at Barrang Lompo Island. The general attitude of people is that "disasters cannot be predicted and prevented" and/or that "it is God's punishment". General problems on livelihood are in the fore ground, such as having enough income and food, to be healthy and to be educated.

Maumere, Flores

In Maumere, Flores the overall attitude seems to be that disasters are up to God and good faith can give protection. Many people still do not know much about natural disasters because of their relatively low level of education. Even if people know of natural disasters, many continue life like before (near the coast), because of their tradition and customs. To some fishermen adjustments would mean more transport cost if they had to move to higher grounds. Moreover, some do not want to leave because there are also their places of worship and burial.

Padang, Sumatra

At Padang there is high awareness and interest of the society and local government in natural disasters is evident. People in this area are psychologically traumatized by catastrophes, with more than 200 earthquakes in 2005. Development of risk maps, evacuation routes, and evacuation routines to some extent are already prepared.

Conclusions

Indonesia is one of the most disaster prone areas in the world. The current disaster risk technology and management is far from being able to respond adequately to disasters, and disaster risk management had played a rather small role in the minds of many people and government/s on national as well as regional and local levels. This, however, has changed since the 26th December 2004 Tsunami. A Tsunami Early Warning System shall now be established with international assistance. An early warning system will, however, remain a big problem on the west coast of Sumatra or the south coast of Java since there is a very short warning time (less than 15 minutes) (in contrast to Hawaii, where it usually takes several hours until the island is hit by a tsunami). Apart from tremendous technical challenges, a lot of work is also needed to link to and organize coastal people. It is recommended that preparation for disasters at the coast should be part of future management plans. However, management structures in Indonesia follow very different organizational lines (disaster risk management is in a different body than regional development planning), which complicates implementation. There is a scope for streamlining, In conclusion: a lot of work still remains to be done!

References

Bernard, P. (2005). Tsunamis im Mittelmeer? *Spektrum der Wissenschaft*, 4/2005.

Bigelow Laboratory for Ocean Sciences (2006). *Tsunami experiment- crosscurricular ideas.* http://www.bigelow.org/virtual/handson/tsunami.html#glossary.

Dudley, Walter C.; Min Lee. (1998). *Tsunami! 2nd ed.* University of Hawaii Press.

Effendi, A.; Nasution, A.; Djarwoto, A.; Murdohardono, D.; Kertapati, E.; Hermawan; Hidajat, R.; Sutawidjaja, I.S.; Jäger, St.; Manhart, A.; Ranke, U.; Rehmann, T.; Dalimin, R.; Sugalang; Weiland, L. (2004). *Mitigation of Geohazards in Indonesia, Status Report on the Project "Civil-society and Inter-municipal Cooperation for Better Urban Services/Mitigation of Geohazards.* GeoRisk Project, Bandung.

Hwang, D.J. (2005) *Hawaii Coastal Hazard Mitigation Handbook.* School of Ocean and Earth Science and Technology, University of Hawaii, Hawaii.

ISORI-ICSARD, BPPT (2005) *Tsunami Effect on Coastal Zone of Nanggroe Aceh Darussalam and North Sumatra Provinces (Draft).*

Krumme, Uwe. (2005) *Die südasiatische Tsunami-Katastrophe–Ökologische Aspekte des Desasters und der Folgenbewältigung [The Southeastasian Tsunami Catastrophe–Ecological Aspects of a Disaster and Coping with Its Impact], Presentation on 8th June 2005 at the Bundesamt für Seeschifffahrt, Hamburg.* Centre for Tropical Marine Ecology (ZMT), Bremen.

Mazda, Yoshihiro, Magi, Michimasa, Kogo, Motohiko and Phan Nguyen Hong (1997) *Mangroves as a coastal protection from waves in the Tong King delta,* Vietnam, Mangroves and Salt Marhes 1: 127-136.

NASA Observatorium (1999) *What to do when they hit.* Accessed on 1.3.2006. http://observe.arc.nasa.gov/nasa/exhibits/tsunami/tsun_hit2.html

Niedek, I.; Frater, H. (Eds) (2004) *Naturkatastrophen,* Springer Verlag, Berlin.

NOAA: *DART 2 System.* Accessed on 20.4.2006. http://nctr.pmel.noaa.gov/Dart/Jpg/DART_II_metric-page.jpg, n.p.

NOAA/PMEL: *Tsunami terminology.* Accessed on 15.3.2006. http://www.pmel.noaa.gov/tsunami-hazard/terms.html

Oak Ridge National Laboratory (2005) *Tsunami Hazard Map of the World.* National Geographic: 4/2005.

Patlis, J.M. *The Role of Law and Legal Institutions in Determining the Sustainability of Integrated Coastal Management Projects in Indonesia.*

Patlis, J.M.; Knight, M.; Benoit, J. (2003) *Integrated Coastal Management in Decentralizing Developing Countries: The General Paradigm, the U.S. Model, and the Indonesian Example.* Ocean Yearbook 17: 380-416.

Pinet, P. (1998) *Invitation to Oceanography.* Web Enhanced edition. Jones and Bartlett.

Research Centre for Disaster Reduction Systems: *December 26, 2004 Earthquake Tsunami Disaster of Indian Ocean.* Accessed on 21. March 2006. http://www. drs.dpri.kyoto-u.ac.jp/sumatradrs, 2005.

Roberts, E. (2005) *Prepare for the Worst: It's Going to Happen,* Marine Scientist No. 10 (1): 20-23.

Schwarzer, K. (2006) *ISATEC Lecture 1: Introduction.*

SOEST: *How is a Tsunami Generated? 1. By an Earthquake.* Accessed on 21st March 2006. http://www.soest.hawaii.edu/tsunami/tsugen.html, 2006

Tomascik, T-; Mah, A.J.; Nontji, A.; Kasim Moosa, M. (1997) *Tsunamis.* In Coral Reefs: Natural Disturbances, The Ecology of the Indonesian Seas, Vol. VII, Part I, Oxford University Press, Chapter 11.

Weserkurier. Nr. 59, 2006

Wood, E. (2005) *Tsunami und Korallenriffe,* Koralle 6(2), Heft 32, April/May, 72-76.

Chapter 3

Weather Forecasting: Atmospheric Modelling

Rattan K. Datta
National Centre for Medium Range Weather Forecasting (NCMRWF),
Ministry of Earth Sciences, Government of India
E-mail: rkdatta_in@yahoo.com

Introduction

Throughout the known history of mankind, man has been forced to adapt himself to the vagaries of nature. It was mankind's fear of the forces of nature that forced man to worship the sun, moon, earth, water, fire and even plants and animals. Out of all the natural phenomena, probably weather and its vagaries impact the human being the most. Individual has always dreamed of being able to understand how the climate system works and from that knowledge develop ways and means to predict perhaps even modify the weather. Intelligentsia from time immemorial have noticed the signs that portend the major weather events and verbally passed on the weather events to successive generation creating an invaluable inventory of weather folklore. Indian literature has wealth of such information. Manusmriti written as earlier as in 7th century A.D. has a number of meteorological references. As a matter of fact, the motto of the India Meteorological Department (IMD) "Aditya Jayate Vrishti" which means– from the sun rain is born is taken from Manusmriti. Brihat Samhita also contains significant knowledge of astronomy and astrology, which was prevalent in ancient India. Kalidas's Meghadoot contains the references which tally with the present structures of Indian monsoon. The panchang makers in India for ages have been preparing the weather forecast based on various positioning of the planetary systems.

Inspite of all the developments in ancient India, the meteorological science, as we know today, is of comparatively recent origin. Much of the progress in Meteorology is related to progress in the fields of Physics and Mathematics, one can list three inventions/discoveries of 16th and 17th century, one each of 18th and 19th century and three of the present century, which have helped the growth of meteorological to its present level. These are: (i) The invention of thermometer by Galileo in 1593, (ii) The invention of barometer by Galileo's student Torricelli in 1643, (iii) Enunciation of gas law by Robert Boyle in 1661 and principle of latent heat later, (iv) The discovery by Benjamin Franklin in 1743 that the weather systems propagate from one place to the other, (v) Invention of telegraph in 1845 by Marconi, (vi) Invention of rawin and radiosonde to observe vertical profile of wind temperature humidity and pressure, (vii) Invention of digital computer in 1940s and (viii) Meteorological satellite in 1961.

Though soon after its invention, barometer became a useful tool because of close relationship between weather and atmospheric pressure, but the first synoptic map of surface pressure is supposed to have been prepared only in 1820. India Meteorological Department (IMD) was established in 1875.

The discovery by Benjamin Franklin and the invention of telegraph brought to the light that weather over any area is influenced not only by conditions over adjacent areas but also those in far distant areas. Worldwide cooperation thus became indispensable in this discipline. Recognising this vital aspect of weather and climate, a nongovernmental body- International Meteorological Organisations (IMO)- was established in 1873 to foster collaboration amongst meteorological services of the world for the rapid and free exchange of meteorological information in the internationally accepted codes. With the passage of time, importance of meteorology grew manifold and International Meteorological Organisations was transformed into an intergovernmental official body named as World Meteorological Organisation (WMO) on 23rd March 1950 as a specialised agency of the UN system.

If one looks at a typical weather chart (Figure 3.1) what one observes is the meandering airflow in wave form, with vortices, troughs and ridges. These meandering eddies propagate retaining identity from one day to the other. It is the regularity of this propagation, which forms the essence of art and science of weather forecasting. The development of science of meteorology from empirical to mathematical modelling can be broadly considered in three epochs:

1. The period 1891-1910, classified as an empirical era. During the period there was a rapid establishment of meteorological services in various countries who developed empirical relations to provide forecasts for warning against Tropical Cyclones, Floods, for agriculture operation, navigation and public in general. The nations found the forecasts useful and observations of great significance to monitor drought etc. this helped in expansions of observational network and development of standard coding/archival procedures. Statistics found interesting applications. As a matter of fact early period statistics is full of examples related to the problems of weather forecasting.
2. The period 1911-1950, may be classified as an era of development of theoretical meteorology. The invention of radiosonde helped in monitoring and understanding the 3-dimensional structure of the atmosphere. The meteorologists started getting deeper understanding into the physics of

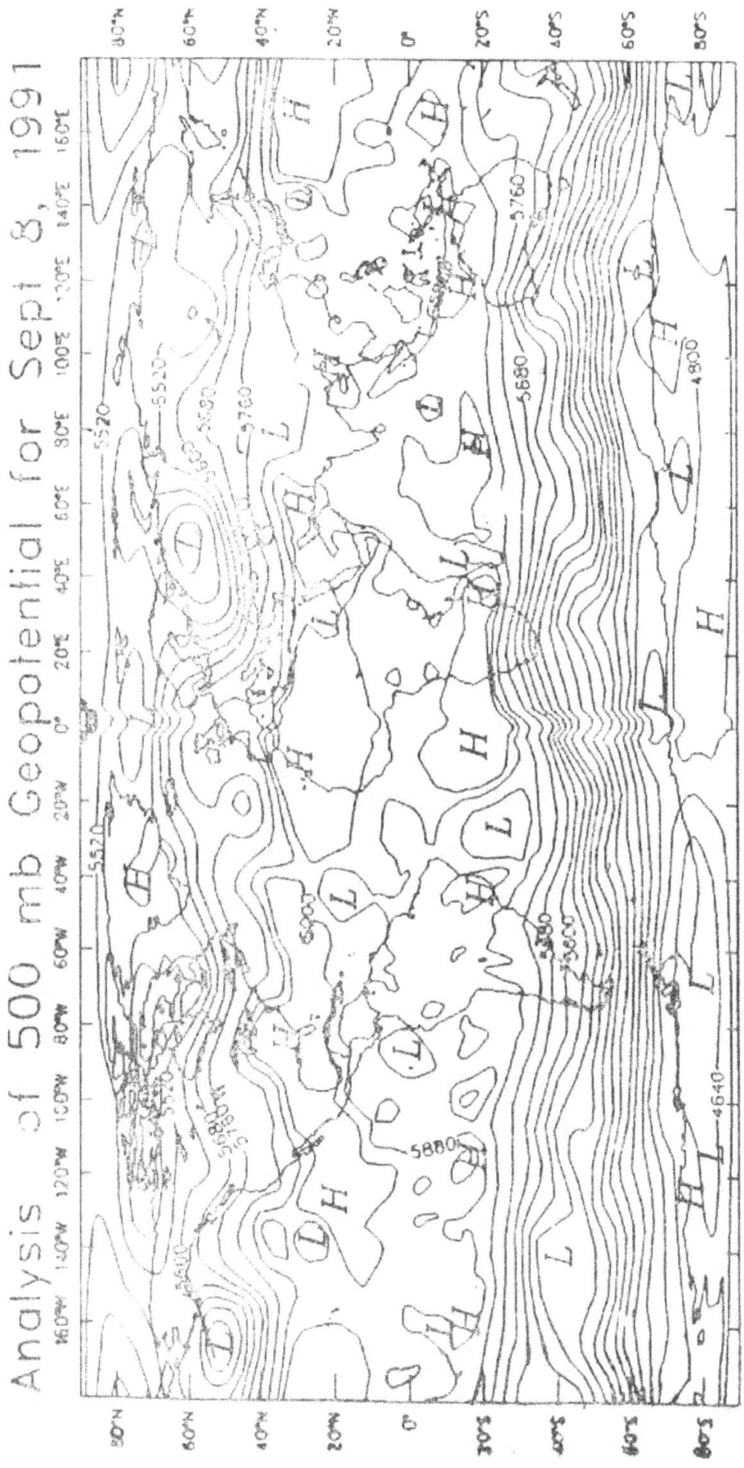

Figure 3.1

atmosphere circulation and learnt to develop mathematical models to represent atmospheric circulation features. Though Helmholtz, as early as in 1858, used the hydrodynamical equation for the meteorological studies and V. Bjerknes in 1904 conceptualised that the weather forecasting is an initial value problem in mathematical physics but the problem was how to solve the non-linear partial differential equations which are not amenable to the analytical solutions? It was during 1911, that a British Meteorologist Lewis Fry Richardson developed solution of partial differential equations through finite different methods of approximation. Though he was not successful in working out accurate weather forecasts by this method but his study laid foundation of the modern technique, well known as 'Numerical Weather Prediction' (NWP). And today, we more or less know the reasons of his failure.

While the mathematical modelling was taking concrete shape, another school of meteorologists was exploiting the use of statistics in modelling the atmosphere. The prominent among them is the work by Gilbert Walker during 1916-1924, when he was Director General of Observatory in India and by Prof. Mahalonabis, meteorologist in IMD.

Thus this period was full of excitement in use of mathematics and statistics in modelling the atmosphere for its better understanding and prediction.

3. The period 1951 onward may be classified as an era of active cooperation amongst meteorologists of various nations under the aegis of World Meteorological Organisation (WMO) and meteorologists with other disciplines of science. All national meteorological services of the world started working on the same platform of World Meteorological Organisation to improve the operational efficiency of weather services. The field became so important because of its social and economical impact that there was remarkable input from the other scientific disciplines, especially in numerical methods, observational tools both surface and space based. Rapid development in computing and communication changed the complete structure of meteorological services. From manual, all processes changed to modelling and automation.

With the above broad introduction in the following section, we briefly present the methods of weather forecasting, limit of predictability and present status in weather forecasting, especially over tropical monsoon area.

Weather Forecasting

The weather forecasting with respect to time are generally divided into three categories:

Short Range Weather Forecasts

These are valid for the next 72 hours or less.

Medium Range Weather Forecasts

Validity for these forecasts is from 3 to 10 days.

Long Range Weather Forecasts

These forecasts are valid beyond 10 days.

With the recent development and the requirements, additional time scales have been added into the system. These are: 'Now-casting' which refer to the forecasts valid for next couple of hours, 'Extended Range Forecasts' referring to the forecasts for the period between 10 to 30 days. The earlier defined Long Range Forecasts now refer to monthly or seasonal forecasts. The attempts are also being made for annual/ decadal or climatic assessment.

It may further be stressed that the forecasts are of two types, namely 'instantaneous' forecasts, which provide the snapshot information for the period of validity. The 'Mean' forecasts on the other hand provide the information on variation of weather elements for the period and the area as a whole compared to the normal value (long time average). The long range and climatic assessment forecasts are in that category because these forecasts are given for the whole period and the country or globe as a whole with respect to the long term mean and do not provide any additional information in terms of the shorter time and space variations within that period. From this consideration, therefore, the mean forecasts are important for large scale and macro-level planning whereas the short and medium range forecasts are very useful for operational level planning.

Technique of Weather Forecasting

The various technique on weather forecasting used could be broadly classified into three categories, namely:

1. Conventional or synoptic method;
2. Statistical methods; and
3. Numerical weather prediction (NWP)

Conventional or Synoptic Methods of Forecasting

Synoptic methods consist in the detailed analysis of current weather report from sufficiently large area with the area of interest in the middle (a few thousand km on either side to capture the propagating system). The current weather patterns after analysis are mentally associated with past analogous situation and forecasts are then prepared based on the assumption that the current situation will evolve following the past analogous situation. Often the selection of the past analogous situations are based on experience and the memory of the human forecaster. With the advent of computer and technique of pattern recognition, it is now possible to carry out this process of fixing the analogous situations more objectively. The recent studies and general experiences show that such methods are good only for 'now-casting' and short range weather forecasting. The significant reason of this limitation is because of the inbuilt chaotic processes within the apparent large scale systematic patterns. It is interesting to indicate here that the 'Theory of Chaos' was also enunciated by meteorologists considering the non-linear processes involved in the atmospheric circulation and their indefiniteness.

Statistical Methods

The statistical methods consist in establishing suitable regression or in certain cases, more sophisticated statistical relationship between the predictant and the predictors.

Normally, the selection of the predictors is based on possible physical relationship between the predictor and predictant. Such techniques have been found to be very useful for the short range and long range forecasting. Although such techniques were also tested for their application in the medium range forecasting but so far success has been very limited.

The statistical techniques have, however, become popular with improved output predictions with NPW models for predicting specific elements. Such techniques are known as Perfect Prog Methods (PPM) and Model Output Statistical (MOS) techniques. As a matter of fact, PPM and MOS have become essential component of NWP system. In these techniques, the predictor selected are the variables got as output from NPW models pertains to situation like hail/no hail, fog, minimum temperature, etc. *i.e.* such parameters which do not form the normal output products from NWP models.

Numerical Weather Prediction (NWP) Technique

The NWP techniques as described earlier consider that the behaviour of the atmosphere can be represented and governed by physical laws which can be expressed by differential equations with the boundary conditions. These equations when expressed as a closed system, are known as atmospheric model. The basic equations which form the atmospheric models are:

1. Equation of horizontal motion; the two dimension Navier-stokes equation;
2. Continuity equation for dry air;
3. Equation representing conservation of moisture;
4. Equation representing the first law of the thermodynamics;
5. Equation for surface pressure tendency
6. Hydrostatistical equation
7. Equation of state

The full description of these equations is given in Appendix 1. The main meteorological parameters used are wind vector, pressure, temperature and humidity.

Since several dynamically important phenomena occur at scales much lower than the normal resolution of the existing models (known as sub-grade scale), we have to include their effect during the solution of the equations. The examples of such physical equations are: turbulent fluxes, cumulus convection, precipitation processes, etc. the formulations which statistically provide the feedback of these processes on the computed parameters are known as parameterizations and their proper formulation is very crucial for the accuracy of the forecast. The orography also play very important role in modifying the atmospheric circulation and associated weather. Incorporation of orography in the model structure is, therefore, very important and require sophisticated techniques.

A simple way of illustrating the philosophy of NWP techniques would be to express the set of equation in the form given below:

$$\frac{d\Phi}{dt} = AF(\Phi) + B,$$

where,

Φ is the variable to be predicted, $F(\Phi)$ is the sum total of non-linear terms based on the initial state of variable Φ, A being coefficient of non-linear interaction and B is the term related to the physical processes. If one refers the set of equations in appendix, one can visualize that all the predictive equations (*i.e.* equations with the term dt/d on L.H.S.) can be put in the above form, so that all the terms on the R.H.S. are functions of known values at time t_0 (initial time). The procedure used in NWP is that the R.H.S. is computed and then through finite difference new value of Φ is computed by the relation:

$$\Phi_0(NEW) = \Phi + \Delta t\frac{d\Phi}{dt} = \Phi(Old) + \Delta t \times [RHS]$$

if integration process is explicit scheme. It is possible to use implicit or semi-implicit scheme to increase the time step Δt. The time step cannot be used arbitrarily but has to consider the wave speed and the size of grid mesh used besides the type of integration scheme used, namely, explicit, implicit, or semi-implicit. If this is not done carefully, the solutions are likely to diverge because of computational instability.

Further taking Φ (*New*) as the initial value, the step is repeated, till the limit of forecasting range. Thus the new values of variables computed become the initial stage for the next step. The solution of the partial differential equations (terms on RHS, refer Appendix 1) can be achieved through (*i*) finite difference representation of the equations, or (*ii*) by expressing the variable in spectral form. The models, which use the first form of representations, use information interpolated at regular meshes in the horizontal domain and is known as grid point model and the second type are known as spectral model.

The domain of a numerical model over which the present state of atmosphere is needed, increases with the increase in the range of forecast period. This is depicted in Figure 3.2. It may be noted that if the prediction is required for period greater than three days, initial state is required for the entire globe. This can be easily shown theoretically by the consideration of propagation of errors.

It would be interesting at this stage to note the order of computing resources required to process all the global meteorological information, analyse and integrate through solving all the equations numerically. Although the atmospheric circulation is a continuum but to solve numerically, it is necessary to discretize the atmosphere in the horizontal and vertical planes. Whatever technique is used, namely, finite differences or spectral, all the meteorological variables are interpolated to discrete

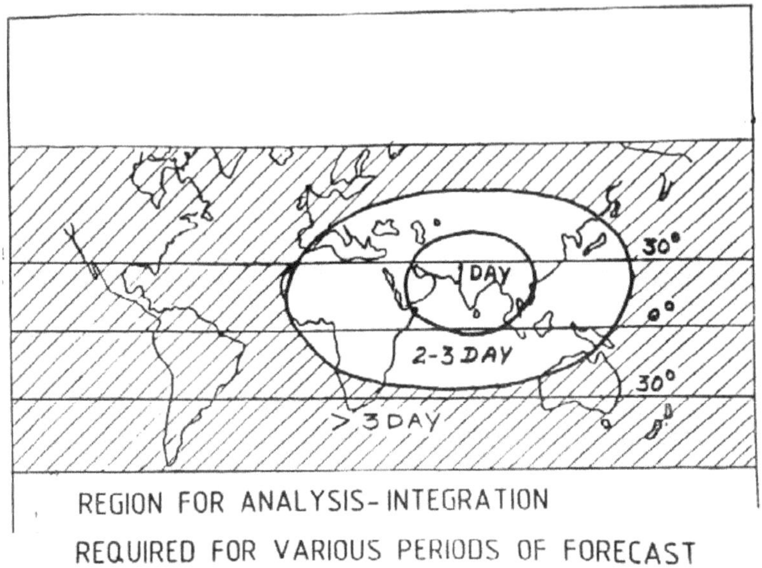

Figure 3.2

points as shown in Figure 3.3. For improved capability of forecasting lower truncation error in solution it is required that horizontal and vertical resolution is as best as the computing power can cope up. For example, for 100 km resolution in horizontal and 20 levels in the vertical would require solution for approx. 20 X 180 X 360 grid point at each time step Δt. all the variables need to be stored in the central memory in case of semi-implicit time integration can be of the order of 12 minutes. It can be easily shown that to produce a 5-10 days forecast by this process would require over 10^{13} operations by computer. We would naturally need to complete the work in few hours, if we want to forewarn the farmers and public in general of the expected weather, this requires a very fast computer which can work with a speed of over Trillion instructions per second. Now a days one talks the same in terms of Tera Floating Point Operation Per Second (TFLOPS).

Predictability and Present State-of-the Art

Predictability indicates the period upto which the forecast variables are well co-related to the realized field and the forecast products have adequate skill for use the various parameters which set the limit of predictability are:

1. Hydrodynamic instability inherent in a non-linear set of governing equations and growth of errors starting from small uncertainties at the initial state;
2. Lack of understanding of physical processes and their inclusion in the atmospheric models; and
3. Lack of precise information of initial state because of data gaps.

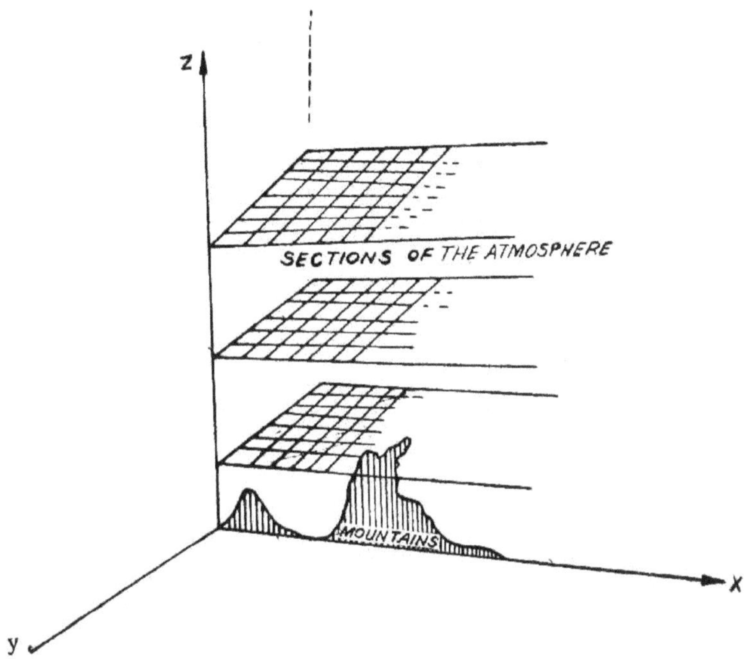

Figure 3.3

The upper limit of predictability may also be considered to have reached when the forecast field is no different than a randomly selected situation in the month of season. From theoretical consideration it is possible to arrive at an upper limit of predictability, which is 14 days in the middle latitude and less than 10 days in the tropical areas for the instantaneous type of weather forecasts. The practical limit however always falls short of the theoretical limit because of the existing data gaps, the resolution of the models and truncation errors because of the finite accuracy of computation. The practical limit today is about 6 to 8 days for the middle latitude areas and is 3 to 5 days in the tropical belts. Naturally, all the efforts of future research are to push the practical limits of predictability as close to the theoretical predictivity as possible.

It may also be interesting to see why weather in tropical areas is less predictable than that in the middle latitude areas in terms of the instantaneous forecast.

As stated earlier, the instantaneous weather forecasts provide snapshot of weather forecast and these are dependent upon the scale of meteorological system, which dominate in running the circulation system. In the tropical areas, the predominant circulation features are controlled by the cumulus scale system, which has short life and therefore short time of predictability.

Furthermore, there are other uncertainties related to the mass wind relationship in the tropical area compared to the middle latitude. The middle latitude systems are governed by large scale synoptic scale circulation and therefore are better predictable. That is why upper limit of predictability is higher in the middle latitude areas than for the tropics. When we consider the practical predictability, again the tropical areas are the regions which are least monitored because of oceans and lack of resources with the countries who cover the tropical zones.

In the case of 'Mean weather forecast' such as long range weather forecast, the tropical regions have better chance of being predictable because of the seasonal bias of the tropical circulation system compared to the transient type of circulation features in the middle latitude.

This aspect has been made use of by the Indian scientists in planning their research strategies.

Forecasting Technique as Evolved in India

It will be useful to discuss this under 3 different categories *i.e.* short range weather forecasting, medium range forecasting and long range forecasting

Short-range Forecasting

Right from the beginning the scientists at the Indian meteorological department have been developing objective techniques to predict the weather based on available meteorological information. The tools of observations have improved and so also the mode of information. The meteorologists have adapted themselves well to the use of modern aid of observation and analysis for providing the short-range weather forecast. The Indian meteorologists today use the synoptic methods, guidance from the NWP products meshed with human skill to issue the short-range weather forecast. There has been considerable improvement during the last over 3 decades so that today the aviators, the water resource managers, the agriculturist and the public in general all look for information with high degree of creditability. The efforts are to make these forecasts more specific in terms of time and space.

Medium Range Forecasting

For efficient management of agricultural operation as well as water reservoirs, more than short range it is really the medium range weather forecast, which are very crucial. Right from early 1960s the Indian meteorologists have been at it with limited success primarily because of lack of adequate tools. One of the significant lessons learnt from monsoon experiment (well known as MONEX-79, as a part of the global experiment) was that medium range weather forecasting is possible through numerical weather prediction techniques. With the fast advancement in the computer technology and space based observational system in terms of Indian National Satellite (INSAT) medium range weather forecasting came between the realm of reality. Noting the advantage of these forecasts for agriculture, and motivated by the advancement in the relevant technology and associated mathematical techniques, Government of India started a project well known as National Centre For Medium Range Weather Forecasting And Development Of Agrometeorological Advisory Services in 1988. The main objective of the project was: development and operationalisation of medium range weather forecasting techniques and integrate these weather forecasts with the state and stage of soil and crop issue weather based agroadvisors to farmers in collaboration with IMD-ICAR and state agricultural universities. The medium range weather forecasting problem itself is a challenging task but more so for Indian area because of uniqueness of our monsoon, special topography and other features of sub-continent. Our hope and endeavour are that our scientific community will adequately meet the challenge.

Long Range Weather Forecasting

The long range weather forecast for the monsoon have been prepared in India in 1891. The technique mainly consists in relating monsoon rainfall with different antecedent meteorological parameters over the monsoon region and the globe. The first parameter found as early as 1891 to be significantly correlated was winter/ spring snow over Himalayas. Since then large number of parameters have been identified which are physically connected with the monsoon circulation. All these years a few parameters selected out of many have been used as leading parameters to issue the long range forecast.

The monsoon circulation is a regional scale climatological features of the atmosphere and is a very important part of the global generation circulation. It is well known that performance of monsoon varies from year to year. The basic cause of these variations has therefore to be linked with meteorological anomalies at various places in the globe. The monsoon circulation is known to be caused by the differential heating of the landmass and ocean. Elevated landmass like the Himalayas and the Tibetan plateau, and the latent heat released by the convective clouds over the South East Asia also play important role in generating monsoon circulation. In 1988 a technique was developed to combine all these known global and regional parameters.

After elaborate screening, 16 parameters were selected. The relationship of monsoon rainfall with individual predictors exhibit a non-linear relationship. To take care of this non-linearity, a curvilinear relationship has been determined by fitting the equation of different degrees. The best fit degree of relationship with all the parameters has been combined and a power regression model developed. The forecast from this power regression model gave encouraging result both during the sample period (1958-1980) on which the model is developed, as well as during the independent test period (1981-88).

This model, well known as Gowariker model gave very useful seasonal forecast for the perfomance of the monsoon in quantitative terms for the years 1988 to 2000. Now IMD has built up a modified version. The parameters need upgradation, as they tend to loose the prediction power. The models thus need continuous upgradation.

Concluding Remarks

Meteorological community of India believe in self-reliance. It is more so because Indian sub-continent has unique features, like two monsoons, two seasons of tropical cyclones, the steepest orography of the Himalayas and typicality of land and sea distribution and thus sharing of experience with meteorologists of developed countries is limited. The development of science of meteorology in tropics, especially Indian monsoon is an opportunity and a challenge. We the mathematicians, physicists, engineers, agriculturists, in short scientists of all disciplines, have to work together to meet this challenge to produce the best tropical forecasts.

APPENDIX 1
MODEL EQUATIONS

$$\frac{\partial u}{\partial t}+m^2\left[\frac{\partial}{\partial x}\left(u\frac{u}{m}\right)+\frac{\partial}{\partial y}\left(u\frac{v}{m}\right)\right]+\frac{\partial}{\partial p}(wu)-fv+mg\frac{\partial z}{\partial x}$$
$$=-g\left(\frac{\partial \tau}{\partial p}\right)_x+F_u \tag{1}$$

$$\frac{\partial v}{\partial t}+m^2\left[\frac{\partial}{\partial x}\left(v\frac{u}{m}\right)+\frac{\partial}{\partial y}\left(v\frac{v}{m}\right)\right]+\frac{\partial}{\partial p}(wv)+fu+mg\frac{\partial z}{\partial y}$$
$$=-g\left(\frac{\partial \vec{\tau}}{\partial p}\right)+F_V \tag{2}$$

$$\frac{\partial z}{\partial t}=-\frac{RT_v}{pg} \tag{3}$$

$$m^2\left[\frac{\partial}{\partial x}\left(\frac{u}{m}\right)+\frac{\partial}{\partial y}\left(\frac{v}{m}\right)\right]+\frac{\partial w}{\partial p}=0 \tag{4}$$

$$\frac{\partial p_s}{\partial t}+mu_s\frac{\partial p_s}{\partial x}+mv_s\frac{\partial p_s}{\partial y}-w_s=0 \tag{5}$$

$$\frac{\partial}{\partial t}(C_pT)+m^2\left[\frac{\partial}{\partial x}\left(C_PT\frac{u}{m}\right)+\frac{\partial}{\partial y}\left(C_PT\frac{v}{m}\right)\right]+$$
$$\frac{\partial}{\partial p}(C_pTw)+g\frac{\partial z}{\partial p}\omega=\frac{dQ}{dt}-g\frac{\partial H}{\partial p}+C_PF_T \tag{6}$$

$$\frac{\partial q}{\partial t}+m^2\left[\frac{\partial}{\partial x}\left(q\frac{u}{m}\right)+\frac{\partial}{\partial y}\left(q\frac{v}{m}\right)\right]+\frac{\partial}{\partial p}(qw)=M-g\frac{\partial w}{\partial p}F_q \tag{7}$$

1. *u, v, w* are the three components of wind vector.
2. *T* is temperature in degree Kelvin.
3. *g* is acceleration due to gravity.
4. *p* is the pressure
5. *f* is the corolis parameter (=2Ω sinΦ, Ω is angular momentum of rotation of earth and Φ, degree latitude).
6. *m* is the map scale factor.
7. C_p is the specific heat of gas/air under constant pressure.
8. *Z* is the geopotential height.
9. F_u, F_v, F_q etc are frictional effect.
10. τ refers to eddy stresses.
11. *Q* is diabeting heating.
12. *H*, sensible heat.

Chapter 4

Global Climate and Regional Environmental Change Affecting Coastal Ecosystems: I. Biogeochemistry During Anthropocene

Tim C. Jennerjahn
Centre for Tropical Marine Ecology, Fahrenheitstrasse 6, 28359 Bremen, Germany
E-mail: tim.jennerjahn@zmt-bremen.de

Introduction

Nowadays more and more scientists use the term 'Anthropocene' for the recent period in the earth's history. It is a very short period of time in the 4.6-billion-year history of our planet, but is characterized by drastic environmental and ecological changes caused by human societies and their activities. Throughout major part of its history elemental cycles on earth have been driven by basic natural control factors, the water and rock cycles and solar radiation (*e.g.* Degens, 1989; Berner and Berner, 1996). The development of life starting about 3.8 billion years ago added another factor affecting elemental cycles which culminated in the activities of humans during the 'Anthropocene'. Although there is debate about its beginning, the late 18th century is seen as the starting point by many (*e.g.* Crutzen and Stoermer, 2000). Coastal zones are of prime importance in this respect, because they are rich in natural resources and provide goods and services to humans for a long time. Consequently, they are among the most-densely populated regions of the world. This, in turn, makes them particularly vulnerable to environmental and ecological changes. Tropical coastal zones receive major part of the annual riverine inputs of freshwater and dissolved and particulate substances. They harbour some of the most productive and diverse ecosystems on earth like coral reefs, seagrasses and mangroves. It has also turned out that tropical regions play a more important role in global climate than previously thought. Taken together, it appears that tropical coastal regions are key for global climate, the global distribution of energy, matter and life and the use of coastal resources, hence in the response to natural and anthropogenic environmental changes.

The Indian Ocean Tsunami on December 26, 2004, was an extreme event reminding the global society of the vulnerability of coastal ecosystems and the many hazards they are facing. It is a complex interplay of three oftenly interlinked groups of factors threatening coastal ecosystems: climate change, human activities and extreme events (Figure 4.1). Although basically natural phenomena like climate change and the occurrence of extreme events can also be linked to the human activities to a large extent.

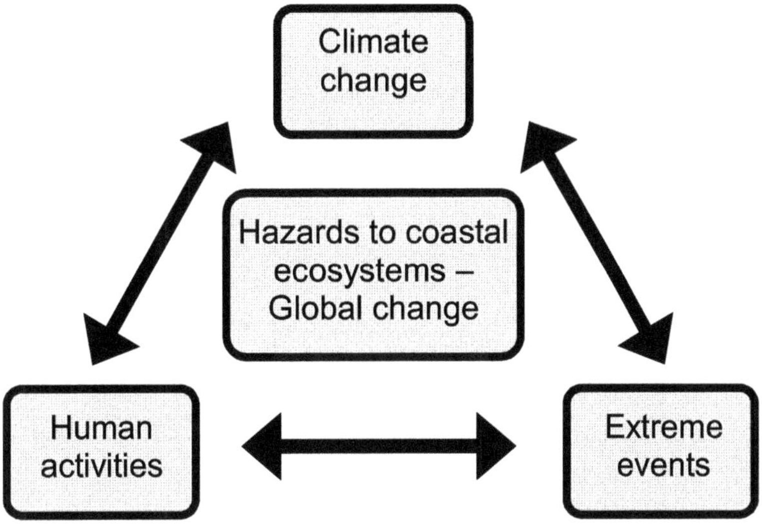

Figure 4.1: Three Major Groups of Interlinked Factors Threatening the Health and Ultimately the Economic Potential of Coastal Ecosystems

Hazards to Coastal Ecosystems

There are numerous hazards to coastal ecosystems summed up in three groups the most important of which are summarized in Table 4.1 and will be discussed in the Table 4.1.

Climate Change

In the geological past climate of our planet varied with a first glaciation occurring about 2.5 billion years ago. The Quaternary period, the past 2 million years, is subdivided into two major epochs, the Pleistocene lasting from 2 million years ago until about 12,000 years ago and from then on the Holocene. The Pleistocene was characterized by major climate oscillations culminating in the so-called Last Glacial Maximum about 20,000 years ago. Climate oscillations of the Quaternary were driven by earth' s orbital parameters, the natural factors which are determining the earth's movement around the sun. At the Last Glacial Maximum temperatures dropped by up to 15°C in polar regions and by 5-8°C in low latitudes. A large volume of water was bound in glaciers and consequently sea level was lower than today. Concurrent with the temperature drop the level of atmospheric carbon dioxide (CO_2) was lower by 80-100 ppm (*e.g.* Bradley, 1999). At the beginning of the industrialisation the CO_2 level was around 280 ppm. From then on it rose to the present-day 380 ppm simultaneously with the man-made carbon emissions. It is common sense that this 100 ppm-rise of atmospheric CO_2 is mainly due to human activities, *e.g.* the burning of fossil fuels. It is in the range of the natural variations of atmospheric CO_2, but added on top of the natural maximum CO_2 levels of 280-300 ppm. Some of the major consequences are changes in temperature, moisture distribution, seasonality, frequency of extreme weather events and sealevel (see Houghton *et al.,* 2001 and www.ipcc.ch).

Table 4.1: Hazards to Coastal Ecosystems According to the Three Groups Displayed in Figure 4.1

1. Climate change	Temperature changes
	Moisture distribution (precipitation)
	Seasonality
	Frequency of extreme weather events
	Sealevel changes
2. Human activities	Land use/cover change (deforestation, use of agrochemicals, conversion)
	Urbanisation (sewage, pollution, sealing)
	Overexploitation of natural resources (fisheries, logging, mining)
	Alteration of river catchment hydrology
	Tourism
	Coastal construction
	Transportation
3. Extreme events	Geohazards (earthquake, tsunami, volcano eruption, landslide)
	Climatic hazards (flood, drought, tropical storms and hurricanes)
	Industrial/other hazards (oil spill, nuclear accident, meteor impact)

Despite the overall trend of global warming there is large regional variability with even cooling in some areas. In general, warming is strongest in high latitudes of the northern hemisphere while some mid-latitude oceanic areas of the southern hemisphere are cooling (Houghton *et al.,* 2001). Consequently, the melting of polar ice caps and glaciers and the thermal expansion of water have led to an ever increasing sealevel rise. In the past decade it amounted to about 3 cm while it was on the order of 1.5-2.0 cm per decade in the past century (WBGU, 2006, see www.wbgu.de). Regional variation in temperature development in combination with vegetation change on land also affects the wind systems and evaporation and precipitation. This, in turn, affects the overall moisture distribution on earth. Despite considerable regional variation precipitation generally increased in mid to high latitudes of both hemispheres while it decreased in the low latitudes which is particularly evident in the expanding desert areas. Attendant with these changes observations indicate an increase in extreme weather events like storms, floods and droughts and changes in the frequency of natural climate phenomena like El Niño (Houghton *et al.*, 2001). These changes in combination with human activities considerably affect the biogeochemistry of coastal ecosystems. Some of the major consequences are, for example, changes in the amount and composition of river fluxes of dissolved nutrients and changes in the river fluxes of suspended sediments.

Ocean acidification due to increase of atmospheric carbon dioxide is another global threat. A doubling of atmospheric carbon dioxide from the preindustrial level of 280 ppm is expected for the middle of the 21st century (Houghton *et al.*, 2001). This has important implications for ocean carbonate chemistry. The increase of atmospheric carbon dioxide leads to an overall increase of inorganic carbon in the surface ocean. The lowered pH then, in turn, reduces calcification. A reduced biogenic carbonate precipitation in the tropics by 14-30 per cent is projected for the middle of the next century which will mainly affect coral reefs, but also other calcifying marine organisms (Kleypas *et al.*, 1999).

Human Activities

Rivers/Estuaries to Coastal Seas

Rivers are the life veins of continents. Through estuaries they are connecting land and ocean and nourish the continents as well as the coastal ocean and the people living in the respective areas. They are of particular ecological and economic importance in the tropics, because about three quarters of the annual global water, sediment and carbon inputs into the ocean are introduced in the tropics (Degens *et al.*, 1991; Syvitski *et al.*, 2005). Coral reefs, seagrass meadows and mangroves are fragile ecosystems of the tropics threatened by environmental changes. In the tropics three quarters of the population live in the coastal region, particularly in the vicinity of river mouths, and economically mainly depend on its resources. Natural and especially anthropogenic environmental changes of the past decades are leading to substantial changes of these fragile ecosystems and the biogeochemical cycles of the coastal ocean.

Because of regional climate, geology and human activities this is particularly important in the indoasian/western Pacific region. About one third of the annual water input and half of the annual sediment input into the ocean are introduced in this region (Syvitski *et al.*, 2005). Moreover, S and SE Asia are among the regions with the strongest human modifications of the coastal zone worldwide (Elvidge *et al.*, 1997) implying that human activities there have the most vigorous effects on the ecology and the elemental cycles of coastal ecosystems.

However, anthropogenically induced environmental changes can have contrasting effects in the tropical humid climate of Asia. Deforestation, extensive agriculture, fertilizer and pesticide use, waste discharge, canalization and destabilization of river beds lead to increased nutrient and sediment loads of rivers.

Damming of the rivers has an opposite effect. Two thirds of the >45,000 large dams operated worldwide are in Asia (The World Commission on Dams, 2000). Mainly used for power generation and irrigation these dams and the respective reservoir lakes also retain nutrients and sediments. Changed amounts and ratios of the essential nutrients Si, N and P entail changes of the plankton community and the biogeochemical cycles (*e.g.* Dudgeon, 2000; Ittekkot *et al.*, 2000; Rosenberg *et al.*, 2000). Toxic algal blooms and anoxic conditions in the water column of coastal seas are examples for the effects of these anthropogenically induced environmental changes. Prominent examples in this respect are the Mississippi–Gulf of Mexico and the Danube–NW Black Sea.

The Mississippi is one of the largest rivers in the world and drains major part of the United States. Intensive use of agrochemicals in the Mississippi catchment led to a drastic increase in the dissolved nitrogen load of the river in the past century. Attendant with an increase of nitrogen fertilizer use in the U.S. from about $3x10^6$ t yr^{-1} in the 1960s to $12x10^6$ t yr^{-1} in the 1990s the nitrate concentration of the river increased from about 40 µM to 120 µM (Goolsby *et al.*, 2000). During that time the dissolved silicon to nitrogen ratio dropped from 4:1 to 1:1. Despite an overall increase of primary productivity including siliceous organisms in the coastal waters of the Gulf of Mexico a shift in the phytoplankton community pattern was observed from heavily silicified diatoms to lightly silicified diatoms to non-biomineralizing phytoplankton. As a consequence of massive algal blooms the decomposition of the very high biomass led to extensive zones (up to 20,000 km^2) of midsummer hypoxia in the Gulf of Mexico (Rabalais *et al.*, 2000). This has further biogeochemical and ecological consequences. Hypoxia deteriorate living conditions for marine organisms and may ultimately lead to fish kills. Also, when oxygen is depleted organic matter decomposition requires other oxidants like, for example, nitrate. The process of denitrification, however, also leads to emission of the greenhouse gas nitrous oxide.

The Danube is Europe's second largest river which has major impoundments, the so-called "Iron Gate" dams, besides numerous other small ones. The "Iron Gate" dams are located about 1000 km upstream of the estuary and were constructed in the early 1970s. Sediments and the dissolved nutrients like nitrogen, phosphorus and silicon are retained in the large reservoir lakes. Downstream in the Romanian lowlands there is little input of silicon which almost exclusively comes from natural sources. However, large inputs of nitrogen and phosphorus from agriculture and urban and industrial wastewater disposal have led to a dramatic change in amount and composition of nutrients finally entering the NW Black Sea (Humborg *et al.*, 1997). Nitrate and phosphate concentrations in the NW Black Sea increased and silicate concentrations decreased from the 1970s to the 1990s leading to a drop in the Si/N ratio from about 45 to 3. Concurrently, phytoplankton more than trebled and its composition changed. While only diatoms and dinoflagellates were observed in the 1960s blooms of calcareous algae as well as a disproportionate increase of the non-biomineralizing dinoflagellates were observed in the NW Black Sea in the 1990s (Humborg *et al.*, 1997). Overexploitation of fish stocks which had increased as a result of increased primary productivity indirectly stimulated the explosive development of jellyfish. Because of a lack of predators the organic matter of dead organisms increased bacterial growth ultimately leading to oxygen-deficient conditions on the NW Black Sea shelf (Lancelot *et al.*, 2002).

More than half of the large river systems of the world are affected by dams or other hydrological alterations (Nilsson *et al.*, 2005). Human activities have also greatly affected the river fluxes of suspended sediments into the ocean. In a recent estimate the additional sediment input into the ocean derived from increased soil erosion due to deforestation and land use change in the 'Anthropocene' was calculated to 2.3 x 10^9 t while the amount of sediments retained in reservoir lakes behind dams amounts to 3.7 x 10^9 t. The net effect is a reduction of sediment fluxes by 1.4 x 10^9 t from 14.0 x 10^9 t in prehuman times to 12.6 x 10^9 t in the 'Anthropocene' (Syvitski *et al.*, 2005). On a global scale, riverine nitrogen and phosphorus inputs into the ocean have trebled due to human activities from the 1970s to the 1990s whereas the input of dissolved silicate almost exclusively stemming from natural sources was significantly reduced (Smith *et al.*, 2003). Changes of the fluvial carbon input into the oceans may have significant consequences for the climate-relevant marine carbon cycle. In contrast to previous knowledge recent investigations have shown that riverine organic carbon can be very old in large as well as in small rivers (Kao and Liu, 1996; Raymond and Bauer, 2001). Balancing of riverine carbon in the global carbon cycle is based on the young age of the organic matter assumed up to now, implying it to be mainly of atmospheric origin. However, "old" carbon results from the erosion of deep soil layers or basement rocks due to human activities in the catchment areas. This riverine carbon therefore does not stem from the atmospheric reservoir but from a carbon reservoir relevant on geological timescales. Rivers introducing such "old" carbon into the ocean hence are a carbon source relevant on man's timescales which are not considered in actual global carbon balances (Liu *et al.*, 2000). Generally it appears that element fluxes at the land–ocean boundary are globally more significant than previously thought, particularly with regard to the large amplitude changes caused by human activities during the 'Anthropocene'.

Mangroves

Mangroves, besides coral reefs and seagrasses belong to the unique ecosystems found in the coastal stretch of tropical regions. Located in the intertidal zone mangroves are the most landward of those three ecosystems. They have a lot of important ecological functions in providing shelter and nursery to many aquatic organisms and stabilizing coasts. With an average productivity of 2,500 mg C m^{-2} d^{-1} and a maximum of up to 7,000 mg C m^{-2} d^{-1} they are among the most productive ecosystems on earth (Bunt, 1992). Generally, the high nutrient demand to maintain this productivity is met by external supply from runoff, rivers and tides and by intensive internal recycling through microbial and benthic activity. There are six major mangrove settings of which the most important for the production and sedimentation of organic matter in the coastal zone are the river- and tide-dominated settings (Woodroffe, 1992). Generally, they receive high inputs of biogenic and abiogenic materials from land and from sea, they have a high turnover of organic matter and they are in permanent exchange with coastal waters. Besides important ecological functions they provide physical protection to the coast due to the intense network of roots and the high accumulation of sediment in areas where allochthonous input is high. This high accumulation rate and the mostly anoxic conditions of sediments make them also an efficient filter for nutrients and pollutants which are being immobilised in the sediments. In such a way they also provide protection to seagrass meadows and coral reefs (Valiela and Cole, 2002).

The ecological functions, however, are endangered by the substantial reduction of mangrove area in the past decades. A recent conservative estimate states a 35 per cent loss of mangrove forest area and 167,000 km^2 remaining (Valiela *et al.*, 2002). The Segara Anakan Lagoon in south central Java, Indonesia, is a typical example in this respect. It is the largest remaining single mangrove area on the densely-populated island of Java. Its natural resources and hence socio-economic goods and services are endangered by degradation and pollution through unsustainable land use practices in the hinterland, overexploitation of resources and pollution from industry and households. The mangrove forest area has been reduced by about 50 per cent in the past three decades, approximately 5,000 ha are left (Yuwono *et al.*, in press). Hydrodynamics of the lagoon are governed by tidal exchange with the Indian Ocean through two outlets in its western and eastern parts and freshwater input of the Citanduy River and a couple of other small rivers in its western part. Deforestation, land use change and fertilizer application in the hinterland of the lagoon have led to increased river inputs of dissolved nutrients and extremely high sediment inputs leading to rapid accretion of land in the lagoon. The shrinking lagoon and changing water quality affect habitats and living conditions of the organisms. For example, consumption of particulate organic matter derived from primary production and river inputs reduces dissolved oxygen oftenly to concentrations <5 mg l^{-1}, a threshold concentration below which many fish species get under stress (Yuwono *et al.*, in press).

On a global scale, the carbon exported from mangroves to the coastal seas may account for up to 11 per cent of the annual total of 429×10^{12} g C exported from rivers and mangroves. Despite this high export and high rates of recycling carbon accumulation in mangrove sediments makes up about 15 per cent of the 157×10^{12} g of organic carbon currently accumulating in modern marine sediments (Berner, 1982; Jennerjahn and Ittekkot, 2002). High biological activity inside the mangroves and the permanent recycling and exchange of nutrients between mangroves and coastal waters are the major factors allowing for high rates of carbon production and accumulation in as well as for export from mangroves. This nutrient seesaw is an important function of mangrove ecosystems in terms of the biogeochemical cycling of elements in the coastal zone, as such fertilizing both ecosystems. The substantial reduction of mangrove areas worldwide in the past decades, however, will probably have consequences for material exchange at the land–ocean boundary and it has yet to be estimated how these changes will affect elemental cycles in the tropical seas.

Coral Reefs

Coral reefs are also among the unique ecosystems of tropical coasts which are occupying about 1.4 million km^2. They are built of biogenic calcium carbonate deposits excreted by colonies of coral polyps and other reef organisms and they harbour the highest biodiversity of any marine ecosystem. Reef-building corals live in a mutualistic symbiosis with unicellular algae. The photosynthetic algae produce organic matter from inorganic nutrients. This organic matter is consumed by the coral. The excretion products, inorganic nutrients, are then being made available for photosynthesis again. The direct coupling of production and remineralisation processes in the coral-zooxanthellae symbiosis is a trophic shortcut that minimizes losses of carbon and nutrients from the system. This recycling process besides other efficient mechanisms to extract organic and inorganic nutrients from otherwise oligotrophic tropical waters make coral reefs one of the most productive marine ecosystems with an annual gross primary production of 1,500-5,000 g C m^{-2} yr^{-1}. They sequester Ca 7 x 10^8 t C yr^{-1} in form of calcium carbonate (*e.g.* Sorokin, 1993; Birkeland, 1997). This makes them an important actor in the global carbon cycle to some extent counteracting the accumulation of carbon dioxide as a greenhouse gas in the atmosphere.

In the past decades coral reefs became more and more threatened through global climate change and human activities impairing their important ecological functions. On a global scale, approximately 20 per cent of coral reefs are already destroyed, 24 per cent are under imminent risk of collapse through human activities and another 26 per cent are under a longer term risk (*e.g.* Bryant *et al.*, 1998; Wilkinson, 2004). Global climate change and human activities are the major causes of the coral reef decline. For example, coral bleaching caused by the extreme El Niño/La Niña event in 1997/98 destroyed about 16 per cent of the world's coral reefs (Wilkinson, 2004).

Increased terrestrial inputs of nutrients and sediments into coastal seas are the most important effects of human activities impairing coral reef ecology and biogeochemistry (*e.g.* Cole, 2003). Deforestation and the subsequent increase in erosion lead to increased river inputs of suspended sediments which upon reaching coral reefs may decrease light penetration and disturb feeding or in the worst case simply bury the reef. The nutrient input into coastal seas increased dramatically during the 'Anthropocene'. It is of particular importance in the tropics where aquaculture and agriculture, oftenly with intensive use of fertilizers, are major land uses and income sources. S and SE Asia is among the regions with maximum river inputs of dissolved nitrogen and phosphorus (Smith *et al.*, 2003) as well as the centre of coral biodiversity. Anthropogenic nutrient loading has been observed in coral reefs (*e.g.* Heikoop *et al.*, 2000). The increased nutrient load, for example, favours phytoplankton growth in the water thereby reducing light penetration for the coral and also promotes growth of other organisms competing with the corals like sponges, worms and molluscs. Overfishing may also result in changes of the biogeochemical cycling of elements in corals. For example, removal of grazing fish may favour growth of macroalgae and bomb and cyanide fishing may partly destroy the reef or poison its organisms and hence alter carbon and nutrient cycling (Wilkinson, 2004). It appears that an efficient nutrient recycling system of tropical coasts is impaired by global climate and regional environmental change which, in turn, has consequences for carbon and nutrient cycling in the tropical coastal ocean.

Coastal Connectivity

Coral reefs, seagrasses and mangroves can be found in close proximity of each other along tropical coasts. In many instances coral reefs provide physical protection by attenuating wave energy to mangroves and seagrasses which are only found in low energy coastal environments. Mangroves by trapping sediments and associated substances act as filters which protect seagrasses and coral reefs from smothering with sediments. However, investigations on the connectivity of these ecosystems with respect to exchange of energy and substances have been started only in recent years. For example, it has been shown that mangroves and saltmarshes may protect seagrasses from land-derived nitrogen (Valiela and Cole, 2002). Oftenly seagrass meadows which are sensitive to nutrient pollution like coral reefs are found in estuarine systems. These, in turn, are the major pathway of the anthropogenically enhanced nutrient load of rivers introduced into the coastal ocean. Coastal wetlands like mangroves and saltmarshes by their high denitrification and burial rates may help to immobilize part of this excess nitrogen and hence reduce the nitrogen input into coastal waters (Valiela and Cole, 2002). The primary production per unit area by seagrasses increases and the loss of seagrass habitat decreases the larger the portion of wetland area of a total estuary is. However, there appear to be threshold values of nitrogen load which can be immobilized in coastal wetlands. From their global set of data on estuaries with seagrasses compiled from the literature Valiela and Cole (2002) found a critical range of nitrogen loads impairing seagrasses of 20–100 kg N ha^{-1} yr^{-1}. Within this range wetlands may intercept between 0 and 100 per cent of the nitrogen load while in areas with loads >100 kg N ha^{-1} yr^{-1} the retention in mangroves and saltmarshes is <10 per cent. In their study Valiela and Cole (2002) found 67 per cent of the sites to receive land-derived inputs of <100 kg N ha^{-1} yr^{-1} suggesting that in two thirds of the investigated areas neighbouring seagrass meadows benefitted from the nitrogen immobilisation capacity of the wetlands.

An interconnection of mangroves and coral reefs that goes beyond simple physical protection has been reported from the Caribbean. Mumby *et al.* (2004) investigated the community structure and biomass of reef fish species in the Mesoamerican reef system of Belize and Mexico. They compared sites with little or no mangroves in their vicinity to sites with extensive mangroves. Reef fish community structure and biomass of reef fish differed significantly between mangrove-rich and mangrove-scarce systems. Despite variations in the magnitude and pattern of biomass enrichment among species a general enrichment of biomass was observed in reef fish from mangrove-rich systems over those from mangrove-scarce systems. In all systems juveniles migrate from seagrass beds to coral reefs, but it appears that in the case of the mangrove-rich systems the mangroves provide an intermediate nursery stage on the way to the adult habitat. It allows juveniles to feed and grow larger and to better withstand the predation pressure which increases at each stage from the juvenile to the adult habitat. As shown in this example from the Caribbean the interaction of mangroves with coral reefs apparently goes beyond the stage of simple physical protection. However, the pathways and fluxes of energy and substances remain to be investigated (Mumby *et al.*, 2004). In such a way it seems that the connectivity of tropical coastal ecosystems is an important factor for their productivity, biodiversity and resilience.

Extreme Events

The extreme events impairing the functions of coastal ecosystems span a range from exclusively naturally induced geohazards to the man-made hazards. Nuclear accidents and particularly oil spills are common extreme events of anthropogenic origin which have important consequences for elemental cycles and the ecology of coastal waters and ecosystems. For example, the radioactive fallout of nuclides released by one of the most catastrophic nuclear accidents ever in Chernobyl in 1986 was detected in the North Sea, the Mediterranean, the Black Sea and other aquatic and terrestrial ecosystems shortly after the accident (*e.g.* Buesseler *et al.*, 1987; Fowler *et al.*, 1987; Kempe and Nies, 1987). The radioactive fallout deposited in coastal waters was rapidly removed from the surface by biological activity. Particle-associated nuclides were incorporated into large and fast sinking aggregates produced by phyto- and zooplankton. In such a way these particles transport the radioactive nuclides to the seafloor where benthic organisms may feed on it. Particularly, the longer-lived nuclides like ^{137}Cs (30 years halflife) and ^{134}Cs (2.1 years halflife) can impose longterm threats to the benthic community. Changes in the benthic community, in turn, can alter element cycling at the sediment-water interface.

The Indian Ocean tsunami of December 2004 was a major geohazard of global concern. With its epicentre located near the northwestern end of the Indonesian island of Sumatra it mainly affected the coasts of S and SE Asia (Titov *et al.*, 2005). Many coastal ecosystems were affected or completely destroyed like, for example, the Banda Aceh region on the Sumtran coast near the epicentre. Observations indicate that coral reefs affected by high anthropogenic nutrient loads and bioerosion are more vulnerable to physical impacts like the tsunami wave of 2004. Recent model simulations now suggest that reef health is one of the factors contributing to the buffering capacity of coral reefs against the impact of tsunamis (Kunkel *et al.*, 2006). The protective role of coastal vegetation, namely mangrove forests, against the tsunami impact has been reported from the Indian coast (Danielsen *et al.*, 2005; Kathiresan and Rajendran, 2005). There villages located behind dense mangroves and tree plantations suffered much less from physical destruction than those located on the open shore. Despite the protective role of mangroves and coral reefs it is conceivable that there are thresholds of the buffering capacity of these coastal systems which remain to be explored. For example, nothing could withstand the physical impact of the tsunami in the Banda Aceh region on Sumatra so close to the epicentre. It is conceivable that element fluxes and transformations at the land-ocean boundary change dramatically entailing further consequences for organisms and foodwebs in the coastal zone when such an event occurs.

Summary and Conclusions

Coastal ecosystems are particularly vulnerable to natural and man-made environmental change because the majority of an everincreasing population lives in the coastal zone and to a large extent depends on its natural resources. The steadily increasing human uses of coastal resources endanger ecosystem health which, in turn, ultimately affects the economic potential of the coastal region. It is of particular importance in the tropics which appear to be key for global climate, the global distribution of energy, matter and life and the use of coastal resources, hence also in the response to natural and man-made environmental changes. The most important hazards to coastal ecosystems can be summed up in three major factors which are often intimately linked: climate change, human activities and extreme events.

Global warming, the rise in atmospheric carbon dioxide and sealevel rise alter circulation and chemistry of atmosphere and ocean, moisture distribution and also the physical settings of coastal ecosystems. Human activities in the coastal zone and also in its hinterland like land use/cover change, river damming, urbanisation, overexploitation of resources, tourism, transportation and coastal construction affect the hydrology of watersheds, weathering and erosion and the inputs of dissolved and particulate substances into rivers. While the former are processes that basically imply gradual changes singular extreme events like, for example, earthquakes and tsunamis, storms and floods or oil spills have immediate consequences for the biogeochemistry and ecology of affected regions. Some of the most important consequences with respect to the biogeochemical cycling of elements are reduced calcification in the coastal ocean and changes in the amount and composition of dissolved and particulate river fluxes. These are entailing further changes for the biogeochemistry and ecology of coastal living resources like eutrophication, red tides, hypoxia, release of greenhouse gases and changes in biodiversity and foodweb structure.

A lack of understanding of the complex interaction of the above-mentioned hazards and its consequences appears to be a major factor for the often dissatisfactory results of coastal management programmes. A sustainable use of coastal resources requires the development of management tools based on an integrated understanding of the dynamics of social-ecological systems which are beyond natural evolution in times of global change.

References

Berner, R.A., 1982. Burial of organic carbon and pyrite sulfur in the modern ocean: Its geochemical and environmental significance. American Journal of Science, 282: 451-473.

Berner, E.K., R.A. Berner, 1996. Global Environment: Water, Air, and Geochemical Cycles. Prentice Hall, Upper Saddle River, New Jersey, 376 p.

Birkeland, C., 1997. Life and death of coral reefs. Chapman and Hall, London, 536 p.

Bradley, R.S., 1999. Paleoclimatology–Reconstructing Climates of the Quaternary. Academic Press, San Diego, 2nd ed., 610 p.

Bryant, D., L. Burke, J. McManus, M. Spalding, 1998. Reefs at Risk. A map-based indicator of threats to the world's coral reefs. World Resources Institute, 56 p.

Buesseler, K.O., H.D. Livingston, S. Honjo, B.J. Hay, S.J. Manganini, E.T. Degens, V. Ittekkot, E. Izdar, T. Konuk, 1987. Chernobyl radionuclides in a Black Sea sediment trap. Nature 329: 825-828.

Bunt, J.S., 1992. How can fragile ecosystems best be conserved? In: Hsü, K.J., J. Thiede (eds.), Use and Misuse of the Seafloor. Wiley and Sons, Chichester, Dahlem Workshop Reports, Environmental Sciences Research Report 11: 229-242.

Cole, J., 2003. Dishing the dirt on coral reefs. Nature 421: 705-706.

Crutzen, P.J., E.F. Stoermer, 2000. The 'Anthropocene'. Global Change Newsletter 41: 17-18.

Danielsen, F., M.K. Sörensen, M.F. Olwig, V. Selvam, F. Parish, N.D. Burgess, T. Hiraishi, V.M. Karunagaran, M.S. Rasmussen, L.B. Hansen, A. Quarto, N. Suryadiputra, 2005. The Asian tsunami: a protective role for coastal vegetation. Science 310: 643.

Degens, E.T., 1989. Perspectives on Biogeochemistry. Springer, Berlin, 423 p.

Degens, E.T., S. Kempe, J.E. Richey, 1991. Summary: Biogeochemistry of major world rivers. In: Degens, E.T., S. Kempe, J.E. Richey (eds.), Biogeochemistry of Major World Rivers. Wiley and Sons, Chichester, SCOPE 42: 323-347.

Dudgeon, D., 2000. Large-scale hydrological changes in tropical Asia: prospects for riverine biodiversity. BioScience 50: 793-806.

Elvidge, C.D., K.E. Baugh, E.A. Kihn, H.W. Kroehl, E.R. Davis. 1997. Mapping city lights with nighttime data from the DMSP operational linescan system. Photogrammetric Engineering and Remote Sensing 63: 727-734.

Fowler, S.W., P. Buat-Menard, Y. Yokoyama, S. Ballestra, E. Holm, H. Van Nguyen, 1987. Rapid removal of Chernobyl fallout from Mediterranean surface waters by biological activity. Nature 329: 56-58.

Goolsby, D.A., W.A. Battaglin, B.T. Aulenbach, R.P. Hooper, 2000. Nitrogen flux and sources in the Mississippi River basin. Science of the Total Environment 248: 75-86.

Heikoop, J.M., M.J. Risk, A.V. Lazier, E.N. Edinger, J. Jompa, G.V. Limmon, J.J. Dunn, D.R. Browne, H.P. Schwarcz, 2000. Nitrogen-15 signals of anthropogenic nutrient loading in reef corals. Marine Pollution Bulletin 40: 628-636.

Houghton, J.T., Y. Ding, D.J. Griggs, M. Noguer, P.J. van der Linden, D. Xiaosu (eds.), 2001. Climate Change 2001. The Scientific Basis. Contribution of Working Group I to the Third Assessment Report of the Intergovernmental Panel on Climate Change (IPCC). Cambridge University Press, Cambridge, 944 p.

Humborg, C., V. Ittekkot, A. Cociasu, B. V. Bodungen, 1997. Effect of Danube River dam on Black Sea biogeochemistry and ecosystem structure. Nature 386: 385-388.

Ittekkot, V., C. Humborg, P. Schäfer, 2000. Hydrological alterations and marine biogeochemistry: a silicate issue? BioScience 50: 776-782.

Jennerjahn, T.C., V. Ittekkot, 2002. Relevance of mangroves for the production and deposition of organic matter along tropical continental margins. Naturwissenschaften 89: 23-30.

Kao, S.-J., K.-K. Liu, 1996. Particulate organic carbon export from a subtropical mountainous river (Lanyang Hsi) in Taiwan. Limnology and Oceanography 41: 1749-1757.

Kathiresan, K., N. Rajendran, 2005. Coastal mangrove forests mitigated tsunami. Estuarine, Coastal and Shelf Science 65: 601-606.

Kempe, S., H. Nies, 1987. Chernobyl nuclide record from a North Sea sediment trap. Nature 329: 828-831.

Kleypas, J.A., R.W. Buddemeier, D. Archer, J.-P. Gattuso, C. Langdon, B.N. Opdyke, 1999. Geochemical consequences of increased atmospheric carbon dioxide on coral reefs. Science 284: 118-120.

Kunkel, C.M., R.W. Hallberg, M. Oppenheimer, 2006. Coral reefs reduce tsunami impact in model simulations. Geophysical Research Letters 33, doi:10.1029/2006GL027892.

Lancelot, C., J.-M. Martin, N. Panin, Y. Zaitsev, 2002. The North-western Black Sea: a pilot site to understand the complex intercation between human activities and the coastal environment. Estuarine, Coastal and Shelf Science 54: 279-283.

Liu, K.-K., L. Atkinson, C.T.A. Chen, S. Gao, J. Hall, R.W. MacDonald, L. Talaue McManus, R. Quinones, 2000. Exploring continental margin carbon fluxes on a global scale. EOS Transactions 81: 641-642.

Mumby, P.J., A.J. Edwards, J.E. Arias-Gonzalez, K.C. Lindeman, P.G. Blackwell, A. Gall, M.I. Gorczynska, A.R. Harborne, C.L. Pescod, H. Renken, C.C. Wabnitz, G. Llewellyn, 2004. Mangroves enhance the biomass of coral reef fish communities in the Caribbean. Nature 427: 533-536.

Nilsson, C., C.A. Reidy, M. Dynesius, C. Revenga, 2005. Fragmentation and flow regulation of the world's large river systems. Science 308: 405-408.

Rabalais, N.N., R.E. Turner, D. Justic, Q. Dortch, W.J. Wiseman Jr., B.K. Sen Gupta, 2000. Gulf of Mexico biological system responses to nutrient changes in the Mississippi River. In: Hobbie, J. (ed.), Estuarine science: a synthetic approach to research and practice, Island Press, Washington, 241-268.

Raymond, P.A., J.E. Bauer, 2001. Riverine export of aged terrestrial organic matter to the North Atlantic Ocean. Nature 409: 497-500.

Rosenberg, D.M., P. McCully, C.M. Pringle, 2000. Global-scale environmental effects of hydrological alterations: introduction. BioScience 50: 746-751.

Smith, S.V., D.P. Swaney, L. Talaue-McManus, J.D. Bartley, P.T. Sandhei, C.J. McLaughlin, V.C. Dupra, C.J. Crossland, R.W. Buddemeier, B.A. Maxwell, F. Wulff, 2003. Humans, hydrology, and the distribution of inorganic nitrogen loading to the ocean. BioScience 53: 235-245.

Sorokin, Y.I., 1993. Coral reef ecology. Springer, Berlin, 463 p.

Syvitski, J.P.M., C.J. Vörösmarty, A.J. Kettner, P. Green, 2005. Impact of humans on the flux of terrestrial sediment to the global coastal ocean. Science 308: 376-380.

The World Commission on Dams, 2000. Dams and Development–A New Framework for Decision-Making. Earthscan Publications, London, 404 p.

Titov, V., A.B. Rabinovich, H.O. Mofjeld, R.E. Thomson, F.I. Gonzalez, 2005. The global reach of the 26 December 2004 Sumatra tsunami. Science 309: 2045-2048.

Valiela, I. and M.L. Cole, 2002. Comparative evidence that salt marshes and mangroves may protect seagrass meadows from land-derived nitrogen loads. Ecosystems 5: 92-102.

Valiela, I., J.L. Bowen and J.K. York, 2002. Mangrove forests: one of the world's threatened major tropical environments. BioScience 51: 807-815.

WBGU (German Advisory Council on Climate Change), 2006. The future oceans– warming up, rising high, turning sour. Special report, Berlin, 110 p.

Wilkinson, C. (ed.), 2004. Status of Coral Reefs of the World: 2004. Global Coral Reef Monitoring Network, Australian Institute of Marine Science, 301 p.

Woodroffe, C.D., 1992. Mangrove sediments and geomorphology. In: Robertson, A.I., D.M. Alongi (eds.), Tropical Mangrove Ecosystems. AGU, Washington, Coastal and Estuarine Studies 41: 7-41.

Yuwono, E., T.C. Jennerjahn, I. Nordhaus, E.A. Riyanto, M.H. Sastranegara, R. Pribadi, 2007. Ecological status of Segara Anakan, Java, Indonesia, a mangrove-fringed lagoon affected by human activities. Asian Journal of Water, Environment and Pollution, in press.

Chapter 5

Global Climate and Regional Environmental Change Affecting Coastal Ecosystems: II. Ecology and Resources of Mangrove Forests

Inga Nordhaus
Zentrum für Marine Tropenökologie, Fahrenheitstraße 6, Bremen, Germany
E-mail: inga.nordhaus@zmt-bremen.de

Introduction

Tropical coastal zones harbour some of the most diverse and productive ecosystems on earth. They have important ecological functions and values and are rich in natural resources which have been used by humans at all times. Mangroves, together with adjacent mud flats, seagrass beds and coral reefs are fragile tropical ecosystems and vulnerable to natural and man-made environmental changes. The anthropogenic pressure on these ecosystems has highly increased during the past decades due to the continual increase of the coastal population in tropical countries which depend on the natural resources of these ecosystems. Large parts of mangrove forests, seagrass meadows and coral reefs have already been degraded or completely destroyed due to human activities. Besides these, coastal ecosystems face the impacts of climate change and extreme natural events, including climatic hazards (*e.g.* storms, floods) and geohazards (*e.g.* tsunamis, earthquakes). These three groups of factors, human activities, climate change, and extreme events, have impaired and will impair the functions and the health of coastal ecosystems and are often intimately connected with each other.

It is now widely accepted that the increase of the atmospheric CO_2 concentration due to human activities contributes to the observed climate change, including an increase of air and water temperature, sea-level rise, changes in precipitation, an increase in extreme weather events and acidification of the oceans (McLeod and Salm, 2006; www.ipcc.ch). Mangrove ecosystems have persisted throughout the Quaternary in many places despite substantial fluctuations of sea level. Nowadays, however, sea-level rise is probably the greatest challenge that mangroves are facing (Field, 1995). While they could migrate more or less freely with changes in sea-level in the past, their landward movement is often restricted by humans and their activities today.

It is difficult to predict the effects of global climate change on the different ecosystems because of their high complexity. Manipulative experiments in the field would be helpful in investigating the ecological response to climate change. This, however, is almost impracticable because of the many interacting control factors and therefore experiments are mostly restricted to the laboratory. These experiments remain on the individual or species level and are often designed to test single variables. Results are thus not sufficient to understand and predict the complex interactions and long-term responses on the ecosystem level. Also, coastal ecosystems are connected to each other via the exchange of dissolved and particulate matter and via migrating organisms. In some places, mangrove systems depend on coral reefs which provide shelter from waves and on the other hand, mangroves prevent corals from being covered by sediment. Natural or man-made environmental changes in one ecosystem thus likely also influence adjacent ecosystems.

The goal of this chapter is to highlight the role of mangroves in this context. The important functions and values of this ecosystem are described and the various impacts of human activities, global climate change and natural hazards are compiled and discussed.

Distribution and Current Status of Mangroves

Mangroves are unique ecosystems in the land-ocean interface of the tropics and subtropics, extending on shallow coastal regions with an average water temperature of at least 20°C during the coldest month. These areas are subject to tidal inundation and are usually restricted to low energy coasts. The local extension of mangrove forests depends on the interaction of several environmental factors, in particular salinity, water temperature, tidal range, precipitation and coastal gradient. The limiting factor for mangroves in higher latitudes is sea surface and/or atmospheric temperature. Today, the worldwide mangrove area accounts for approximately 170.000 km^2 (Valiela *et al.*, 2001). The largest mangrove areas are located in Indonesia (with a quarter of the world total, approx. 42 500 km^2), Brazil (13 800 km^2), Nigeria (10 500 km^2), and Australia (10 000 km^2) (Robertson and Duke, 1990; Kjerfve and Lacerda, 1993; Spalding *et al.*, 1997; Saenger and Bellan 1995, cited in Valiela *et al.*, 2001).

The worldwide decline of mangrove forests is immense and was estimated to at least 35 per cent within the last 20 years (Valiela *et al.*, 2001). Average loss rates are higher for American (38 per cent) and Asian mangroves (36 per cent) than for Australian mangroves (14 per cent). The worldwide areal loss of mangroves is about 3000 km^2 each year since the early 1980s, corresponding to an overall loss rate of 2.1 per cent per year. Loss rates of mangrove forests are even higher than those estimated for tropical rain forests and coral reefs (Valiela *et al.*, 2001). However, losses vary highly between locations. Mangrove destruction in some Asian countries over the last 20 years approaches 50–80 per cent regionally, such as in Java, Sulawesi, Sumatra and the Philippines (Wolanski *et al.*, 2000).

Functions and Values

Mangroves have important functions and values. They are crucial for coastal protection (waves, storms, erosion) which has recently been confirmed by investigations after the Indian Ocean tsunami in December 2004 (Danielsen *et al.*, 2005; Kathiresan and Rajendran, 2005). Due to their root system, mangroves function as sediment traps and stabilize the shoreline. Mangrove sediments are known to retain pollutants from wastewater, giving them a fundamental filter function (*e.g.* Tam and Wong, 1999). They prevent that harmful substances derived from human activities reach adjacent coral reefs and seagrass beds.

Mangroves are highly productive ecosystems, reaching an average production of 2.5 g C m^{-2} d^{-1} (Bunt, 1992). They thus retain a large amount of CO_2. At the higher end of the range, the net photosynthetic production of a 22 year-old *Rhizophora apiculata* forest in Malaysia was estimated to be 56 t C ha^{-1} yr^{-1} (Clough *et al.*, 1997). The ecological functions of mangroves are manifold. Mangroves are well known to serve as nursery sites, feeding grounds and protection areas for many fish species, invertebrates, mammals and birds (Robertson and Duke, 1987; Little *et al.*, 1988; Robertson and Blaber, 1992; Sasekumar *et al.*, 1992; Krumme, 2003). As the juveniles of many economically important fish and shrimp species are reliant on the food sources of the mangroves, fishery yields in the coastal zone depend on the existence of this ecosystem (*e.g.* Pauly and Ingles, 1986). In addition, benthic organisms living in or on the muddy sediment in mangrove forests provide a significant food source for local people (*e.g.* large mangrove crabs; Glaser and Diele, 2004).

Nutrient recycling in the mangrove habitat is highly promoted by mangrove crabs (*e.g.* Nordhaus and Wolff, 2007). Mangrove litter plays a key role as nutrient supplier in the trophic chain of this initially nutrient-poor ecosystem. Litter-consuming mangrove crabs can consume the bulk of mangrove litter, *e.g.* up to 81 per cent in North Brazil (Nordhaus *et al.*, 2006), 79 per cent in Australia (Robertson and Daniel, 1989), and 44 per cent in South Africa (Emmerson and Mc Gwynne, 1992). Thus, through litter consumption these crabs help to retain nutrients in the mangrove habitat which would otherwise be exported by the tides. However, a part of litter material is always exported by the tides and provides organic matter for primary and secondary production in estuaries and the coastal waters. Although several studies emphasize the relevance of mangroves for the deposition of organic matter along tropical continental margins (*e.g.* Jennerjahn and Ittekkot, 2002), stable isotope studies have also shown that this carbon source is only important for secondary production in areas close to the mangrove forests whereof it is unlikely to contribute to offshore food webs (*e.g.* Rodelli *et al.*, 1984; Loneragan *et al.*, 1997).

Due to the construction of burrows, mangrove crabs as well as polychaete worms promote the stirring and mixing of the sediment (bioturbation). Their burrows can form a labyrinth of interconnected tubes through which water can flow (Ridd, 1996).

This flow may possibly provide an extremely efficient pathway for the transport of nutrients, oxygen or other dissolved substances, across the swamp-bed interface (Ridd, 1996). Thus, flushing of burrows by the tides provides an efficient mechanism for the exchange of these substances (Heron and Ridd, 2003). Self-constructed burrows of large mangrove crabs can have a depth of more than two meters (Schories *et al.*, 2003). It appears that burrowing by crabs affects soil aeration which in turn can affect the productivity and reproductive output of *Rhizophora* trees (Smith III *et al.*, 1991). Meiofaunal abundance was significantly higher within crustacean burrows than in adjacent sediments in tropical tidal flats (Dittmann, 1996). Burrows are a favourable habitat, as they offer a cooler environment than the sediment surface and thus provide refuge from desiccation.

Biodiversity of mangrove flora and fauna is highest in the Indo-Pacific region. The number of mangrove tree species is 39 in Indo-Malaysia, 35 in Australasia but only 7 in the Eastern Pacific and 5 in West Africa (Ricklefs and Latham, 1993). Mollusk diversity for instance is also higher in Indo-Malaysia (211 species) compared to the West-Atlantic-Caribbean region (124 species) (Ricklefs and Latham, 1993). Mangroves along arid coastlines and high latitude mangroves generally have fewer species than tropical mangroves.

Mangrove resources have always been used by local communities. The traditional use of mangroves includes extraction of wood for cooking and heating, for building houses and huts, for constructing fish traps, boats, furniture, and bridges, and for producing tannins, charcoal, medicines, and alcohol (Alongi, 2002 and references therein). The annual economic value of mangroves, estimated by the costs of the products and services they provide, has been estimated to be US$200,000–$900,000 per km^2 (Wells *et al.*, 2006).

Impact of Human Disturbances

Local people use the fishery resources of mangroves (fish, shrimps, crabs and mollusks) and cultivate fish and crustaceans on a small-scale. However, these activities are potentially sustainable (Alongi, 2002). By contrast, the commercial exploitation of mangroves is unsustainable and causes losses of mangrove habitats. This includes logging for wood products and housing, modifications of natural waterways for bridges and levees, conversion to aquaculture ponds and agriculture land, release of toxins and pathogens, and urbanization (Alongi, 2002 and references therein). Mangrove losses were also caused by conversion to salt flats for salt products (Hamilton and Snedaker, 1984), interception of freshwater, *e.g.* in Bangladesh (Saenger *et al.*, 1983), and the use of herbicides in Vietnam during wartime. Including all losses caused by the various human activities, a total area of 36,000 km^2 has been destroyed, corresponding to 64 per cent of the overall mangrove loss during the last two decades (Valiela *et al.*, 2001).

Conversion to Aquaculture Ponds

In particular the conversion of mangrove forests to aquaculture ponds and large-scale forest use caused the remarkable recent loss of this habitat. Shrimp and fish culture are responsible for 52 per cent of the worldwide mangrove decline that is attributed to human activities (Valiela *et al.*, 2001). For Southeast-Asia a destruction of 50-80 per cent due to pond culture has been reported (Wolanski *et al.*, 2000). For instance, around half of the 279,000 ha of Philippine mangroves that were destroyed between 1951 and 1988 were converted into ponds mainly for milkfish, but also for shrimp (Primavera, 2000; Primavera, 2005). Only a fifth of the 500,000 ha of Philippine mangroves at the turn of the 20th century remained, whereas brackish water ponds have increased almost fourfold, from 61,000 ha in 1940 to 230,000 ha at present. Aquaculture provides more than one-third of the yield of the world´s fishery. Total aquaculture production in 2003 was 54.8 million metric tons, valued at US$67.3 billion. More than 90 per cent of this output comes from Asia, where aquaculture has its origin (Primavera, 2005).

The construction of ponds not only destroys the mangrove forest in the particular area but often also kills the flora and fauna in the areas surrounding the ponds, because of major changes in drainage conditions, changes in inundation frequency, changes in nutrient availability, and toxicity of runoff waters from the ponds in areas of acid sulphate soils. Serious problems occur in highly pyrite sulphur soils because the efficiency of phosphate fertilizers is inhibited, resulting in inadequate growth of algae on which fishponds depend for high yields. Consequences are a high acidity and high aluminium concentrations which will finally kill the fish (Saenger *et al.*, 1983).

Indirect ecological effects are also manifold. Intensive shrimp farming demands intensive use of fish meal as shrimp food, which puts pressure on offshore stocks (Primavera, 1991). Because the conversion of food to shrimp is incomplete, considerable amounts of organic matter and nutrients pass through the ponds, and the effluent may cause or exacerbate the eutrophication of downstream estuaries and mangrove forests (Wolanski *et al.*, 2000). Due to the problems of eutrophication, accumulation of toxins, acidification, and crop diseases mariculture ponds have a short life span, usually comprising 5-10 years. After the ponds have been abandoned, owners convert new areas of mangrove into ponds (Wolanski *et al.*, 2000). This shifting cultivation pattern accelerates loss of mangroves, because the rate of recovery of abandoned ponds to new mangrove forests is much slower than the rate of habitat loss (Primavera, 1991).

In addition, shrimp culture demands supplies of juveniles which have led to a reduction of natural stocks (Sasekumar *et al.*, 1992). Other environmental effects include the loss of bycatch during the collection of wild seed used for stocking in ponds and of adult broodstock for hatcheries, introduction of exotic species, spread of parasites and diseases, interactions of escapees from ponds with wild populations, misuse of chemicals and antibiotics, salinization of soil and water, and coastal pollution (Primavera, 2005). Often, saltwater contamination by adjacent shrimp ponds makes agricultural land barren and the only option of poor farmers with no capital for aquaculture is to sell their land (Primavera, 1997).

Timber Production

Other forest uses (mainly industrial timber and wood–chip production) account for a mangrove loss of 26 per cent of the total loss ascribable to human activities

(Valiela *et al.*, 2001). Timber exploitation on a large scale is based on the cash return from the sale of timber for chips and pulp. Indonesia is one of the most important countries in terms of timber production. Large-scale mangrove logging occurs for instance in Kalimantan, Sulawesi and North Sumatra, resulting in a timber production in Indonesia of at least 250 000 m^3/year (Saenger *et al.*, 1983).

The main consequences of large-scale logging are a complete loss of large mangrove areas where natural regeneration is usually unsuccessful, a change in the floristic composition which affects the mangrove´s function as a breeding and feeding area for prawns, fish and crabs, and long-term detrimental effects due to soil compaction caused by the construction of rail lines. In the long term, timber productivity will decline as a result of change in floristic composition and a higher proportion of non-commercial species. In addition, debris is left behind due to the clearfelling process which can act as an effective barrier in the dispersal of propagules into the water (Saenger *et al.*, 1983).

Urbanization

The destruction of mangroves and the conversion of the land for domestic and industrial development are a great threat to mangroves. The most common forms of conversion are the building of houses and residential development, coastal tourist facilities, industry and port development (Saenger *et al.*, 1983). Urban development includes the construction of roads, harbours, airports and dams. For instance, 10–12 per cent of the land surface of what is now the country of Singapore was covered by mangroves in 1922. This area had been reduced to 6.2 per cent in 1969 and to 3 per cent in 1979. Besides the direct destruction of mangrove forests, city development is likely to strongly modify the quantity and quality of freshwater flow in mangroves. As a consequence, seawater penetration will be higher upstream and closer to the surface, resulting in a change of the zonation of flora and fauna. The reduction of freshwater input caused by dams and barrages results in downstream effects on coastal vegetation and nearshore fisheries. Dominant effects are the progressive increase in dry-season soil salinity and a gradual displacement of mangrove species by others more tolerant of higher salinities. Also, urban and industrial development causes the disposal of liquid and solid wastes to the environment, including toxic substances (Saenger *et al.*, 1983).

Oil Pollution

Main oil transport routes frequently pass coasts which are vegetated with mangroves. Oil spills repeatedly caused serious damage to mangrove flora and fauna throughout the world. The main problem is that mangrove forests are excellent traps for drifting oil slicks. These slicks cover the aerial root system and accumulate in the soil (Saenger *et al.*, 1983). Effects on the ecosystem can be divided into immediate and long-term effects. Immediate effects include the physical smothering on the plant surfaces responsible for gas exchange (bark, prop roots, pneumatophores) with oil. Mangrove trees can be killed within 48-72 hours. Long-term effects comprise the chronic poisoning of mangrove trees and fauna by the toxic components of retained oil. Surviving mangroves show chronic stress, reduced productivity, and lower leaf litter production rates. Trees are then highly susceptible to any additional perturbation or stress (Saenger *et al.*, 1983). For example, the germination of *Aegiceras corniculatum* and *Acanthus ilicifolius* in mangrove sediments was totally inhibited by lubricating oil. Early growth including height, leaf number and biomass of the oil seedlings of *Bruguiera gymnorrhiza* was significantly reduced (Zhang *et al.*, 2006). It was suggested that lubricating oil causes oxidation stress and lipid membrane damages leading to this growth reduction. Other studies reported the occurrence of mutant propagules of *Rhizophora mangle* (Klekowski *et al.*, 1994) and *Avicennia marina* (Duke and Watkinson, 2002) growing in oil contaminated sediments. These were termed ´albinos´ since they lacked chlorophyll and normal green coloration, leaving them white, yellow or red. The frequency of mutations in *Rhizophora mangle* was positively correlated with the concentrations of polycyclic aromatic hydrocarbons in the underlying sediment and with acute and chronic petroleum pollution (Klekowski *et al.*, 1994). The mutation was considered lethal since newly established albino seedlings appeared unable to survive more than a few months (Duke and Watkinson, 2002). The exposure of mangrove seedlings to fumes of the volatile fraction of light Arabian crude oil disturbs the normal functions of stomata and the salt glands of the leaves, leading to significantly reduced salt secretion and transpiration and a significant decrease in the stomatal resistance to gas diffusion (Youssef and Ghanem, 2002).

Benthic communities are also sensitive to oil spills. An investigation on the impact of oil pollution on macrofauna near an oil platform in Norway showed a reduction of diversity of benthic macrofauna and an increased dominance of opportunistic species within 500 to 1000 m off the discharge source (Gray *et al.*, 1990). After the oil spill from the grounded MV river Princess, a significant decrease (60 per cent) in the number of intertidal benthic species in a beach at Goa was detected (Ingole *et al.*, 2006). However, responses to oil pollution are species-specific. Some species, such as several amphipods, are sensitive to oil and are considered good indicators of oil pollution; by contrast, polychaetes appear to be resistant to high levels of hydrocarbons in sediment (Gómez Gesteira and Dauvin, 2000). Organisms that accumulate contaminants but are resistant to their toxicity can also be used as bioindicators, which has been shown for mussels (Cossa, 1989). The mangrove crab *Ucides cordatus* has proven to be a suitable bioindicator for oil contamination in mangroves as it is resilient to even high concentrations of polycyclic aromatic hydrocarbons and it accumulates these substances in the hepatopancreas (Haddad Nudi *et al.*, in press).

Liquid Waste Disposal and Heavy Metals

Mangroves also suffer from other chemical contamination, in particular effluents from agriculture, industry and households. Many organic substances which are discharged into the marine environment as a waste product are not toxic per se but the total load may exceed the carrying capacity of an aquatic system (Saenger *et al.*, 1983). Liquid waste disposal can cause problems in waters adjacent to mangrove forests. Nutrient input from sewage stimulates phytoplankton growth. When phytoplankton sinks to bottom, bacterial decomposition is induced, leading to the reduction of oxygen in bottom waters. This can result in the development of hypoxia in an estuary and cause the death of bottom-dwelling organisms.

Heavy metals introduced into the marine environment by industrial, mining and domestic activities are incorporated into sediments (Everaarts *et al.*, 1989). Benthic macrofauna occurring in sediments with increased metal levels may, to varying degrees, accumulate some of these metals in their tissue, resulting in high concentrations even in habitats with low metal concentrations (Everaarts *et al.*, 1989). The introduction into the food chain at very low levels and rates can result in a bioaccumulation of these compounds or elements which may reach concentrations that become toxic or lethal (Saenger *et al.*, 1983). Studies on the concentration levels of heavy metals in benthic animals from the Gulf of Thailand and Malaysia indicated considerable contamination levels with lead and cadmium in bivalves and some crustaceans (Menasveta and Cheevaparanapiwat, 1981; Phillips and Muttarasin, 1985; Everaarts and Swennen, 1987). However, the uptake pattern for each investigated element is species-specific (Boon *et al.*, 1989; Everaarts *et al.*, 1989; Rees *et al.*, 1999) and calls for in-depth studies to get a detailed picture of the impacts of contaminants. The direct effect of harmful substances may also remain undetected and only be revealed by the disruption of a particular food chain, or break of a basic recycling or regeneration process (Saenger *et al.*, 1983).

Sedimentation

River fluxes of suspended sediments have been increased due to enhanced soil erosion as a result of deforestation and land use changes. It has been calculated that the additional sediment transport by global rivers due to these human activities is 2.3 x 109 t compared to prehuman conditions. While the average sediment load that reaches the oceans has been reduced because of sediment retention within reservoir lakes behind dams, sediment fluxes to coastal regions have locally highly increased, *e.g.* in Indonesia (Syvitski *et al.*, 2005). For instance, sediment input by rivers due to unsustainable land use practices in the hinterland is an important stress factor for the vegetation and the benthic epifauna in mangrove areas of the Segara Anakan Lagoon, Java, Indonesia (Yuwono *et al.*, in press). The cutting of mangroves also causes higher erosion and thus higher sediment loads entering the lagoon. This most likely leads to a reduction of benthic populations, especially of filter-feeding mussels. In this highly dynamic lagoon system, vast growing pioneer plants, *e.g. Derris trifoliata* have established in areas with high sedimentation rates and areas where the mangrove vegetation had been cut down. It is likely that the capacity of the lagoon as a nursery ground and thus its ecological and economical importance is constantly diminishing due to sedimentation.

Conversion to Salt Ponds

The conversion of mangroves for salt production causes losses of lesser magnitude, but this activity can be important locally (Valiela *et al.*, 2001). A highly intensive operation of salt ponds has been reported for instance from India, Bangladesh, Benin, El Salvador, and Australia. Climatic constraints on solar salt production require the implementation of salt flats in arid to semiarid zones where the mangrove resource is already climatically restricted. The operation of salt processing in marginal climatic zones often requires the further utilization of limited mangrove timber resources for fuel wood to accelerate brine evaporation (Hamilton and Snedaker, 1984).

The construction of salt ponds require a complete eradication of the trees and shrubs, levelling and diking of the land, construction of a flooding canal system, and intensive mechanical compaction of the soil surface (Saenger *et al.*, 1983). Therefore, serious consequences can be observed, including an increase of the sediment salt content, drastically changed physical and chemical soil conditions, and increased soil compactness by continuous trampling, creating local, barren, poorly drained depressions flooded only during the rainy season. These changes will not allow re-colonization from adjacent mangrove stands. In almost all known cases the conversion process has been irreversible (Saenger *et al.*, 1983; Hamilton and Snedaker, 1984). A study in West Africa showed that local mangrove woody species were replaced by common herbaceous salt-tolerant pioneer plants.

Impact of Global Climate Change on Mangroves

Rise in Atmospheric CO_2 and Mangroves

Since plants depend on CO_2 for photosynthesis, enhanced CO_2 concentrations resulted in higher assimilation and growth rates in two mangrove tree species during experiments (Ball *et al.*, 1997). At higher CO_2 levels, stomatal conductance and water loss of trees via transpiration are reduced while CO_2 uptake levels are maintained (Snedaker and Araújo, 1998). This results in an increase in water use efficiency and varying effects on transpiration rate and growth, depending on other circumstances (Hogarth, 1999). Due to the increase in water use efficiency and thus decreased water loss, mangroves in arid regions may benefit (Ball and Munns, 1992).

For instance, *Rhizophora stylosa* and *Rhizophora apiculata* in Australia showed enhanced photosynthesis and average growth rates with increased CO_2 concentration, but only when grown at a low salinity level (125 mM NaCl) (Ball *et al.*, 1997). It was suggested that a rise in atmospheric CO_2 might result in the expansion of these species into areas of greater aridity, thus increasing species diversity in those regions. In contrast to several tree species, *Laguncularia racemosa* responded with a significant decrease in net primary productivity to elevated CO_2 concentrations (Snedaker and Araújo, 1998). It was postulated that a global rise in atmospheric CO_2 may result in a competitive disadvantage of *L. racemosa* in mixed mangrove communities. Thus, the species specific response to elevated CO_2 may cause an alteration of the mangrove community composition in certain locations.

Increased CO_2 concentrations may also promote maturation. In an experiment with doubled CO_2 levels, *Rhizophora mangle* seedlings showed a significant increase in biomass, total stem length, branching activity, and total leaf area. The initiation of reproduction was recorded after only one year compared to normally two years (Farnsworth *et al.*, 1996). However, Ellison and Farnsworth (1996) predict that the enhanced growth of mangroves due to global rise of CO_2 will likely not be enough to compensate for the negative impacts of sea-level rise.

Some long-term responses to CO_2 elevation in terrestrial tree species are surprising. The initial increase in photosynthesis was not maintained and compensating mechanisms were detected, resulting in a reduction of the photosynthesis rate towards the previous level. Over longer periods, alterations in plant morphology may reduce the impact of elevated CO_2. The investigation of herbarium specimens revealed a decrease in stomata density since the Industrial Revolution, which correlates closely with CO_2 increase. Therefore, a given level of photosynthesis will be maintained with fewer stomata if the CO_2 level increases (Graves and Reavey, 1996).

One indirect impact on mangroves of increased temperature and CO_2 is the degradation of coral reefs caused by mass bleaching and impaired growth (Hoegh-Guldberg, 1999). Damage to coral reefs may adversely impact mangrove systems that depend on the reefs which provide shelter from wave action.

Global Warming and Mangroves

The geographical distribution of mangroves as well as photosynthesis and production rates are dependent on temperature. Also, water loss, transpiration, and salt loss are affected by temperature (Pernetta, 1993). Mangroves grow over a wide range of temperatures. They occur in climates with average minimum air and soil temperatures of about 15°C and 20°C, respectively, during the coldest months and average maximum air and soil temperatures around 35°C and 30°C, respectively, during the warmest months (Clough, 1992). Photosynthesis and leaf production processes decrease at lower and higher temperatures. At leaf temperatures of 38-40°C photosynthesis processes cease (Clough *et al.*, 1982). However, since the diurnal changes of temperatures that mangroves experience within a day are higher than the projected increases in atmospheric and sea temperature, it is unlikely that global warming will cause a great difference to mangrove productivity (Field, 1995; Hogarth, 1999).

It was suggested that the geographical distribution may be affected and mangroves may extend northward and southward with increasing temperatures (Field, 1995). Since the temperature tolerance of mangrove species varies, it is likely that species spread to different extents which may result in a change of species composition at certain locations (Hogarth, 1999). However, other scientists argue that extreme cold events will probably limit the poleward expansion of mangroves (Woodroffe and Grindrod, 1991; Snedaker, 1995). For instance, mangroves extended northward along the east coast of the United States in the 1970s, but several frost periods caused mangrove death and a reduction in the north range limit (Snedaker, 1995).

Increased air and water temperatures are likely to benefit mangroves at the north- and southward limits of distribution. Species diversity and litter production are likely to increase and trees will grow higher (Edwards, 1995). It is possible that the seasonal patterns of reproduction will also change with increasing temperature. Global warming may have some effect on other aspects, *e.g.* it can cause the acceleration of litter decomposition (Hogarth, 1999) and accelerate nutrient cycling due to increased production rates of bacteria.

It is likely that the geographical distribution and the physiology of mangrove fauna is also affected by increasing temperatures, *e.g.* a poleward expansion of mangrove crab distribution is likely to occur. The length of inactive periods in certain locations may be reduced due to a shorter cold season. This could lead to an increased litter consumption by mangrove crabs and thus accelerate nutrient cycling. However, since long-term experiments are still lacking, these projections remain speculative.

Sea-level Rise and Mangroves

Sea-level rise is probably the most important factor influencing the future distribution of mangroves (Field, 1995). However, uncertainties exist about the global sea-level rise as well as about regional rates of sea-level rise as they are affected by tectonic movements that can cause land subsidence or uplift. Humans also contribute to land subsidence, *e.g.* through coastal development that causes deficits in the sediment budget, shipping channels that cause bank erosion, groundwater or oil extraction that causes submergence, and dredging and mining that causes losses of land (McLeod and Salm, 2006).

Geological records indicate that previous sea-level fluctuations presented both crises and opportunities for mangrove communities, and they have survived or even expanded in several refuges (Field, 1995). Mangroves may adapt to sea-level rise if it occurs slowly enough and adequate space exists to expand landwards (Ellison and Stoddart, 1991). Therefore, the net balance between sea-level rise and sediment or peat accumulation will be important for the fate of mangroves. Mangroves produce deposits from decaying litter fall and root growth and by trapping allochthonous sediment which gives them the opportunity to adjust to a rising sea level (McLeod and Salm, 2006). Thus, mangroves can survive despite sea-level rise if the rate of sediment accumulation is sufficient to keep up with sea-level rise and if migration is not blocked by local conditions, such as infrastructure and topography. For instance, in western Jamaica, mangroves were able to survive because the sedimentation rate exceeded the rate of sea-level rise during the Holocene (Hendry and Digerfeldt, 1989). Mangroves may also adapt to changes in sea-level by growing upward in place. However, if migration or growth cannot occur fast enough to adjust to the rising sea level in particular regions, then mangroves will probably become progressively smaller and may disappear.

Besides sediment supply, tidal range is another important factor that controls mangrove response to sea-level rise. Mangrove communities in macrotidal, sediment rich areas, *e.g.*, mangrove communities in northern Australia have a greater chance to survive sea-level rise than those in micro-tidal areas with low allochthonous sediment input, *e.g.*, mangroves in Caribbean islands (Parkinson *et al.*, 1994; Semeniuk, 1994; Woodroffe, 1995). Mangrove communities in carbonate islands are considered extremely vulnerable to sea-level rise since allochthonous sediment input is often limited. Riverine mangroves with high allochthonous sediment input are more likely to survive sea-level rise than those with low external inputs (Woodroffe, 1990; Pernetta, 1993).

According to Ellison and Stoddart (1991) many island mangroves will probably not persist. Through sediment accretion, low island mangroves should be able to keep pace with sea-level rise of up to 8-9 cm/100 years, but at rates of over 12 cm/100 years they could probably not survive. In the low-lying island mangroves in Bermuda, the rate of peat accretion has been 8.5–10.6 cm/100 years over the last 2000 years, which is less than the rate of sea-level rise of 28 ± 18 cm/100 years in this century (Ellison, 1993). Retreat and erosion of the seaward margin at Hungry Bay, Bermuda, caused the loss of large mangrove areas within the last few hundred years. The study indicates that low island mangroves will experience problems with the rates of sea-level rise predicted for the next 50 years (Ellison, 1993).

However, mangrove systems in Florida have persisted and are even expanding in some areas despite sea-level rise has been about 20 cm over the last 147 years (Snedaker, 1995). Therefore, it is difficult to predict the effects of sea-level rise on the mangroves in particular regions due to the high complexity of this ecosystem. In order to understand the impact of sea-level rise it is essential to take into account other factors, such as the substrate type, coastal processes, local tectonics, availability of freshwater and sediment, salinity of soil and groundwater, and frequency and intensity of cyclonic storms (Semeniuk, 1994; Blasco *et al.*, 1996). In addition, it is important to note that too much sediment input (*e.g.*, resulting from poor agricultural practices) can bury the pneumatophores and kill mangroves (Ellison and Stoddart, 1991).

Since mangrove trees and animals show a zonation of species up and down the river within estuaries related to the shore profile, soil type, and the salinity gradient, it is likely that sea-level rise would cause a shifting of zonation patterns in an upriver direction. Different species may be able to colonize new areas at different speeds, resulting in species-specific survival rates (Pernetta, 1993; Semeniuk, 1994). For instance, in Caribbean mangroves prop roots of *Rhizophora mangle* stand higher above mean sea-level than the short aerial roots of *Avicennia germinans*. It was suggested that rapid sea-level rise may lead to local extinctions of *A. germinans* but have an insignificant effect on *R. mangle* (Vicente *et al.*, 1989 cited in McLeod and Salm 2006). Ellison and Stoddart (1991) also suggested that *Rhizophora* trees are better positioned to survive rising sea-level due to higher peat accumulation rates than *Bruguiera* and *Excoecaria*.

Changes in Precipitation and Mangroves

Depending on location, precipitation may decrease or increase as a result of climate change. Increased precipitation results in decreased soil salinity which can cause an increase of mangrove growth rates in some species (Field, 1995). Ellison (2000) suggested an increase of mangrove area, tree height, diversity of mangrove zones and growth rates in areas of enhanced rainfall in Australia. In addition, mangroves may migrate and outcompete saltmarsh vegetation in areas of reduced salinity levels (Harty, 2004; Rogers *et al.*, 2005).

Decreased precipitation can cause a reduced freshwater input to mangroves and thus an increase of soil salinity which in turn may result in increased salt in mangrove tissues and decreased mangrove productivity, growth, and seedling survival. Mangroves are short, of narrower margins, and interrupted by salt flats in areas of lower rainfall mainly due to salt stress (Ellison, 2000). More salt tolerant species are thus favoured, and a change of species composition is likely. Decreased precipitation in arid areas together with an increase in evaporation may result in a reduction of the mangrove area, decrease in diversity, and loss of the landward zone to unvegetated hypersaline flats (Snedaker, 1995).

Impact of Natural Hazards on Mangroves

Storms and Floods

It is likely that tropical storms will be more frequent and more severe as a consequence of climate change, representing an additional threat to mangroves (Field, 1995). In most cases, the impact of typhoons or hurricanes is highly destructive (Hogarth, 1999). Hurricane Andrew caused a heavy damage to about 150 km^2 of mangroves in Florida in 1992. About 60 per cent of the trees were either uprooted or broken and 86 per cent of the trees showed defoliation. Many trees that survived the hurricane died subsequently, *e.g.* one year later. Although the affected areas were rapidly recolonized by seedlings, high mortality and slow growth of these seedlings point to long-term effects and slow recovery from the damages (McCoy *et al.*, 1996). Large storm impacts that caused the destruction of mangrove forests were also recorded in Honduras (Cahoon *et al.*, 2003) and other countries (Jiménez *et al.*, 1985 and references therein).

Storm surges can also flood mangroves and, when combined with sea-level rise, lead to mangrove destruction, *e.g.* in Florida (Lollar, 2004). Flooding of the aerial roots can result in decreased oxygen concentrations in the trees, leading to death. Hurricanes may deteriorate the impacts of sea-level rise by uprooting trees, leaving sediment vulnerable to erosion (McLeod and Salm, 2006). In Tanzania, flooding and heavy rains during the 1998 El Niño also caused the death of large mangroves areas (Erftemeijer and Hamerlynck, 2005). Flooding, caused by increased rainfall, storms, or sea-level rise may cause a reduced productivity, photosynthesis, and survival of mangrove trees (Ellison, 2000).

A significant change in community structure was recorded due to species specific destruction of mangrove trees caused by Hurricane Andrew. The hurricane heavily damaged large *Laguncularia racemosa* trees, but less affected smaller *Rhizophora mangle* trees. It was observed that some species are favoured at the expense of others. *Rhizophora* is more resilient when exposed to winds, *Laguncularia* is a good coloniser and *Avicennia* the most cold-tolerant (Hogarth, 1999). In addition, a higher mortality can be expected for species with larger mean tree size (Baldwin *et al.*, 1995; McCoy *et al.*, 1996), thus an increase in hurricane intensity may result in a decrease in the average tree height in Florida mangroves (Lugo and Snedaker, 1974). The regeneration process of mangrove forests in Florida after the hurricane Andrew was determined by the complex interactions between resprouting capability, seedling survival, post-hurricane seedling recruitment, and colonization by herbaceous vegetation. This resulted in a forest structure different from the pre-hurricane structure (Baldwin *et al.*, 2001). Irregular events, such as storms, flooding and coldness can therefore be important in shaping community structure.

Tsunamis

The tsunami of December 26, 2004 in the Indian Ocean is the latest and most devastating natural disaster of global concern which jolted interest in the value of mangrove forests in protecting vulnerable coastal communities (Williams, 2005). Its epicentre was located near the north-western end of the island of Sumatra, Indonesia (Titov *et al.*, 2005) and caused economic and ecological harm in 13 Asian and African countries (Kathiresan and Rajendran, 2005). Several studies conducted after the 2004 tsunami suggest that mangrove forests had protected humans and their inhabitations against the tsunami-induced waves whereas the damage was magnified in areas with degraded coastal ecosystems (Danielsen *et al.*, 2005; Kathiresan and Rajendran, 2005; Williams, 2005; Stone, 2006).

At Cuddalore, Tamil Nadu, India, the tsunami destroyed parts of the village at the river mouth and completely destroyed villages on the coast which were not protected by vegetation (Danielsen *et al.*, 2005). By contrast, areas with mangroves and tree shelterbelts were significantly less damaged than other areas. Villages behind the mangrove suffered no destruction and those located within *Casuarina* plantations experienced only partial damage (Danielsen *et al.*, 2005). There, mangroves and *Casuarina* plantations apparently mitigated the waves and protected shorelines against damage.

Another study conducted at Parangipettai, Tamil Nadu, India, found a significant negative correlation between the human death toll and the distance of human inhabitations from sea, the elevation from mean sea-level and the area of mangroves and other coastal vegetation (Kathiresan and Rajendran, 2005). It was recommended to construct human inhabitations more than 1 km from the shoreline in elevated places, behind dense mangroves or other coastal vegetation. After the tsunami, most of the coastal vegetation has been affected in the study area with shedding of leaves or browning of canopies, but mangroves tolerated the waves without showing any apparent damage. It was observed that the affected parts of the forests were mostly seedlings, forests with low tree density, narrow strips of forests and the aerial roots that were clogged with salt deposits (Kathiresan and Rajendran, 2005). It was concluded that mangroves are the most suitable species to attenuate the effects of huge tidal waves (Kathiresan and Rajendran, 2005).

In Indonesia, the epicentre of the tsunami was closer to Simeuleu Island than to Sumatra. However, due to the protection by mangrove forests in a very good condition the human death rate was significantly lower on this small island (Kathiresan and Rajendran, 2005). Also, mangroves stretching thousands of kilometers along the coast of Sunderbans were effective barriers against the waves (Kathiresan and Rajendran, 2005). A study conducted in Sri Lanka revealed that at most mangrove fringes near the water edge took all the energy and were damaged. Behind this fringe, true mangrove representatives, such as *Sonneratia* spp., *Rhizophara* spp. or *Bruguiera* spp., stood firm against the ocean surge whereas forests dominated by less typical mangrove associates were severely damaged (Dahdouh-Guebas *et al.*, 2005). It was thus concluded that even though a coastal area might superficially seem to be protected by a mangrove forest, that habitat could be cryptically degraded and not offer the desired protection. In addition, mangroves are likely to play a critical role in storm protection, but with the constraint that this depends on the quality of the mangrove forest (Dahdouh-Guebas *et al.*, 2005).

It remains to be seen whether humans learnt their lesson from the catastrophic effects of the recent tsunami. Mangroves are resilient and may be restored which has been shown for instance in East-Africa (Kairo *et al.*, 2001). Conserving or replanting coastal mangroves and greenbelts should buffer communities from future tsunami events (Dahdouh-Guebas *et al.*, 2005; Danielsen *et al.*, 2005).

Summary and Conclusions

Mangroves are unique ecosystems along the tropical and subtropical coasts. They are highly productive forests with important ecological and economic functions and values. The anthropogenic pressure on this ecosystem has highly increased during the past decades and resulted in the destruction of large mangrove areas. Additional threats to mangroves are the continuous change of the global climate and the occurrence of extreme climatic hazards and geohazards.

The living resources of mangroves have always been used by local communities. Whereas the traditional use of timber and fishery resources has mostly been sustainable, the commercial exploitation during the last decades was not. Taking natural and man-made causes of mangrove destruction together, the worldwide decline of mangrove forests was estimated to at least 35 per cent within the last 20 years (Valiela *et al.*, 2001). Mangrove destruction is particularly high in some Asian countries where the main causes are the conversion of mangrove areas to aquaculture ponds and large-scale timber production. Other human activities that contribute to the continuous degradation of mangroves are the consequences of urbanization, the conversion of mangroves to agriculture land and salt ponds, waste disposal to the aquatic habitats, including toxic substances, and enhanced sediment fluxes to coastal regions due to human activities.

Mangroves are likely to be early indicators of the effects of climate change. Sea-level rise is probably the greatest challenge that mangroves are facing (Field, 1995). Mangroves may survive in areas where the rate of sediment accretion is sufficient to keep up with sea-level rise and where their landward movement is not restricted by humans. Mangrove forests in macro-tidal sediment rich areas have probably better chances to survive sea-level rise than those in micro-tidal areas with low allochthonuos sediment input (Parkinson *et al.*, 1994; Woodroffe, 1995). The global rise in atmospheric CO_2 and temperature, sea-level rise and changes in precipitation will probably lead to changes in the distribution and zonation patterns of mangrove flora and fauna, their productivity, biodiversity and species composition. However, due to the high complexity of this ecosystem, effects of global climate change on a particular mangrove forest are difficult to test experimentally and are thus difficult to predict.

Tropical storms can be highly destructive to mangroves like, for instance, in Florida and the Caribbean (Jiménez *et al.*, 1985; McCoy *et al.*, 1996). They will most likely be more frequent and more severe in the future as a result of climate change (Field, 1995). Anyway, mangroves have proven to protect coastal zones and human settlements from storms and floods. In particular the Indian Ocean tsunami in 2004 reminded the global community of this protection value since damages were often much higher in degraded coastal regions than in villages which were located behind mangrove forests (*e.g.* Kathiresan and Rajendran, 2005).

Therefore, it is of great interest to rehabilitate and reforest degraded mangrove forests and coastal areas which had been converted for other purposes, in particular aquaculture ponds. Mangrove greenbelts can provide coastal protection from storms and erosion und mitigate the impacts of climate change but would also improve the resilience of regions and enhance the socioeconomic value of presently degraded areas. A lack of understanding of the complex interactions between the different natural and man-made impacts on mangroves is a major reason for the failure of previous coastal management programs. In-depth scientific studies are required to ascertain areas that are suitable for restoration programs. Successful mangrove restoration programs in several countries have shown that the efforts are worthwhile.

References

Alongi, D.M., 2002. Present state and future of the world's mangrove forests. Environmental Conservation 29, 331-349.

Baldwin, A., Egnotovich, M., Ford, M., Platt, W., 2001. Regeneration in fringe mangrove forests damaged by Hurricane Andrew. Plant Ecology 157, 149-162.

Baldwin, A.H., Platt, W.J., Gathen, K.L., Lessmann, J.M., Rauch, T.J., 1995. Hurricane damage and regeneration in fringe mangrove forests of southeast Florida, USA. Journal of Coastal Research 21, 169-183.

Ball, M.C., Cochrane, M.J., Rawson, H.M., 1997. Growth and water use of the mangroves *Rhizophora apiculata* and *R. stylosa* in response to salinity and humidity under ambient and elevated concentrations of atmospheric CO_2. Plant, Cell and Environment 20, 1158-1166.

Ball, M.C., Munns, R., 1992. Plant responses to salinity under elevated atmospheric concentrations of CO_2. Australian Journal of Botany 40, 515-525.

Blasco, F., Saenger, P., Janodet, E., 1996. Mangroves as indicators of coastal change. Catena 27, 167-178.

Boon, J.P., Everaarts, J.M., Kastoro, W., Razak, H., Sumanta, I., Sumarno, Nelissen, P.H., Stefels, J., Hillebrand, M.T.J., 1989. Cyclic organochlorines in epibenthic organisms from coastal waters around East Java. Netherlands Journal of Sea Research 23, 427-439.

Bunt, J.S., 1992. How can fragile marine ecosystems best be conserved? In: K.J., H., Thiede, J. (Eds.), Use and misuse of the seafloor. John Wiley and Sons, Berlin, pp. 229-242.

Cahoon, D., R., Hensel, P., Rybczyk, J., McKee, K.L., Proffitt, C.E., Perez, B.C., 2003. Mass tree mortality leads to mangrove peat collapse at Bay Islands, Honduras after Hurricane Mitch. Journal of Ecology 91, 1093-1105.

Clough, B.F., 1992. Primary Productivity and Growth of mangrove Forests. In: Robertson, A.I., Alongi, D.M. (Eds.), Tropical Mangrove Ecosystem. American Geophysical Union, Washington DC, pp. 225-249.

Clough, B.F., Andrews, T.J., Cowan, I.R., 1982. Primary productivity of mangroves. In: Clough, B.F. ed., Mangrove Ecosystems in Australia–structure function and management. AIMS with ANU press, Canberra, Australia.

Clough, B.F., Ong, J.E., Gong, W.K., 1997. Estimating leaf area index and photosynthetic production in canopies of the mangrove *Rhizophora apiculata*. Marine Ecology Progress Series 159, 285-292.

Cossa, D., 1989. A review of the use of *Mytilus* spp. as quantitative indicators of cadmium and mercury contamination in coastal waters. Oceanologica acta 12, 417-432.

Dahdouh-Guebas, F., Jayatissa, L.P., Di Nitto, D., Bosire, J.O., Lo Seen, D., Koedam, N., 2005. How effective were mangroves as a defence against the recent tsunami? Current Biology 15, 443-447.

Danielsen, F., Sørensen, M.K., Olwig, M.F., Selvam, V., Parish, F., Burgess, N.D., Hiraishi, T., Karunagaran, V.M., Rasmussen, M.S., Hansen, L.B., Quarto, A., Suryadiputra, N., 2005. The Asian Tsunami: A Protective Role for Coastal Vegetation. Science 310, 643.

Dittmann, S., 1996. Effects of macrobenthic burrows on infaunal communities in tropical tidal flats. Mar. Ecol. Prog. Ser. 134, 119-130.

Duke, N.C., Watkinson, A.J., 2002. Chlorophyll-deficient propagules of *Avicennia marina* and apparent longer term deterioration of mangrove fitness in oil-polluted sediments. Marine Pollution Bulletin 44, 1269-1276.

Edwards, A.J., 1995. Impact of Climatic Change on Coral Reefs, Mangroves, and Tropical Seagrass Ecosystems. In: Eisma, D. ed., Climate Change, Impact on Coastal Habitation. Lewis Publisher, Texel, The Netherlands, pp. 209-234.

Ellison, A.M., Farnsworth, E.J., 1996. Anthropogenic disturbance of Caribbean mangrove ecosystems: Past impacts, present trends, and future predictions. Biotropica 28, 549-565.

Ellison, J.C., 1993. Mangrove Retreat with Rising Sea-level, Bermuda. Estuarine, Coastal and Shelf Science 37, 75-87.

Ellison, J.C., 2000. How South Pacific mangroves may respond to predicted climate change and sea-level rise. In: Gillespie, A., Burns, W. (Eds.), Climate change in the South Pacific: Impacts and responses in Australia, New Zealand, and small island states. Kluwer Academic Publishers, Dordrecht, pp. 289-301.

Ellison, J.C., Stoddart, D.R., 1991. Mangrove ecosystem collapse during predicted sea-level rise: Holocene analogues and implications. Journal of Coastal Research 7, 151-165.

Emmerson, W.D., Mc Gwynne, L.E., 1992. Feeding and assimilation of mangrove leaves by the crab *Sesarma meinerti* de Man in relation to leaf-litter production in

Mgazana, a warm-temperate southern African mangrove swamp. J. Exp. Mar. Biol. Ecol. 157, 41-53.

Erftemeijer, P.L.A., Hamerlynck, O., 2005. Die-back of the mangrove *Heritiera littoralis* dryland, in the Rufiji delta (Tanzania) following El Niño floods. Journal of Coastal Research 42, 228-235.

Everaarts, J.M., Boon, J.P., Kastoro, W., Fischer, C.V., Razak, H., Sumanta, I., 1989. Copper, zinc and cadmium in benthic organisms from the Java Sea and estuarine and coastal areas around East Java. Netherlands Journal of Sea Research 23, 415-426.

Everaarts, J.M., Swennen, C., 1987. Heavy metals (Zn, Cu, Cd, Pb) in some benthic invertebrate species and in sediment from three coastal areas in Thailand and Malaysia. J. Sci. Soc. Thailand 13, 189-203.

Farnsworth, E.J., Ellison, A.M., Gong, W.K., 1996. Elevated CO_2 alters anatomy, physiology, growth, and reproduction of red mangrove (*Rhizophora mangle* L). Oecologia 108, 599-609.

Field, C.D., 1995. Impact of expected climate change on mangroves. Hydrobiologia 295, 75-81.

Glaser, M., Diele, K., 2004. Asymmetric outcomes: assessing central aspects of the biological, economic and social sustainability of a mangrove crab fishery, *Ucides cordatus* (Ocypodidae), in North Brazil. Ecological economics 49, 361-373.

Gómez Gesteira, J.L., Dauvin, J.C., 2000. Amphipods are Good Bioindicators of the Impact of Oil Spills on Soft-Bottom Macrobenthic Communities. Marine Pollution Bulletin 40, 1017-1027.

Graves, J.D., Reavey, D., 1996. Global environmental change: plants, animals and communities. Longman, Essex, England, 226 pp.

Gray, J.S., Clarke, K.R., Warwick, R.M., G., H., 1990. Detection of initial effects of pollution on marine benthos: an example from the Ekofisk and Eldfisk oilfields, North Sea. Marine Ecoloy Progress Series 66, 285-299.

Haddad Nudi, A., De Luca Rebello Wagener, A., Francioni, E., De Lemos Scofield, A., Sette, C.B., Veiga, A., in press. Validation of *Ucides cordatus* as a bioindicator of oil contamination and bioavailability in mangroves by evaluating sediment and crab PAH records. Environment International.

Hamilton, L.S., Snedaker, S.C., 1984. Handbook for Mangrove Area Management, Environment and Policy Institute, IUCN, UNESCO, UN, UNEP.

Harty, C., 2004. Planning strategies for mangrove and saltmarsh changes in Southeast Australia. Coastal management 32, 405-415.

Hendry, M., Digerfeldt, G., 1989. Palaeogeography and palaeoenvironments of a tropical coastal wetland and offshore shelf during Holocene submergence, Jamaica. Palaeogeography, Palaeoclimatology, Palaeoecology 73, 1-10.

Heron, S.F., Ridd, P.V., 2003. The effect of water density variations on the tidal flushing of animal burrows. Estuarine, Coastal and Shelf Science 58, 137-145.

Hoegh-Guldberg, O., 1999. Climate change, coral bleaching and the future of the world's coral reefs. Marine Freshwater Research 50, 839-866.

Hogarth, P.J., 1999. The Biology of Mangroves. Oxford University Press.

Ingole, B., Sivadas, S., Goltekar, R., Clemente, S., Nanajkar, M., Sawant, R., D'Silva, C., Sarkar, A., Ansari, Z., 2006. Ecotoxicological effect of grounded *MV River Princess* on the intertidal benthic organisms off Goa. Environmental International 32, 284-291.

Jennerjahn, T.C., Ittekkot, V., 2002. Relevance of mangroves for the production and deposition of organic matter along tropical continental margins. Naturwissenschaften 89, 23-30.

Jiménez, J.A., Lugo, A.E., Cintrón, G., 1985. Tree mortality in mangrove forests. Biotropica 17, 177-185.

Kairo, J.G., Dahdouh-Guebas, F., Bosire, J., Koedam, N., 2001. Restoration and management of mangrove systems — a lesson for and from the East African region. South African Journal of Botany 67, 383-389.

Kathiresan, K., Rajendran, N., 2005. Coastal mangrove forests mitigated tsunami. Estuarine, Coastal and Shelf Science 65, 601-606.

Kjerfve, B., Lacerda, L.D., 1993. Mangroves of Brazil. In: Lacerda, L.D. ed., Conservation and sustainable utilization of mangrove forests in Latin America and African regions. Part 1: Latin America. Mangrove Ecosystems Technical Reports, pp. 245-272.

Klekowski, E.J., Jr., Corredor, J.E., Morell, J.M., Del Castillo, C.A., 1994. Petroleum pollution and mutation in mangroves. Marine Pollution Bulletin 28, 166-169.

Krumme, U., 2003. Tidal and diel dynamics in a nursery area: patterns in fish migration in a mangrove in North Brazil. [PhD thesis], University of Bremen.

Little, M.C., Reay, P.J., Grove, S.J., 1988. The fish community of an East African mangrove creek. J. Fish. Biol. 32, 729-747.

Lollar, K., 2004. Mangrove Die-Off Being Studied, http://www.flmnh.ufl.edu/Fish/southflorida/news/mangrovedieoff2004.html.

Loneragan, N.R., Bunn, S.E., Kellaway, D.M., 1997. Are mangroves and seagrasses sources of organic carbon for penaeid prawns in a tropical Australian estuary? A multi stable-isotope study. Mar. Biol. 130, 289-300.

Lugo, A.E., Snedaker, S.C., 1974. The ecology of mangroves. Ann. Rev. Ecol. Syst. 5, 39-64.

McCoy, E.D., Mushinsky, H.R., Johnson, D., Meshaka Jr., W.E., 1996. Mangrove damage caused by hurricane Andrew on the southwestern coast of Florida. Bulletin of Marine Science 59, 1-8.

McLeod, E., Salm, R.V., 2006. Managing mangroves for resilience to climate change: Gland, Switzerland, IUCN, The World Conservation Union and The Nature Conservancy, p. 64 pp.

Menasveta, P., Cheevaparanapiwat, V., 1981. Heavy metals, organochlorine pesticides and PCB´s in green mussels, mullets and sediments of river mouth in Thailand. Mar. Pollution Bulletin 12, 19-25.

Nordhaus, I., Wolff, M., 2007. Feeding ecology of the mangrove crab *Ucides cordatus* (Ocypodidae): Food choice, food quality and assimilation efficiency. Marine Biology, DOI 10.1007/s00227-006-0597-5.

Nordhaus, I., Wolff, M., Diele, K., 2006. Litter processing and population food intake of the mangrove crab *Ucides cordatus* in a high intertidal forest in northern Brazil. Estuarine Coastal Shelf Science 67, 239-250.

Parkinson, R.W., DeLaune, R.D., White, J.R., 1994. Holocene sea-level rise and the fate of mangrove forests within the wider Caribbean Region. Journal of Coastal Research 10, 1077-1086.

Pauly, D., Ingles, J., 1986. The relationship between shrimp yields and intertidal vegetation (mangrove) areas: a reassessment, IOC/FAO Workshop on Recruitment in Tropical Coastal Demersal Communities, Volume 44: Ciudad del Carmen, Campeche, Mexico, IOC/FAO Workshop on Recruitment in Tropical Coastal Demersal Communities, p. 277-283.

Pernetta, J.C., 1993. Mangrove forests, climate change and sea level rise: Hydrobiological influences on community structure and survival, with examples from the Indo-West Pacific. IUCN, Gland, Switzerland, 46 pp.

Phillips, D.J.H., Muttarasin, K., 1985. Trace metals in bivalve molluscs from Thailand. Mar. Environm. Res. 15, 215-234.

Primavera, J.H., 1991. Intensive prawn farming in the Phillipines: Ecological, social, and economic implication. Ambio 20, 28-33.

Primavera, J.H., 1997. Socio-economic impacts of shrimp culture. Aquaculture Research 28, 815-827.

Primavera, J.H., 2000. Development and conservation of Philippine mangroves: institutional issues. Ecological Economics 35, 91-106.

Primavera, J.H., 2005. Mangroves, fishponds and the quest for sustainability. Science 310, 57-59.

Rees, J.G., Setiapermana, D., Sharp, V.A., Weeks, J.M., Williams, T.M., 1999. Evaluation of the impacts of land-based contaminants on the benthic faunas of Jakarta Bay, Indonesia. Oceanologica Acta 22, 627-640.

Ricklefs, R.E., Latham, R.E., 1993. Global patterns of diversity in mangrove floras. In: Ricklefs, R.E., Schluter, D. (Eds.), Species Diversity in Ecological Communities - Historical and Geographical Perspectives. The University of Chicago Press, Chicago, pp. 215-229.

Ridd, P.V., 1996. Flow Through Animal Burrows in Mangrove Creeks. Estuarine, Coastal and Shelf Science 43, 617-625.

Robertson, A.I., Blaber, S.J.M., 1992. Plankton, epibenthos and fish communities. In: Robertson, A.I., Alongi, D.M. (Eds.), Tropical Mangrove Ecosystems. American Geophysical Union, Washington, DC, pp. 173-249.

Robertson, A.I., Daniel, P.A., 1989. The influence of crabs on litter processing in high intertidal mangrove forests in tropical Australia. Oecologia 78, 191-198.

Robertson, A.I., Duke, N.C., 1987. Mangroves as nursery sites: comparisons of the abundance and species composition of fish and crustaceans in mangroves and other nearshore habitats in tropical Australia. Marine Biology 96, 193-205.

Robertson, A.I., Duke, N.C., 1990. Recruitment, Growth and Residence Time of Fishes in a Tropical Australian Mangrove System. Estuarine, Coastal and Shelf Science 31, 723-743.

Rodelli, M.R., Gearing, J.N., Gearing, P.J., Marshall, N., Sasekumar, A., 1984. Stable isotope ratio as a tracer of mangrove carbon in Malaysian ecosystems. Oecologia 61, 326-333.

Rogers, K., Saintilan, N., Heijns, H., 2005. Mangrove Encroachment of Salt Marsh in Western Port Bay, Victoria: The Role of Sedimentation, Subsidence and Sea Level Rise. Estuaries 28, 551-559.

Saenger, P., Hegerl, E.J., Davie, J.D.S., 1983. Global Status of Mangrove Ecosystems. The Environmentalist 3, Supplement No. 3, 88 pp.

Sasekumar, A., Chong, V.C., Leh, M.U., D´Cruz, R., 1992. Mangroves as a habitat for fish and prawns. Hydrobiologia 247, 195-207.

Schories, D., Barletta-Bergan, A., Barletta, M., Krumme, U., Mehlig, U., Rademaker, V., 2003. The keystone role of leaf-removing crabs in mangrove forests of North Brazil. Wetlands Ecology and Management 11, 243-255.

Semeniuk, 1994. Sea-Level Rise and Tidal Wetlands. Journal of Coastal Research 10, 1051-1076.

Smith III, T.J., Boto, K.G., Frusher, S.D., Giddens, R.L., 1991. Keystone species and mangrove forest dynamics: the influence of burrowing by crabs on soil nutrient status and forest productivity. Estuarine, Coastal and Shelf Science 33, 419-432.

Snedaker, S.C., 1995. Mangroves and climate change in the Florida and Caribbean region: Scenarios and hypotheses. Hydrobiologia 295, 43-49.

Snedaker, S.C., Araújo, R.J., 1998. Stomatal conductance and gas exchange in four species of Caribbean mangroves exposed to ambient and increased CO_2. Marine and Freshwater Research 49, 325-327.

Spalding, M., Blasco, F., Field, C.D., 1997. World mangrove atlas. The International Society for Mangrove Ecosystems, Okinawa, Japan.

Stone, R., 2006. A rescue effort for tsunami-ravaged mangrove forests. Science 314, 404.

Syvitski, J.P.M., Vörösmarty, C.J., Kettner, A.J., Green, P., 2005. Impact of Humans on the Flux of Terrestrial Sediment to the Global Coastal Ocean. Science 308, 376-380.

Tam, N.F.Y., Wong, Y.S., 1999. Mangrove soils in removing pollutants from municipal wastewater of different salinities. Journal of environmental quality 28, 556-564.

Titov, V., Rabinovich, A.B., Mofjeld, H.O., Thomson, R.E., González, F.I., 2005. The Global Reach of the 26 December 2004 Sumatra Tsunami. Science 309, 2045-2048.

Valiela, I., Bowen, J.L., York, J.K., 2001. Mangrove forests: One of the world´s threatened major tropical environments. BioScience 51, 807-815.

Wells, S., Ravilious, C., Corcoran, E., 2006. In the front line: Shoreline protection and other ecosystem services from mangroves and coral reefs: Cambridge, UK, UNEP World Conservation Monitoring Centre, 33 pp.

Williams, N., 2005. Tsunami insight to mangrove value. Current Biology 15, R73.

Wolanski, E., Spagnol, S., Thomas, S., Moore, K., Alongi, D.M., Trott, L., Davidson, A., 2000. Modelling and Visualizing the Fate of Shrimp Pond Effluent in a Mangrove-fringed Tidal Creek. Estuarine, Coastal and Shelf Science 50, 85-97.

Woodroffe, C., D., 1990. The impact of sea-level rise on mangrove shorelines. Progress in Physical Geography 14, 483-502.

Woodroffe, C., D., Grindrod, J., 1991. Mangrove biogeography: the role of Quaternary environmental sea-level change. Journal of Biogeography 18, 479-492.

Woodroffe, C.D., 1995. Response of tide-dominated mangrove shorelines in northern Australia to anticipated sea-level rise. Earth Surface Processes and Landforms 20, 65-85.

Youssef, T., Ghanem, A., 2002. Salt secretion and stomatal behaviour in *Avicennia marina* seedlings fumigated with the volatile fraction of light Arabian crude oil. Environmental Pollution 116, 215-223.

Yuwono, E., Jennerjahn, T.C., Nordhaus, I., Riyanto, E.A., Sastranegara, M.H., Pribadi, R., in press. Ecological status of Segara Anakan, Java, Indonsia, a mangrove-fringed lagoon affected by human activities. Asian Journal of Water, Environment and Pollution.

Zhang, C.G., Leung, K.K., Wong, Y.S., Tam, N.F.Y., 2006. Germination, growth and physiological responses of mangrove plant (*Bruguiera gymnorrhiza*) to lubricating oil pollution. Environmental and experimental Botany 60, 127-136.

Chapter 6

Application of Geographical Information System (GIS) and Mathematical Modelling in Coastal Area and Hazard Management

R.S. Kankara

Ministry of Earth Sciences, ICMAM-Project Directorate, NIOT campus, Pallikaranai, Chennai, India

E-mail: brs@icman.gov.in

ABSTRACT

The coastal zone is under pressure on account of increased anthropogenic activity on the coast and vulnerable to coastal hazards such as tsunami, storms, oil spill, erosion, coastal flooding etc. It is necessary to protect the coastal resources from natural and man induced hazards to ensure sustainable development of coastal zone. GIS and mathematical modelling provides a scientific platform to handle, analyse and interprete the data and information. These tools are being increasingly used for coastal zone and hazard management. The application of GIS and modelling can improve the quality and power of analysis of natural hazards assessments, guide development activities and assist planners in the selection of mitigation measures and in the implementation of emergency preparedness and response action. This chapter discusses the applications of GIS and modelling in the management of coastal areas and ecosystems.

Keywords: GIS, Mathematical modelling, Ecosystem, ICMAM, Coastal, Management.

Introduction

India has a long coastline of over 7500 km, supporting a wide variety of coastal activities such as human settlement, industries, ports, harbours, navigation, fishing, tourism, etc., and comprising various types of coastal habitats such as mangroves, coral reefs, seagrass beds, beaches, lagoons, etc. The marine environment and coastal zone play a major role in terms of economy and ecology. Natural processes and human interventions cause changes in the physical, chemical and ecological status of the coastal zone resulting in degradation and/or loss of amenities and habitats. The ecosystems in the coastal zone are highly productive with high biological diversity and they are the nursery habitats for many species. The coastal features such as mangroves, sand dunes, sea grasses and coral reefs form the natural defence against tsunami, storms, floods and erosion.

The concept of Integrated Coastal and Marine Area Management (ICMAM) takes into consideration various natural coastal processes and impacts of human activities through scientific understanding to develop of management strategies to address the issue of resource use conflicts and to control the impacts of human intervention on the environment. To ensure effective and efficient management of coastal areas, it is necessary to develop accurate, up-to-date and comprehensive scientific databases on coastal resources and infrastructure facilities. In this regards, the application of modern scientific tools such as mathematical modelling, remote sensing, GIS and GPS are extremely valuable in understanding the dynamic behaviour of coastal and marine environment, development of databases and risk or impact assessment from coastal hazards in the integrated manner and derive management action plans.

Issues for Coastal Zone Management

The major issues related to coastal and marine area management are listed below:

Assessment of Coastal Ecosystems and Resources

1. Generation of baseline information, conservation and restoration of coastal ecosystems viz. mangroves, coral reefs, seagrass beds, etc.
2. Reclamation of wetland for agricultural and industrial purposes
3. Exploration and sustainable use of living resources

Shoreline Protection

1. Identification of vulnerable areas including eroded areas and developmental activities
2. Planning and implementation of coastal protection work (erosion, flood protection, salt water intrusion, etc.)
3. Impact of engineering structures and dams on coastal processes of erosion, deposition and sediment transport
4. Longshore sediment transport

Coastal Development and Pollution

1. Appropriate site selection for industries, landfall points, aquaculture, recreational activities, etc.
2. Assessment of conditions in regulation zones, areas under construction setback–lines, megacities, etc.
3. Non-point and point pollution.

Coastal Hazards Identification, Forecasting, and Impacts

1. Geo Reference mapping for coastal resources and infrastructures
2. Coastal Hazard monitoring and modelling Hazard risk assessment
3. Hazard warnings and Emergency response plans for natural disasters such as cyclones, sea level rise, or anthropogenic activities such as oil spills.

Coastal Hazards in India

India is a large country and prone to a number of coastal hazards. Among all the natural hazards that country faces, erosion and coastal flooding are the most frequent and often devastating. Indian coastline is prone to very severe cyclonic formations in the Arabian Sea and Bay of Bengal. Threat of oil spill is also major cause of concern.

Coastal areas are the most preferred locations for human settlement. 25 per cent of India's population lives within 50 km of the coastline. The coastal population is the seafaring community, of which fishermen form a sizeable part. About 0.5 million fishermen are exposed to the vagaries of nature in the offshore areas every day and that well over 3500 villages are exposed to the hazards that coastal areas have to confront. The recent tsunami showed that certain areas become far more vulnerable than other neighbouring areas due to differences in physical environment. Coastal topographic slope and the absence of "bio-shields" are some of the factors that determine vulnerability to coastal hazards.

Status of Coastal Ecosystems in India

India's 7,500 km long coastline is dotted with populated coastal cities and ecologically sensitive coastal ecosystems. Ecological and geo-morphological characteristics of the main coastal states of India are summarised in Table 6.1. Brief overviews of various coastal ecosystems found in Indian coastal areas are described here.

Coastal Wetlands

Along the Indian coastline, the brackishwater areas including marshes, backwaters, mangroves, inter- and sub-tidal measures about 14,16,300 hectares. These areas act as feeding and nursery grounds for a variety of commercially important fish, prawn and crabs, media for inland transportation, fishing etc.

Mangroves

In 1987, India had about 6740 km^2 area of mangrove, which is 7 per cent of world's mangroves. The extent of the mangrove has reduced since than as reported by Nayak (1992). In Table 6.2 mangrove extent in 1997 has been compared with that in 1987. By 1990s Mangroves are degraded and destroyed due to conversion of these areas for agriculture, aquaculture on the East Coast and industrial purposes on the West Coast.

Table 6.1: Ecological and Geomorphological Characteristics of the Main Coastal States of India

State	*Characteristic Features*	*Ecologically Sensitive Areas*
	Characteristics of East Coast	
Tamil Nadu	Coastal length–1026 km Narrow belt of sand dunes, low lying beach, plains mostly formed by rivers. Two major, two intermediate and two minor ports.	Tuticorin: Chank and pearl beds Coral Reefs at Gulf of Mannar Wildlife sanctuary at Vedaranyam Mangroves at Pichavaram and Muthupet
Andhra Pradesh	Coastal length–1014 km. The coastline is smooth and long with inundations, deltaic coast Krishna and Godavari, marshy, muddy coasts, Port at Vishakapatnam.	Pulicate Lake: Largest lagoon Mangroves at Krishna and Godavari delta Wildlife sanctuary at Kolleru lake 14 Sanctuaries
Orissa	Coastal length–450 km Coast is depositional, formed by Mahanadi, Brahmani and Baitarani delta. Port at Paradip	Wildlife sanctuary at Bhitarkanika Mangroves at Mahanadi delta Largest lagoon–Chilka lake
West Bengal	Coastal length–200 km Ganga and Brahmaputra river systems create large intertidal, deltaic mass. Hugly mouth is uneven formed by massive sedimentation, coast sand riffed with innumerable tidal creeks and estuaries	Sundarbans: Largest mangrove forests, swamps and backwaters
	Characteristics of Island Coast	
Andaman and Nicobar Islands	Coastal length–1962 km	Mangroves and coral reefs, 16 national parks and 94 wildlife sanctuaries
Lakshadweep	Coastal length–132 km 36 archipelago islands consisting of 12 atolls, 3 reefs and 5 submerged banks	Coral reefs
	Characteristics of West Coast	
Kerala	Coastal length–720 km Rocky coastal belt broken by small bays, creeks and fringed with islands. No major river, major port at Mumbai.	Marine sanctuary at Malvan

Mangroves are also important as "Coastal Stabilizers" and "Shelter belt areas". These formations protect the coasts and the landward areas from erosion and cyclonic destructions to some extent. Apart from these the mangrove forests of India have importance from a wildlife, recreation and education point of view. "Project Tiger" of Sunderbans and "Crocodile Sanctuary" in the Mahanadi delta are examples of such activities. Mangroves are found along the islands, major deltas, estuaries and backwaters.

Coral Reefs

Around India, coral reef formations are found in the Palk Bay, Gulf of Mannar, Gulf of Kachch, Central West coast of India, Lakshadweep atolls, and Andaman-

Nicobar Islands. Both the coral atoll and the fringing coral reefs are of utmost significance in Indian waters. A few species of corals have recently been reported from the Malvan (Maharashtra) coast. 32 genera from Minicoy Islands, 34 genera from Palk Bay and Gulf of Mannar, 25 genera from Andaman Islands, 9 genera from Lakshadweep and 3 genera from Nicobar Islands have also been reported. 342 species belonging to 76 genera from the seas around India have been described(Subramanian 2005).

Table 6.2: State-wise Distribution and Reduction in of Mangroves Area in Coastal India

State/Union Territory	*Government of India, 1987*	*Government of India, 1997*
West Bengal (Sundarbans)	420	212.3
Andaman and Nicobar Islands	119	96.6
Maharashtra	33	12.4
Gujarat	26	99.1
Andhra Pradesh	20	38.3
Tamil Nadu	15	2.1
Orissa	15	21.1
Karnataka	6	0.3
Goa	20	0.5
Kerala	Sparse	Nil
Total	674	482.7

Marine National Parks, Marine Sanctuaries

On account of their high bio-diversity, the Gulf of Mannar and Wandoor (Andaman) have been declared as Marine National Parks and Malvan coast (Maharashtra), Gulf of Kachchh, Jamnagar as Marine Sanctuaries. There are a number of other specialized ecosystems which exhibit a large variety of marine life and they include Chilika and Pulicat Lakes, Point Calimere, etc.

GIS and Modelling Application in Integrated Coastal Zone and Hazard Management

Mathematical modelling of coastal environment is an effective tool which provides a synoptic view of coastal processes like hydrodynamics, transport (salinity and sediment regime), oil spill etc (Walsra *et al.*, 2001). Geographical Information System (GIS) is another powerful set of tools for collecting, storing, retrieving at will, transforming, and displaying spatial data from the real world for a particular set of purposes (Burrough, 1992). The integration of GIS and modelling synthesize the larger data sets from a much wider range. It also provides flexibility to select relevant criteria that might not be achieved by manual methods. The GIS and modelling is widely being used to address following issues:

1. Inventories of Resource Distribution, location of ports, recreation areas, water bodies, land use features, biological organisms etc.
2. Monitoring of shoreline due to seasonal accretion/erosional problems. Inundation Mapping for protection against coastal hazards such as floods, storm surges, Tsunami.
3. Prediction the movement of the oil spills and risk assessment.
4. Designing appropriate solutions of coastal environmental problem by integrating model outputs especially on sea level changes circulation pattern, wave propagation and water quality prediction in coastal areas etc.

With an increase in the perception towards spreading a culture of prevention in the coastal zone management scenario, considerable emphasis is now being placed on research and development activities in the area of GIS and modelling application for coastal hazard preparedness and prevention. This has brought in a significant positive change towards the adoptions of integrated approach for coastal and marine area management concept. The integration of spatial data is a critical theme in the development and utilization of marine habitat modelling. In India, GIS based modelling is being extensively used in the following coastal and marine related applications.

GIS Application in Coastal Ecosystem Management

The coastal areas contain a rich resource base characterised by highly productive and extensive coral reefs, mangrove forests, seagrass beds, estuaries and lagoons. As stated earlier India is bestowed with 4460 km^2 area of mangroves, 2230 km^2 of coral reefs and dense beds of seagrass. In some coastal areas, mass nesting of turtles, considered as endangered, is observed. Critical habitats are those areas which are vital to the survival of a species in some phase of its life cycle or to the survival of a community, because of the ecological processes which occur within it. Critical habitats are rich in biodiversity and serve as feeding, nesting, breeding and nursery areas for marine organisms including endangered organisms. They are unique and support ecosystem stability and rich fishery resources and are susceptible to degradation due to natural processes and human perturbations.

To manage and protect the critical habitats, through ecosystem approach, Critical Habitat Information System (CHIS) is developed using Geographical Information System (GIS), remote sensing and database by ICMAM Project Directorate of Ministry of Earth Sciences for 11 critical habitats along our coastline (Figure 6.1).

Mangroves: Sundarbans, Coringa, Pichavaram, Gulf of Khambat

Coral Reefs: Gulf of Mannar, Kadmat, Malvan, Gulf of Kachchh

Areas of Biodiversity: Islands off Karwar, Cochin

Turtle Nesting: Gahirmatha.

Application of GIS as a tool for assessing the status of the critical habitats, has revealed that more than natural causes, the human perturbation has resulted in considerable damage to these ecosystems (Usha *et al.*, 2003). Large scale changes in areal extent of mangroves have been observed. The major causes for mangrove degradation have been identified as felling of mangroves for timber and fuel, cattle grazing, insufficient tidal flushing, conversion for aquaculture, etc. In case of coral reefs, some of the causes for degradation are quarrying of live and dead corals, dredging for industrial purposes, etc. Extensive areas of seagrass beds are being disturbed by intensive trawling and stakenet fishing. Commercially important seaweeds and shrimp seeds are overexploited by local community, which cause ecological imbalance and reduced sustainability of fishery resources. Uncontrolled mechanised fishing and human interference have led to decline in sea turtle population and shifting of nesting grounds (Usha *et al.*, 2003).

The Figure 6.2, given below indicates the extent of distribution of live corals of the Mandapam Group of Islands in the Gulf of Mannar. The overlay facility enables super imposing of depth contours around these islands. This enables the Resource Manager to understand the extent of distribution of live corals against the depth. Such an inventory based mapping is very useful to exercise preservation, conservation and regulatory measures in the coral reef areas.

Use of Modelling and GIS in Habitat Management: A Case Study

Gulf of Kachch (Gulf), a large macro-tidal water-body located in West Coast of India, occupies an area of 7350 km^2. The Gulf abounds in marine wealth with 37 ecologically rich islands, coral reef patches, mangroves, about 1500 km^2 mudflats, sandy and rocky beaches. The southern coast of Gulf has been declared as Marine sanctuary and Marine National Park. Gulf produces about 28 per cent salt of the nation's supply. Approximately 0.3–0.35 million m^3 sea bed material is being dredged and dumped in adjoining areas every year by port sector for navigational purposes. Gulf is hub of oil handling facilities. About 40 per cent of the oil is being imported through the facilities. Fishing is an important activity in the region. Gulf has suffered a general deterioration of its environment, leading to a strong feeling that the Gulf ecosystem especially corals, mangroves and fisheries are under stress due to development projects (Figure 6.3).

ICMAM concept was used to study and to suggest suitable solutions.

GIS and modelling tools were used to understand the impact of development activities in Gulf of Kachch to maintain a sustainable balance between economics and environment. Extensive data were collected to develop hydrodynamic models, which explain the role played by tide, bathymetry and currents in transporting the dissolved and particulate materials in the GOK region. (ICMAM-PD, 2004).

Two-dimensional hydrodynamic model (MIKE- 21) was used to study the general circulation, residual circulation, tracer studies, dissolved oxygen, salinity and oil spill in Gulf of Kachchh. Analysis of residual currents revealed the formation of

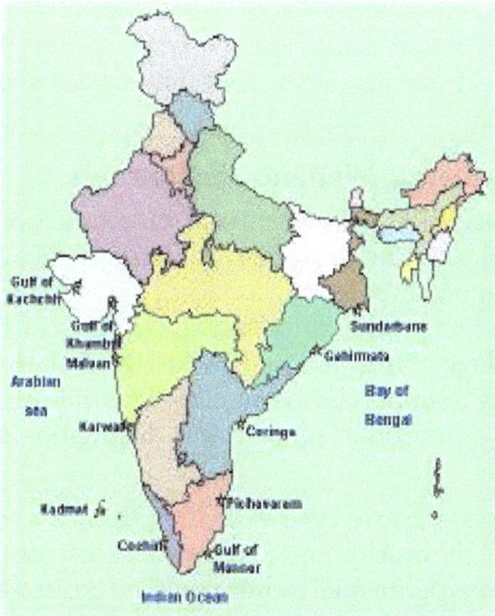

Figure 6.1: Map Showing the Location of CHIS Developed by ICMAM-PD

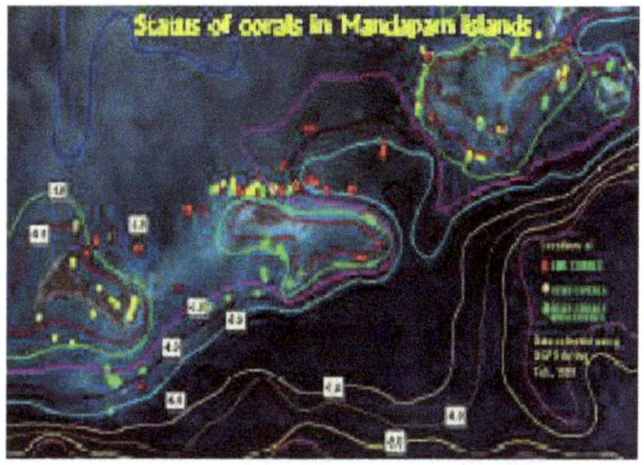

Figure 6.2: Live Coral Points Overlaid on Depth Contours in GIS

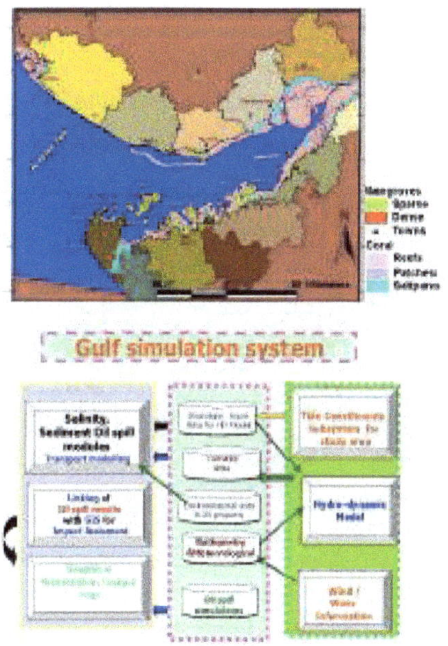

Figure 6.3: A GIS Map of Gulf of Kachchh and Flow Chart for Modelling

eddies in outer Gulf and the mixing capacity of central and outer Gulf (Kankara *et al.*, 2006). The Salinity modelling results give a synoptic view of salinity transport in Gulf. The salinity was found always high (~40 psu) in inner and shallow regions in comparison with the deeper channels of the area. Modelling studies have indicated that the discharge of bitterns have locally increased the salinity by 0.5 ppt, even though their characteristics were not traceable beyond 5 km (Kankara *et al.*, 2003). Such marginal increase of salinity over the years also played a role in increasing salinity along with other factors. The net salt flux was of order of 1.33 x 1013 towards the outer Gulf or open ocean. Dissolved Oxygen modelling is useful tool to estimated the availability and demand of oxygen, which is a indicator for health of water. It provides a clear clue about assimilation capacity of coastal/estuarine water (Kankara, 2006). An oil spill model was developed to simulation oil spill trajectories and to estimate the time travel to reach the oil at shore in different meteorological and hydrodynamic conditions (Kankara and Subramanian 2001). The risk of Gulf environment from oil spill was estimated by integrating GIS to oil spill model to locate and estimate the resources at risk and to set the priorities of the resources for protection (Kankara and Subramanian 2006).

Modelling and GIS Application in Coastal Wetland Restoration

Modelling is very useful tool for analysing the performance of engineering solutions to a coastal problem, prior to implementing the solution in field. Chilika, a "Ramsar Site" and largest brackish water lagoon in of India was facing the problems of siltation, shrinkage of area, decrease in fish productivity and an overall loss of biodiversity. Lagoon is an unique assemblage of fresh water, brackish and marine ecosytem and a hot-spot of biodiversity which support the Socio-economic mileau of about 0.2 million fishermen living in and around the lagoon and a destination of more than one million migratory birds.

Considering the sensitive lagoon eco-system, extensive modelling studies were carried out using MIKE-21 model to find out the best options to restore its ecosystem. The modelling simulation suggested that an experimental mouth could be created to increase the sea and lagoon interaction, so that the salinity flux into the lagoon would be improved by 40 per cent and the tidal flux would be improved by 45 per cent. This would rejuvenate the Lagoon ecosystem and the desired level of salinity *i.e.*, 15 ppt during summer in northern sector can be achieved (Ghosh *et al.*, 1995, Nayak *et al.*, 1997).

Accordingly, the opening to the sea was also accomplished opposite to the village Sipakuda on 23rd September 2000 (Figure 6.4). The opening of the artificial mouth is an historic and successful intervention in the history of the restoration of the ecosystem of the wetlands in India. The gradual reduction in the salinity from the lagoon mouth to the lagoon interior after the opening of the mouth is providing the desirable sense of direction for the euryhaline forms to enter the lagoon from the sea. This is facilitating the auto recruitment of the fish, prawn and crab juvenile in to the lagoon and catch has been increased tremendously (Figure 6.5).

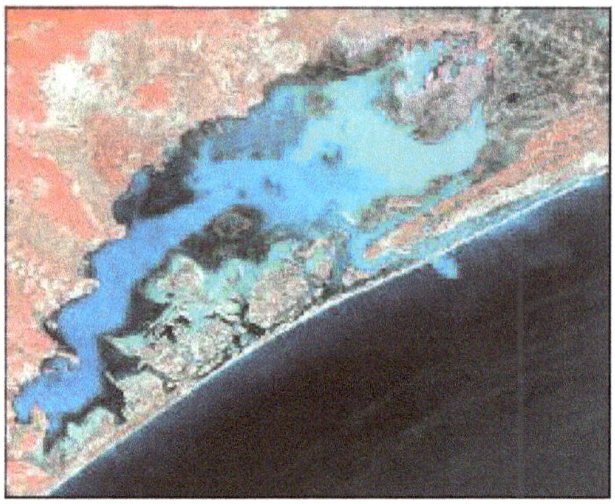

Figure 6.4: Chilika Lagoon and New Mouth

GIS in Shoreline Protection Studies

In world, many areas are being eroded and threaten the life and property of local population. The economic and human costs of the coastal erosion are growing as more people migrate towards coast. One of the major requirements of planning coastal protection work is to understand coastal processes of erosion, deposition, and sediment-transport, flooding and sea level-changes, which continuously modify the shoreline. The historical and functional approaches to study shoreline changes along with various landforms help in deciphering the coastal processes operating in an area (Nayak, 2000). Mathematical models provide proper understanding the oceanographic processes and their impact on coastline. Impact of development activities can be studied using a shoreline model. GIS has been extremely useful in studying the shoreline changes particularly in areas where the seasonal reversal of littoral drift occur. Shoreline-change mapping (1967-68, 1985-89, 1990-92 periods) for the entire Indian coast has been carried out using LANDSAT MSS/TM and IRS LISS II data on 1:250,000 and 1:50,000 scale (Nayak *et al.*, 1992). Erosion has been observed north of Visakhapatnam, Paradip, and Ennore, north of Madras, near Nagapattiam and Kanyakumari ports on the East Coast of India while deposition has been observed south of these ports.

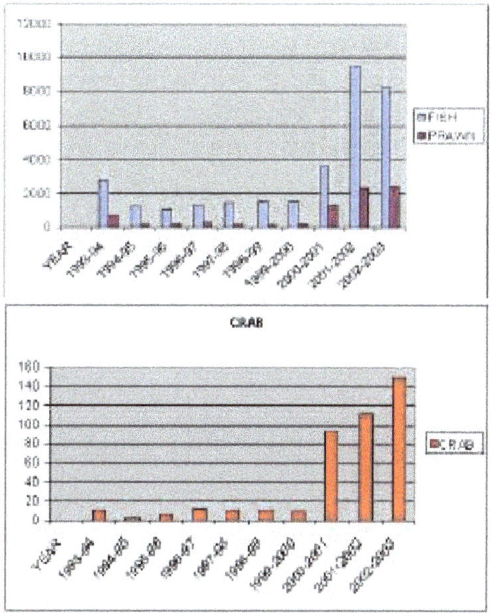

Figure 6.5: Yearwise Fish and Crab Landing in Chilika Lagoon from 1993-2003

This is prominent in the coastline especially where the near shore obstruction like breakwater of ports and harbours etc. exist. Classical example of the effect of obstruction of littoral drift by construction of breakwater leading to erosion of the coastline is found at Chennai, Tamil Nadu (Figure 6.6). This construction of breakwaters resulted in arresting the northerly drift causing accretion on the southern side and erosion on the northern side. Construction of an outer arm in subsequent years and construction of fisheries harbour

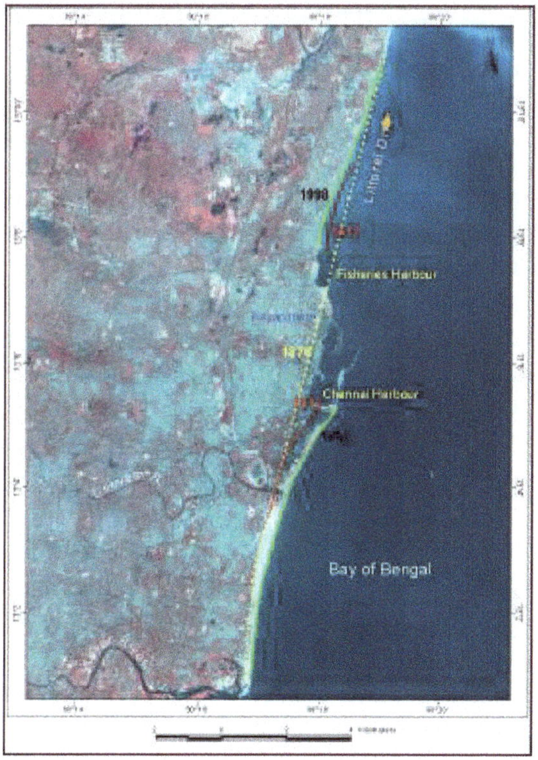

Figure 6.6: Shoreline Changes Along Chennai Port

further north of the port triggered severe erosion along the northern. A stretch of about 5 km north of the fisheries harbour has been eroded and it is estimated that an average width of about 75 m of the beach has been lost in the last few years. The only brighter side is the accumulation of sand on the southern side, which is now known as the Marina beach having width of nearly 1 km (ICMAM-PD 2002).

GIS and Modelling in Surge Prediction and Coastal Flooding

In Indian, the study of numerical storm surge prediction was pioneered in 1972 by scientists in IMD and IIT Delhi. Lata *et al.* (2002) developed storm surge forecasting model for East Coast of India. The model was integrated to GIS resource map for validating the coastal flooding signature of a cyclone prone low-lying area along Nellore to Machalipatnam of Andhra coast. The extent of inundation of due to 3.1 m surge height was estimated to be about 1207 km^2 with varying distance from the coast. Such an output is helpful in planning several preventive and remedial measures, when such calamities occur.

Modelling and GIS in Studying Effect of Oil Spills

Oil spill management is a multi agency task and should coordinated through a national agency. GIS and mathematical modelling is used extensively all over the world to understand the movement of the oil spills especially close to the ecologically sensitive areas like coral reefs and mangroves. Coast Guard is responsible for combating oil spill pollution in Maritime zones of India in case of oil spills and has a capability to deal with a major oil spill of in Exclusive Economic Zone of India. There are three response centres located at Mumbai, Chennai and Port Blair equipped with response equipment and limited capabilities exist at Kochi (Kerala) and Vadinar (Gujarat). Computer models are able to make predictions about the trajectory and fate of spilled oil and to generate the likelihood of sensitive resources being threatened and associated time scales using historical weather data. Modelling simulations can provide clear idea about oil movement and enhance the decisions concerning strategy development and the identification of necessary response capability.

In India, Ministry of Earth Sciences (MoES) has developed an Oil Spill Modelling and Prediction System for Gulf of Kachchh (Figure 6.7). Now efforts are on to develop a online system for entire Coastline. This system will be useful in identifying the pollution source and taking preventive measures like deployment of oil containing devices like booms to protect these resources.

GIS and Modelling Application in Protection of Vulnerable Areas for Tsunami Hazard

In order to evolve strategies for protection of vulnerable areas from coastal hazards, it is important to understand the susceptibility of the coastal areas to coastal hazards like tsunami. Spatial data on the factors that aid in the propagation of the tsunami waves in the coastal areas such as landuse, geomorphology, land elevation, bathymetry, extent of inundation and socio-economics are necessary to prepare vulnerability maps for this area. GIS is appropriate tool to handle such data sets. These data are input in mathematical models to predict the vulnerability of the coast using different source parameters. Then inundation maps can be prepared using GIS overlaid with all other relevant data on critical infrastructure and socio-economics. These maps could be used for planning purposes in case of coastal hazards in future and to understand the impact tsunami for different scenarios.

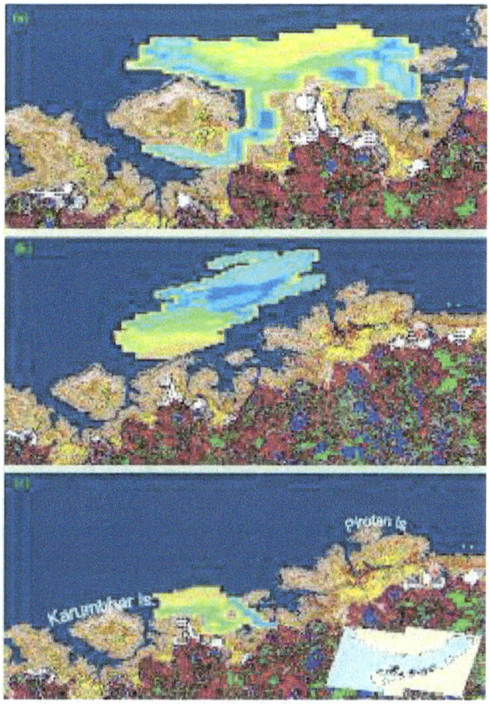

Figure 6.7: Oil Spill in Gulf of Kachchh for Different Wind Conditions

Preparedness to Deal with Coastal Hazards in India

The Ministry of Earth Sciences (MoES) in association with Department of Science and Technology (DST), Department of Space (DOS), Council of Scientific and Industrial Research (CSIR), and academia, has initiated several scientific programs such as ocean observation system, coastal zone management, ocean information system, and tsunami warning and predictions systems. These programs provide a vital information in decision-making processes for issues related to coastal Hazard management. The scientific tools such as mathematical modelling, remote sensing, GIS and GPS are being used for generating inventory on coastal habitats, shoreline condition, condition assessment of protected areas, forecasting of ocean and coastal sea state and potential fisheries zone to user community and industry.

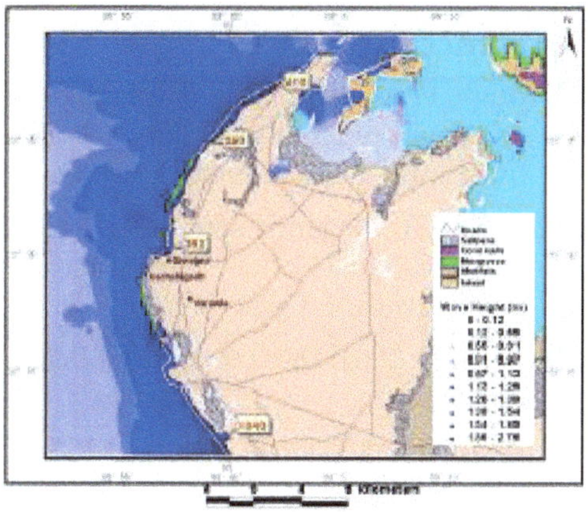

Figure 6.8: A Typical Inundation Map for Dwarka Region for 1945 Makran Tsunami

An Early Warning System for Tsunami and Storm Surges in Indian Ocean is being setup at INCOIS, Hyderabad. The system comprising the Near real time determination of Earthquake parameters and a comprehensive real time Ocean observational network will be placed by from October 2007. The main features of the system are:

1. Developing numerical models for Tsunami and Storm Surges with all associated data inputs.
2. Generating Coastal inundation and vulnerability maps. Setting up a dedicated Tsunami Warning
3. Centre (including Storm Surge) in India and operating it on 24 x 7 basis for generation of timely advisories.
4. Capacity building, education and training of all stakeholders.

Conclusion

The present paper shows a few examples of application of mathematical modelling and GIS for Coastal area and hazard Management. These are the appropriate tools to assess the impact of developmental activities and natural hazards. The use of modelling and GIS in India for coastal and marine applications is increasing day by day. The Coastal Managers must explore the possibilities to use these tools for effective management of various phenomena like erosion, oil spill, resource inventory etc in an innovative manner. The marine application of GIS is being challenging, demands greater R&D on application of this tool in the coming years.

Acknowledgement

I expresses my sincere gratitude and thanks to Dr. P. S. Goel, Secretary, Ministry of Earth Sciences, Govt. of India and Dr. B. R. Subramanian Project Director, ICMAM-PD, Chennai for their constant encouragement and allowing me to participate in this workshop. This chapter has been prepared based upon the studies supported by ICMAM PD/MoES under various projects.

References

Burrough, P. A. 1992. 'Spatial data quality and error analysis issues: GIS functions and environmental modelling', Second International Conference/Workshop on Integrating GIS and Environmental Modeling, Breckenridge, CO, U.S.A.

Ghosh L.K., T. Nagendra and R.S. Kankara. 1997. Salinity conservation in Chilika Lake, National Conference on Harbour and Ocean Engineering INCHOE-1997, Trivendrum, in Dec 1997.

ICMAM-PD. 2004. Draft report of ICMAM plan for Gulf of Kachchh, ICMAM-PD, Chennai.

ICMAM-PD. 2002. Draft report of ICMAM plan for Gulf Chennai, ICMAM-PD, Chennai.

Kankara, R. S. 2006. An Integrated Approach to manage Gulf of Kachchh Habitat (un-published Ph.D thesis), Institute of Ocean Management, Anna University, Chennai

Kankara, R.S. and B.R Subramanian. 2001, Simulation of the movement and fate of oil pollutants in Gulf of Kachchh, HYDRO 2001,at CWPRS, Pune.133-141.

Kankara, R. S., B. R. Subramanian and V. Sampath. 2003, A study on tidal circulation and salinity regime in Gulf of Kachchh- Measurement and Modelling, Proc. of International Conference on Coastal and Ocean Technology (COT-2003), NIOT, Chennai,279- 292

Kankara, R. S., B. R. Subramanian and V. Sampath, 2006. Validation of a Deterministic Hydrodynamic Model in ICZM framework for Gulf of Kachchh, India. Journal of Coastal Research 2006 (In press).

Kankara, R. S. and B. R. Subramanian. 2006. Oil Spill Sensitivity Analysis and Risk Assessment using Integrated Modelling for Gulf of Kachchh. India. Journal of Coastal Research 2006 (In press).

Latha, D., *et al.* 2002. A forecasting model for Storm surges and estimation of coastal inundation along the East Coast of India; Vol. 2, 2000, NIOT, Chennai, India.

Nayak, B. U., L. K. Ghosh, S. K. Roy and R. S. Kankara. 1998. Hydrodynamics and Salinity in the Chilika Lagoon. International workshop for "Sustainable development of Chilika Lake" organised CDA Bhubneswar 1998 .

Nayak, S. *et al.* 1992. Coastal Environment. *Scientific Note.* Space Applications Centre, Ahmedabad.RSAM/SAC/COM/SN/11/92.114 p.

Nayak, S. 2000. Critical issues in coastal zone management and role of remote sensing. In Subtle Issues in Coastal Management, Indian Institute of Remote Sensing, Dehradun. Pp. 77- 98

Nayak, S. 2000. Quaternary geomorphology of the Gujarat coast–an overview. Ind. Jour. of Geomorphology, 5 (1&2): pp.9-22

Subramanian, B. R., 2005. Coastal Zone and its Significance. In Coastal Zone Management training Mannual, ICMAM-PD, Chennai.

Tune Usha, S. Sundaramoorthy, T. Shunmugaraj, GVM Gupta and R. S. Kankara. 2003. Application of GIS in Management of Coastal Critical Habitat, Proc. of International Conference on Coastal and Ocean Technology (COT-2003), NIOT, Chennai,15-24.

Walstra, D.J.R *et al.* 1999. Evaluation of hydrodynamic area model based on the COAST3D data at Teignmouth 1999, COAST3D publication No. D4.

Chapter 7

Coastal Ecosystem, Rehabilitation, Hazards, Vulnerability and Management in India

A.L. Ramanathan[1], S.K. Varshney[2] and S. Chidhambaram[3]

[1]*School of Environmental Sciences, Jawaharlal Nehru University, Delhi*
E-mail: alrjnu@hotmail.com

[2]*Department of Science and Technology, New Mehrauli Road, New Delhi*
E-mail: skvdst@nic.in

[3]*Department of Earth Sciences, Annamalai University, Chidambaram, India*

ABSTRACT

Coastal ecosystems in India is under stress due to the ever increasing population density, over exploitation of natural resources, fresh water shortage, pollution and other anthropogenic activities. They are also under severe threat from natural disasters like storm surge, flood, cyclone, high tide, wave erosion and recently tsunami. Therefore studies on saline water intrusion, degradation of water quality, vulnerability of species in coastal ecosystems are important issues in our country. There have been a few studies in these aspects to find out strategies to over come these problems. There is an urgent need to address immediate environmental issues including climate change impact, salinity problems, aqua culture impacts, etc. In order to reduce vulnerability in the coastal areas, especially from the natural hazards, it is important to establish a regional early warning system. Development of sustainable dunes and greenbelts supported by sound scientific research work may reduce the long term burden of these coastal ecosystems and its communities without ignoring the huge energy and natural dynamics of coastal systems.

Keywords: *Coastal ecosystem, Hazards, Salinity, Management, Pollution, Vulnerabilities.*

Introduction

Coastal ecosystems all along the globe are being affected and influenced by the forces of global environmental changes, resulting in vulnerability, hazards, risks and insecurity, not only of the coasts but also of the coastal communities. The Indian coastal system, with a stretch of ~7500 km^2 coastline and an exclusive economic zone of ~2 million km^2, has been at the receiving end of many natural upheavals and extreme events. Biogeophysical impacts of these changes, its influence on the goods and services from the coastal ecosystem and the larger socio-economic and political ramifications are intertwined in a complex web of interactions and thus crucial to understand and address the uncertainty associated with coastal disaster preparedness in India. The tsunami of December 2004 and subsequent challenges of restoration and rehabilitation, underscores the urgency to build upon the existing knowledge for planned adaptation options, measures and policies in the context of sustainable development and equity.

Throughout the world, natural disasters become one of the most serious problems at low-lying coastal areas, which are strongly affected by flooding or by active processes of shoreline erosion and sedimentation. This also has serious consequences for local communities and tourists. Functions and values of the coastal system have been degraded, and public safety and economy have been impacted.

To reduce the vulnerability of coastal communities from natural hazards, various steps could, establishment of a regional early warning system, and applying construction setbacks, greenbelts and other no-build areas in each nation, supported by sound scientific data base. It is not possible to sustain any development in coastal areas if the huge energy and natural dynamics of coastal systems are ignored. Many of the adverse social and economic impacts of the tsunami have occurred because people have been made more vulnerable to natural hazards through poor planning and the ineffective management of coastal development. The technical and financial constraint should not stand in the way of letting the natural dynamics of the natural coastal process. The long-term benefits that can be gained by working with the natural processes those create and maintain healthy ecosystems and a flow of social and economic benefits to humankind far outweigh the cost. It is more crucial of a time, when human population is increasingly concentrated along coastlines, sea level is rising and long established weather patterns are changing. We therefore know that hazards created by storms, the reshaping of coastlines by processes of erosion and accretion, instabilities produced by new patterns of land use as well as such relatively rare occurrences as cyclones, tsunami, etc., will together make shorelines increasingly hazard-prone.

Tsunami Impact on Coastal Ecosystems

Palanivelu *et al.* (2006) have studied the ground water quality assessment in the tsunami-affected coastal areas of Chennai, located in the 250 km north of the study area. They have studied the TDS values and observed that at few locations away from sea showed an increase of TDS from May to September 2005. The increase may be due to insufficient rainfall in Chennai during this period till September, which in turn shows that the southern part of the coastal Chennai is highly contaminated compared to the northern part. TDS levels observed after tsunami were within the range as observed during September 2004. Thus their study reveals that the recorded TDS values over time indicate that there is no major impact of the tsunami on water quality. They concluded that the groundwater quality has deteriorated due to lack of sufficient rainfall leading to sea-water intrusion that is reflected in high TDS and chloride content of the samples.

Muralideran *et al.* (2005) studied the imprint of 26 December 2004 Sumatra earthquake in aquifers of Hyderabad granite pluton.They explained the earthquake had induced hydrological changes caused by changes in hydrostatic pressure due to earthquake-induced changes in crustal volumetric strain. The effects before and after the Sumatra earthquake were clearly registered in hydrographs of wells located over Hyderabad granite pluton. They presented the micro-level impressions registered in hydrograph of a deep bore-well located in the National Geophysical Research Institute (NGRI) campus, Hyderabad, supported with records of other bore wells located more than ~100 km away from the NGRI well site. The signature of tsunami on water level was well registered by a quiescent prelude followed by turbulence. The quiet period between 08:44:40 and 09:19:20 is the first of its kind in the hydrograph record so far gathered. It coincides with recession of sea observed prior to the tsunami. Perturbations observed between 09:19:20 and 10:00:40 h coincide with the hit time of tsunami on the east coast. According to the tidal records of Survey of India and its notification, the tsunami that hit at Visakhapatnam by 09:10:00 h has considered for analysis as it is near Hyderabad compared with other tidal stations. The disturbance observed on hydrograph lags nearly by 10 min with the hit time of the tsunami. They observed rise in water level from 25 December 2004 to 2 January 2005, and a decline beyond 2 January during non monsoon period, collectively indicates the effect of the Sumatra main earthquake and aftershocks. The rise in water level is about 90 mm. In order to confirm their observation, continuous water-level data of forty wells in Mehaboobnagar district, Andhra Pradesh were also collected from State Groundwater Department and analyzed. Similar rise in water level during that period was also observed by them, in only four out of forty bore-wells near the coast.

Kanakasabai and Rajendran (2005) reported the trend of micrometeorological parameters during tsunami in east coast of India. They measured the micrometeorological parameters continuously at Portonova on the East coast of India (with the help of a 30 m height meteorological tower). The parameters like wind speed, wind direction, temperature and solar radiation were recorded during the time of Tsunami. The wind speeds measured at 10m, 17m and 30m heights show a decreasing trend for three days (25th, 26th, 27th December 2004). Likewise the temperature also shows a decreasing trend on the day of Tsunami. The solar radiation was steadily increasing without any modulation on the tsunami day, which was not observed on the previous day, or the next day.

World Health Organization (2005) report after tsunami observed that destruction brought by the tsunami (mainly in India) on piped-in water supplies, groundwater pumps, the saline water intrusion of shallow wells and surface sources, rendered the pre-tsunami supplies throughout the affected area completely unusable. Despite evidence of certain emergency measures like the household water treatment, including chlorination, filtration, solar disinfection and combined flocculation/disinfection, did not play a significant role in the initial phases of the tsunami response with the possible exception of boiling. The report felt that much of the drinking water response to the Indian Ocean tsunami focused on providing a sufficient quantity of water, with perhaps less focus on quality. Further the report observed that surveillance of water quality was occasionally hampered by confusion over responsibility as well as a lack of equipment and trained personnel. There is evidence in certain areas that water quality, both at the source and especially at the household level, was compromised by high levels of faecal pathogens.

Similar impact studies after the 26th December 2004 Tsunami was reported by Minerals and Geoscience Department of Malaysia in 2005. Selective surveys on groundwater contaminated by saline water were carried out immediately after tsunami. They have monitored the groundwater wells and dug wells in areas inundated by the tsunami in the State of Perak, Pulau Pinang and Kedah. The affected areas were demarked but those observed wells are not in use for daily consumption as piped water supply is available in those areas. The groundwater from the monitoring wells in Perak indicated no contamination from the salt water. In Kedah where groundwater sampling was carried out in Langkawi and Kuala Muda as well as in areas of Pulau Pinang which were significantly affected by tsunami indicated that dug wells which were not protected from the inflow of sea water into the wells, were contaminated by the sea water. Parameters such as sodium, chloride, total dissolved solids were recorded to be much higher. Conductivity readings were also high. These wells were also contaminated by organic materials as well as bacteria. Coliform counts were high in some wells. Some wells were flushed before sampling and these wells indicated that there were no contaminations to the shallow aquifer by the sea water or any other contaminants. They also observed other effects in the area due to tsunami and earth quake, such as sinkholes have occurred in the areas underlain by limestone outside the inundated area, most likely caused by earthquake. Some former mining lakes in the limestone area have been reported to have shown bubbly effect immediately after the big earthquake.

All these studies focused on the importance of ground water quality and its role in drinking water supply to tsunami affected regions in Asia. The authors have conducted a detailed study on this with support from DST and UNSECO- India to get an insight into the ground water quality deterioration and the processes responsible for the hydro chemical changes immediately after tsunami. The study also focuses on its evolution over a period of time, which will reflect the present situation of ground water quality in these affected regions and help in providing guidelines for future for these coastal communities.

We have identified the tsunami inundated region ranges in these regions from 0.5 to 1 km from the coast in Tamil Nadu (Table 7.1).From our detailed study we have inferred the processes responsible for change in water quality after tsunami in the shallow ground waters of this region (Figure 7.1) which have been controlling the hydrochemical changes of ground water after tsunami. Our study identified this process as responsible for the present day ground water quality used for drinking water. After Tsunami, sea water entered the water table through the open wells or tube wells and the entrapped water got infiltered into the water table of the coastal alluvium during January 2005. Later high temperature in the summer months might have resulted in the formation of salt precipitates near the surface or in pore spaces and subsequent dilution in the end of March 2005 by sparse rain. After precipitation and dissolution, salt leached out from surface to the shallow groundwater zones enhancing the EC and TDS of the groundwater during August 2005. Thus the impact of direct infiltration or direct mixing of tsunami waters on these ground waters in coastal areas of Tamil Nadu has relatively lesser effect than the subsequent precipitation and dissolution of salts formed by the entrapped sea water which happened after the tsunami event.

Table 7.1: Run Up and Inundation During Tsunami in the Tamil Nadu Region

Sl.No.	*Location*	*Distance from Coast in km*	*Inundation Distance (km)*	*Run Up (m)*
1.	Thirumullaivasal	0.32	1.02	2.7
2.	Meimokkaral	0.52	0.75	2.2
3.	Pumbuhar	0.2	0.32	3
4.	Vazhuthalaikudi	2.86	3.24	2.9
5.	Palayar	0.64	0.78	2.6
6.	Tandavankulam	1.65	2.97	1.16
7.	Kulayar	1.05	2.67	1.98
8.	T.S. Pettai	1.97	3.1	2.1
9.	Portonovo	1.12	1.68	2.9
10.	Kottaiyamedu	0.32	0.72	3
11.	Nayakarpalayam	0.6	1.56	2.7

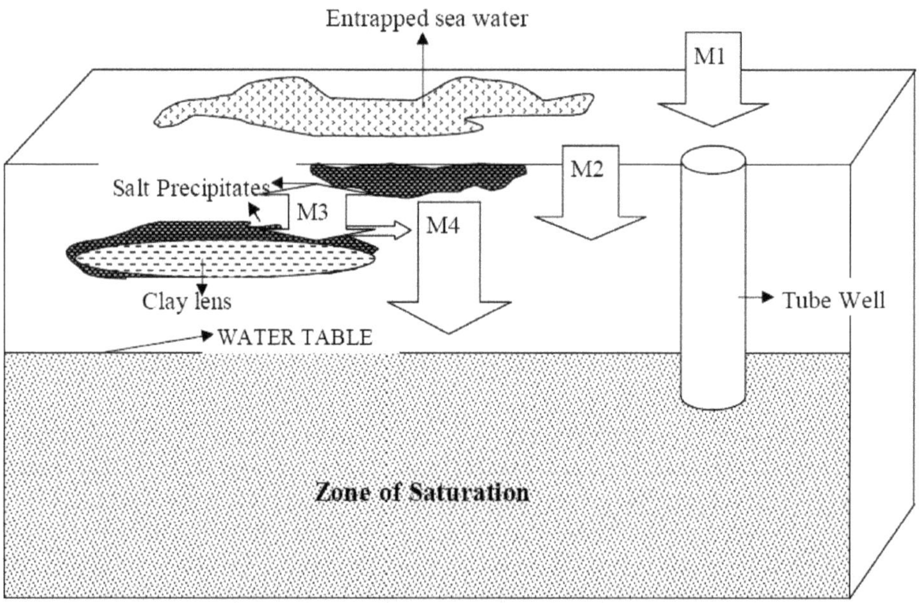

Figure 7.1: Diagram Depicting the Factors Responsible for the Water Quality Change in Ground waters After the Tsunami in Cuddalore–Pichavaram Region of Tamil Nadu

Managing the Impact of Tsunami

The nations of the region and several international organizations are working together to develop an effective tsunami early warning system that will reach the entire regional community, particularly the most vulnerable groups. What is known about past and future coastal change can be applied to define a reference line showing where the shoreline is anticipated to be, for example, by 2050. Detailed aerial photographs could be prepared reach by reach for all of the region's shoreline, showing conditions as they were before and after the tsunami. A reference line could be drawn on such photographs showing the mean high water mark anticipated by the 2050 median projection for a sea level rise of 30cm made by the Intergovernmental Panel on Climate Change (IPCC). This reference line should be modified by other tectonic and coastal data affecting anticipated coastal change in specific areas. Where information is available, data on the anticipated effect of historical trends in erosion and accretion to 2050 and the inland extent of flooding in past storms should also be integrated into estimates of the future position of shorelines.

A recommended construction setback line should be established by each government as a set distance and/or elevation inland of the reference line. The area seaward of the setback line should be designated as a strictly enforced "no-build" zone. It is essential that such setbacks are incorporated into the existing regulatory system and are applied equitably to the wealthy and the poor.

On low-lying shorelines with little topographical relief, practical disaster preparedness plans should be developed and tested that feature speedy evacuation of people to protected shelters. The width of the no-build zone determined by the setback should be greater in as-yet-undeveloped shores than in already urbanized areas. Designate setback lines with permanent on-site markers and enforce them uniformly as a regulatory measure. Exceptions for building structures seaward of the setback line should be granted only where required to support such water-dependent activities as fishing and navigation (not tourist facilities or permanent settlements). Where such exceptions are granted, structures should be temporary or built to withstand flooding by strengthened structural members and elevated first floors that permit flood waters to flow through unimpeded. Attention should be given to the impact of such structures on adjacent coastal areas, and mitigation actions taken.

Avoid or minimize involuntary resettlement in accordance with the Guiding Principles on Internal Displacement presented to the UN Commission on Human Rights and the General Assembly. A "no build zone" applied to a settled coastline may have severe consequences for those deprived of land tenure or rights of residence. Where relocation is judged to be in the best long-term interest of those affected, provide adequate compensation for land and property. Also, establish channels for grievance resolution at the appropriate levels of government. Adopt appropriate building codes for all structures, including seismic codes in earthquake prone areas. Provide potable water, sewage collection and treatment, and adequate drainage systems in all new and reconstructed communities. Favor standardized, modular systems with interchangeable components to achieve cost savings and reduce future maintenance costs. Encourage the use of local labor in all phases of the reconstruction process thereby generating employment and enhancing the marketable skills of the poor. Enhance the ability of the natural system to act as a bioshield to protect people and their livelihoods by conserving, managing and restoring wetlands, mangroves, spawning areas, seagrass beds and coral reefs; and by seeking alternative. Natural barriers to flooding and coastal erosion, such as coral reefs, near-shore rock outcrops, sandbars, and sand dunes should be protected from construction activity and uses that compromise their structural integrity. They reduce, absorb and redirect waves and floodwaters. Wetlands, lagoons, river estuaries, and reefs are essential to sustaining fisheries, public health and the many livelihoods that support coastal populations. They contribute to a healthy and aesthetically pleasing environment for a seaside holiday. A portion of the funds for rehabilitation should therefore be assigned to protect and restore these habitats. Reconstruction will require thousands of cubic meters of sand for cement and for fill, and building materials of every description. Traditionally, many of these materials have been taken from the coast itself. When sand is mined from beaches, dunes and coastal rivers, mangroves are cut for timber, and wetlands filled as building sites coastal settlements become more vulnerable to hazards of every description.

Conduct rapid assessments that involve local people in the identification of natural areas important to fisheries production, the recycling of wastes, shoreline stabilization and scenic quality, including coastal wetlands and mangroves, seagrass beds, and coral reefs. The aerial photographs and maps used for establishing setback lines can be used in this process of identifying critical areas. Incorporate these natural features and habitats into a designated coastal bioshield that maximizes the protection from coastal hazards and the associated benefits provided by these natural features. Adopt measures to protect bio-shields from activities that compromise their natural qualities. Protect them from future disturbance and, where feasible, restore them. Where feasible, plant trees seaward of the setback line to form a greenbelt that buffers the shore from waves, floods and erosion.

1. Prohibit the mining of sand, coral and stone from coastal waters within the 20-meter depth contour.
2. Regulate sand mining from rivers.
3. Declare wetlands and mangroves as off limits for harvest of wood.
4. Prohibit the filling of wetlands and estuaries.
5. Promote design that is cost-effective, appropriate and consistent with best practice and placement of infrastructure away from hazard and resource areas, and favoring innovative and soft engineering solutions to coastal erosion control.
6. The reconstruction is an unprecedented opportunity to relocate communities away from hazardous and unhealthy areas, rectify badly designed infrastructure and services, and reduce previous inequities in their availability and distribution.
7. Place arterial roads, railroads and other transportation infrastructure well inside the setback line, and site access-ways perpendicular to the coast.

8. Limit investments in erosion control to those situations where pre-existing infrastructure or settlements make it cost-effective, and where it is considered environmentally justifiable; favor soft solutions (placement of sand, planting vegetation) over hard solutions (breakwaters, groins, shoreline armoring).
9. Identify natural barriers to flooding and coastal erosion, specifically coral reefs, near-shore rock outcrops, sand bars, and sand dunes; protect them from construction activity and uses that compromise their structural integrity.

Adopt ecosystem based management measures; promote sustainable fisheries management in over-fished areas, and encourage low impact aquaculture. The rehabilitation of hundreds of kilometers of shoreline should generate many opportunities for more diversified and more sustainable livelihoods. A primary concern must be the future prospects of communities dependent upon fishing. These same communities contain a high proportion of the region's poorest people. A recent statement prepared by World Fish points out those coastal fisheries in Asia were severely depleted and over-fished before the tsunami. Too many boats taking too many fish had in some areas reduced fish stocks to less than 10 per cent of their original levels and destroyed or degraded the habitats upon which these potentially renewable resources depend. A trend toward the use of damaging gear and the use of increasingly destructive fishing methods-such as small mesh nets that take juveniles-has made the situation progressively worse. The tsunami has only added to the problem. While assisting fishers by replacing equipment and rebuilding boats, ensure that less destructive and more sustainable fishing practices are adopted. Assist fishers who do not wish to return to fishing by developing alternative livelihoods. This will contribute to reducing fishing effort and restoring natural resources.

Promote employment-intensive fisheries operations that contribute directly to poverty alleviation and food security.

Climate Change and ICZM

Climate Change is often mentioned as one subject of an Integrated Coastal Zone Management (ICZM). Having Climate Change in mind as a global phenomenon, which has impacts not only but especially on coastal regions it could be recognized as an indispensable challenge for an *Integrated* Coastal Zone Management. Impacts of Climate Change and sea level rise might be serious in vulnerable areas like coastal wetlands, coral reefs or small islands and it could be increased by inappropriate land-use and development in the coastal zone. This is why conventions like The Ramsar Convention on Wetlands (Resolution VIII.4, 2002) recognize ICZM as an important method to deal with Climate Change impacts. Therefore one needs as detailed as possible idea of what might happen due to Climate Change in the considered region.

A growing body of literature in the past two decades has examined climate change as the most important issue in global environment, and analysed vulnerability and biodiversity loss arising out of the same. The lack of formal methodologies in the area of climate change–vulnerability relationship poses a big challenge and also an opportunity to continue research in this area. Study by Intergovernmental Panel on Climate Change (IPCC), links vulnerability with climatic change, and point out that the vulnerability of a region depends to a great extent on its wealth, and that poverty limits adaptive capabilities (IPCC, 2001). Further, they argue that socio-economic systems typically are more vulnerable in developing countries where economic and institutional circumstances are less favourable. Also a common theme in the climate change impacts and vulnerability literature is the idea that countries, regions, economic sectors and social groups differ in their degree of vulnerability to climate change (Bohle *et al.*, 1994). This is due partly to the fact that changes in climatic patterns are uneven and are also not evenly distributed around the globe. Though vulnerability differs substantially across regions, it is recognized that "even within regions… impacts, adaptive capacity and vulnerability will vary" (IPCC, 2001). In the Indian scenario it is likely that there will be an increase in the frequency of heavy rainfall events in South and Southeast Asia. The average temperature change is predicted to be in the range of 2.33° C to 4.78° C with a doubling in CO_2 concentrations (Watson *et al.*, 1998). Most of the other studies try to measure the vulnerability of a region to specific events like sea level rise, changes in temperature, rainfall etc. The analysis of the pattern of vulnerability and human security of the people living on the eastern coast of India and examining the relationship, if any, between climate-change and vulnerability for the people living in the eastern coastal districts of India on micro level is the need of the hour.

The phenomenon of climate change has a number of physical consequences. Desertification, sea level rise, and change in climate extremes are a few examples associated with physical consequences of climate change. These physical consequences in turn can be instrumental in triggering of social consequences like conflicts which increase the vulnerability of the people and pose a threat to human security. Societal characteristic plays a major role in increasing vulnerability of the people to climate change. Characteristics like low income levels define the level of poverty and can prevent long-term provisioning at the household level. Other characteristics those are associated with increasing the vulnerability of people in long run like poor infrastructure lack of human capital skills for system planning and management, population pressure etc.

Vulnerable regions and populations are there throughout the globe. Some regions and people are more vulnerable to climate change than the others and hence require immediate attention. In this sense vulnerability not only includes the poor but also those people who risk falling below the poverty line due to shocks at a future point of time. Associated with the shocks is the destruction of income and livelihood opportunities which directly contribute to the vulnerability of the household. Therefore some of the factors that contribute in enhancing the vulnerability of the people can be the following:

1. Lack of access to resources (economic vulnerability)
2. Disintegration of social patterns (social vulnerability)
3. Degradation of environment and inability to protect (organizational vulnerability)

4. Lack of access to information and knowledge (educational vulnerability)
5. Lack of public awareness (attitudinal and motivational vulnerability)
6. Lack of access to political power and representation (political vulnerability)
7. Beliefs and customs (cultural vulnerability)

India is one of the most important countries in the world with regard to the environment. With a large and growing population, India's emissions of greenhouse gases are increasing. Indian population continues to depend upon agriculture for food and livelihood (67 per cent, 1995 estimate) makes the Indian people vulnerable to climate variation and change. The climate change and its potential impacts on agriculture are addressed by several studies in recent times. There are a number of studies in the agricultural sector that signify the effect of climate change. Seshu and Cady (1984) have estimated a decrease in rice yield at the rate of 0.71 ton/ha with an increase in minimum temperature from 18°C to 19°C and a decrease of 0.41 ton/ha with a temperature increase from 22°C to 23°C. Sinha and Swaminathan (1991) show that a 2°C increase in mean air temperature could decrease rice yield by about 0.75 ton/hectare in the high yield areas and by about 0.06 ton/hectare in the low yield coastal regions until and unless new varieties are brought up. Also, a 0.5°C increase in winter temperature would reduce wheat crop duration by seven days and reduce yield by 0.45 ton/hectare. Additionally an increase in winter temperature of 0.5°C would thereby result in a 10 per cent reduction in wheat production in the high yield states of Punjab, Haryana and Uttar Pradesh. The study by Achanta (1993) concluded that the impact on rice production would be positive in the absence of nutrient and water limitations. Rao and Sinha (1994) in their crop-simulation study have estimated that under a doubling carbon dioxide climate change scenario, the wheat yields could decrease by 28 per cent–68 per cent without considering the carbon dioxide fertilization effects. Aggarawal and Sinha (1994) showed that in North India, a 2°C increase would reduce yields in most places. These are model generated conjectures or scenerios for general guidance only.

Climate change could represent an additional stress on ecological and socioeconomic systems that are already facing tremendous pressures due to rapid urbanization, industrialization and economic development in a developing country like India. With its huge and growing population, a 7500 km long densely populated and low-lying coastline, and an economy that is closely tied to its natural resource base, India is considerably vulnerable to the impacts of climate change (DOD, 2002). Despite the rapid growth of India's industrial and service sectors over the past decade, agriculture continues to dominate India's economy. Among a population of almost 1 billion people, approximately 68 per cent are directly or indirectly involved in the agricultural sector. Because the majority of Indian agriculture is rainfed, climatic changes that alter temperature and precipitation patterns may pose serious threats to agricultural production. Scenarios generated by computer models show that India could experience warmer and wetter conditions as a result of climate change, including an increase in the frequency and intensity of heavy rains (Watson *et al.*, 1996). The net impacts of climate change on agricultural output in India are uncertain, yet specific regions and certain groups of farmers, particularly those farming on marginal, rainfed lands, are likely to suffer significant damages as the result of climate change (Dinar *et al.*, 1998, cited in Selvaraj *et al.*, 2002). West coast agricultural regions, including Gujarat, Maharashtra and Karnataka, are expected to be among the most negatively affected by climate change (Sanghi *et al.*, 1998, cited in Selvaraj *et al.*, 2002).

With respect to infrastructure provision, it can be seen that in areas where investments in agricultural infrastructure have lagged, such as Maharashtra and Madhya Pradesh, rates of growth in agricultural productivity and poverty reduction have also lagged (Datt and Ravillion 1995). Climate change may further exacerbate these regional differences, because regions with limited irrigation infrastructure are also the areas where agriculture is most vulnerable to climate variability and change (Rao 1994). With regards to India it can be said that the Eastern Coast is more vulnerable than the Western Coast with respect to the frequency of occurrence of extreme events like cyclones and depressions (Patwardhan *et al.*, 2003; Patwardhan and Narayanan, 2003). Therefore, in this paper we concentrate on the vulnerability scenario in the eastern coastal districts of India. Within the eastern coast the districts in Orissa and Andhra Pradesh are the most vulnerable in terms of exposure to storms, super storms and depressions. Vulnerability is often reflected in the condition of the economic system as well as the socioeconomic characteristics of the population living in that system. This paper attempts to construct a picture of socioeconomic context of vulnerability by focusing on indicators that measure both the state of development of the region as well as its capacity to progress further. The first aspect is reflected through agricultural and industrial development, while the second through infrastructure and others. In this study, the climate change impacts are examined from agriculture, infrastructure and demographic characteristics. The analysis is carried out at the district level. Vulnerability of a particular district is measured by the frequency of occurrence of extreme events, in this case the occurrence of cyclones, storms and depressions. The study aims to build a vulnerability index and rank the various coastal districts on the eastern coast in terms of their performance on the index. The index tries to capture a comprehensive scale of vulnerability by including many indicators that serve as proxies. Specifically, the paper looks at four different sources of vulnerability: viz., the climatic factors, demographic factors, agricultural factors and occupational factors. Also a composite index has been created for the infrastructure development for the districts in question. The results from the index on vulnerability, infrastructure, agricultural development and demographic patterns are compared to capture the true nature of vulnerability of the people living in these areas.

The vulnerability index tries to capture a more comprehensive scale of vulnerability. This is done by including many indicators that serve as proxies to look at different aspects of vulnerability. In other words we assume that vulnerability can arise out of a variety of factors. In particular we look at four different sources of vulnerability. This includes the climatic factors, demographic factors, and agricultural factors and occupational factors which are trivial in determining the overall vulnerability of an area. The idea is to prepare an index to map the vulnerability among the various coastal districts of the eastern coast of India and rank the districts in terms of vulnerability.

Disasters along the Indian coast are crucial given that fact that about 25 per cent of India's population lives within 50 km of the coastline (NRSA, 2005) and that disasters are channeled and distributed in the form of risk within society to political, social and economic practices and institutions (Oliver-Smith 2004). Thus risk is defined as "a measure of the expected losses due to hazard event of a particular magnitude occurring in a given area over a specific time period" (Peduzzi, *et al.*, 2001). It is crucial to understand that disasters are not only natural events that cause them, but are also product of the social, political, and economic environment and thus are a complex mix of natural hazards and human action (Blaikie *et al.*, 1994).

Impact of Sea Level Rise (SLR)

The phenomenal increase of the concentration of greenhouse gases (GHG) in the atmosphere is causing enhancement of atmospheric temperature resulting in a sea level rise. Sea level change is also affected by other factors like geological subsidence, sedimentation etc. SLR could also increase the risk of floods. A rise in sea level would lead to increase tidal influence upstream and also increase salinity in open bays, because the increased cross section would slow the average speed at which fresh water flow to the oceans. Some of other adverse impacts of SLR in the coastal areas are: Inundation, shoreline erosion, coastal flooding, salinity of estuaries and aquifers, modify tidal range in rivers and bays, alter frequency and severity of storms. All the above impacts are threat to ecological balance and to coastal infrastructures. Sea level would induce accelerated erosion too. The primary reason being that with an increase in the sea level, a large area of the beach is wetted and the turbulence generated within the surf zone is brought more land ward. The beach profile itself thus moves land ward, resulting in a loss of beach space. Any increase in the mean sea level would results in increased tidal prism (the volume of water carried into bays of lagoons during a tidal cycle). This increase is a direct result of the increase in the platform area due to the inundation and shoreline retreat. Any increase in the volume of sea water entering an estuary or lagoons would adversely affect the peripheral agricultural crops and lands too (Rekha Nianthi 2005).

Indian coastline is inhabited by more than 100 million people in nine coastal states (West Bengal, Orissa, Andhra Pradesh and Tamil Nadu on the east coast, and Kerala, Karnataka, Goa, Maharashtra and Gujrat on the west coast), two Union Territories (Pondichery and Daman and Diu) and two groups of islands. There were about 65 coastal districts in these nine states, occupying a total area of around 379,610 km^2. With nearly 250 million people, or around 20 per cent of the population, living within a distance of 50 km of the shoreline, the population density of coastal areas (432 persons per km^2) is significantly higher than that of the country as a whole (256 persons per km^2) (Census 2001). There has been an ever increasing pressure due to migration from inland regions, reflected in the fact that three of the four major Indian metropolitan areas are located in the costal region (Mumbai, Kolkata and Chennai) and out of 35 urban agglomerations with a million plus population 18 are situated in the coastal states. There has been large scale migration from neighboring countries of Bangladesh and Sri Lanka as well. The very recent political turmoil in Sri Lanka has forced people to move towards the Tamil Nadu coast. This herd of incoming migrants not only spirals up the demand over natural resources, leading to unsustainable practices of resource exploitation but also destabilizes the existing social harmony, challenges the resource utilization institutions and constrains the social capital.Western coastline has a wide continental shelf having an area of about 0.31 million km^2 which is marked by backwaters and mud flats. East coast consists of Tamil Nadu coast, Andhra coast, Orissa coast and West Bengal coast, which is flat and deltaic and rich in mangrove forests covering an area of about 1430 km^2. Mangroves are located all along estuarine areas, deltas, tidal creeks, mud flats, salt marshes and extend to about 6740 km^2 (about 7 per cent of world's mangrove areas). Major estuarine areas located along the Indian coasts extend to about 2.6 million hectares (Gauda and Panigrahy, 1999).

Coral reefs are predominant on small islands in Gulf of Kutch, Gulf of Mannar in Tamil Nadu and on Lakshadweeep and Andaman and Nicobar groups of islands. Ecosystems such as coral reefs, mangroves, estuaries and deltas are rich in biodiversity which play a crucial role in fishery production besides protecting the coastal zones from erosion by wave action. A total number of 12 major and 187 minor sea ports located in coastal zones are economic engines of International and National Trade and Commerce in India. The 12 major ports carry about three fourths of the total traffic, with Visakhapatnam as the top traffic handler. The total traffic carried by both the Indian major and minor ports during 2004-05 was estimated to be well over 500 million tonnes.

Climatic variations are likely to impact and exacerbate the existing coastal zone problems in India, like erosion, flooding, subsidence, deterioration of coastal ecosystems, such as mangroves, coral reefs and salinization. The risks span a wide spectrum, tropical cyclones, sea-level rise, changes in temperature and precipitation and recently tsunami posing serious threats and thus adding on to the vulnerability of the coastal social-ecological system. Apart from this, various technological and development initiatives undertaken along the coast have been found to be vulnerable and thus further contributing to the overall economic vulnerability and threat to life. The Konkan Railway (Nagrajan *et al.*, 2000) and the Sethusamudram Projects are cases in point. India in general and the coastal regions in particular are considerably vulnerable to the impacts of climate change (DOD, 2002). The Eastern Coast is more vulnerable than the Western Coast with respect to the frequency of occurrence of extreme events like cyclones and depressions (Patwardhan *et al.*, 2003). Rise in sea level is likely to have its impact the following ways (NATCOM, 2004):

1. Land loss and population displacement
2. Increased flooding of low-lying costal areas
3. Agricultural impacts (like, loss of yield and employment) resulting from inundation, salinization, and land loss.
4. Impacts on coastal aquaculture
5. Impacts on coastal tourism, particularly the erosion of sand beaches

Cyclic hazard indices and exposure indices based on Housing Index and Population Index ranks the 52 costal districts belonging to 8 states.The top five districts which had the maximum exposure in terms of housing were: Valsad, Uttar Kannada, Ratnagiri, Ganjam and Kozikode. Four of these lie on the west coast (cyclone incidence is lesser on the west coast when compared to the east coast). This implies vulnerability of west coast (submergence type) to sea level rise. The top five districts with high exposure levels considering exposure in terms of density of population were: South 24 Parganas (a densely populated coastal district of West Bengal along the east-coast of India), industrialized towns like Surat, place of tourist attraction Sindhudurg and towns in river floodplains and deltas West and East Godavari Krishna. Greater Mumbai and Chennai have the highest exposure levels (urban). The exposure levels have increased over the years in terms of the absolute number of people, density of population and size of the housing stock. (Source: India's Initial Communication to the United Nations Framework Convention on Climate Change, *Ministry of Environment and Forests, Government of India, 2004*)

Socio-economic transformations resulting from these changes and the political decisions taken to mitigate and adapt have brought forth unforeseen threats and challenges to survival and livelihood security and in turn have their impacts on larger scale security issues of state and region as a whole. With this background, the following three diverse issues are analyzed and discussed using the approaches to environmental security: Resource Conflicts in the Chilika Lagoon: From Violence to Environmental Movement. The Sethusamudram Canal Project (SSCP): Strategic

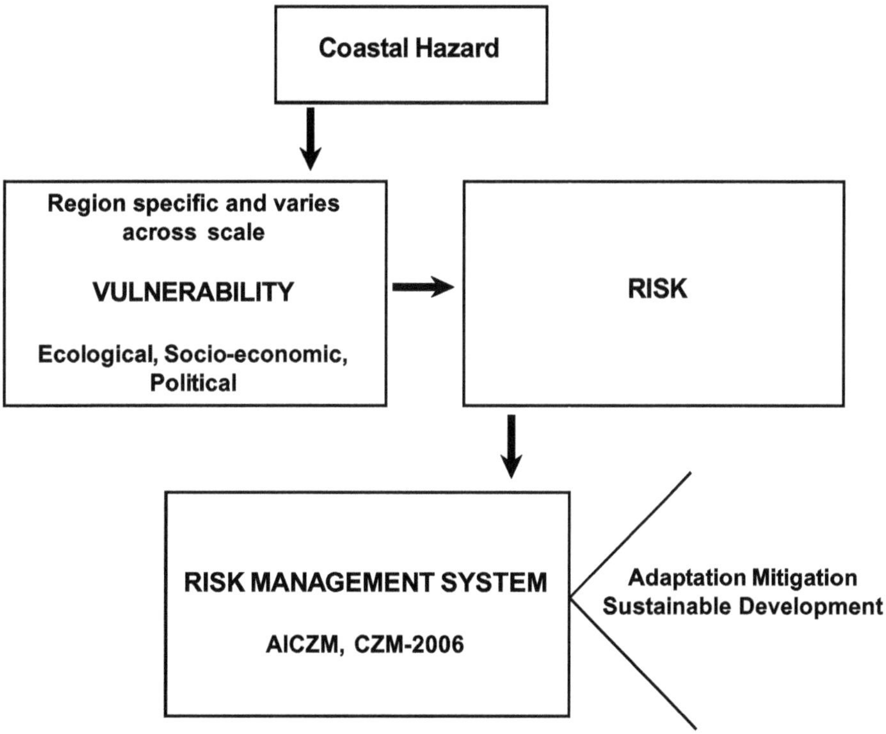

Figure 7.2: A Conceptual Framework for Risk Management in the Indian Coastal System

Security and Beyond. The Sir Creek dispute between Indian and Pakistan in the Rann of Kutch: Prospects for Environmental Peacemaking.

This highlights the need for multidimensionality of hazards and associated risks and the degree of uncertainties and rehabilitation associated with appropriate response strategies. A cursory look at the past developments along the Indian coastline, from tsunami to recent oil spill, hints towards the degree of coastal hazard and vulnerability across time and space. Institutional mechanisms and management strategies, particularly that of the Coastal Zone Regulation notification (1991) and the very recent recommendations of M.S. Swaminathan Committee (2006), are measures geared up for coastal hazard preparedness and mitigation. However, few studies have been carried out in to these aspects which are interdisciplinary in real sense of the term. Secondly, the existing gap between science and policy has to be bridged through effective communication and cross fertilization of ideas in order to evolve an adaptive and integrated coastal zone management (AICZM) strategy. A primary stepping stone towards this endeavor is generation of knowledge through research partnership across disciplines and establishment of an epistemic community at the national level. A conceptual framework for risk management for Indian coastal regions is given in the Figure 7.2.

References

Achanta, A. N. 1993. "An assessment of the potential impact of global warming on Indian rice production" In Achanta A N (ed. *The Climate Change Agenda: An Indian Perspective*. New Delhi: Tata Energy Research Institute.

Aggarwal, P. K. and S. K. Sinha. 1993. "Effect of probable increase in carbon dioxide and temperature on wheat yields in India", *Journal of Agricultural Meteorology* 48(5. 811-814.

Alcamo J., Krol M., Leemans R. 1995. Stabilizing greenhouse gases: Global and Regional Consequences- Results from the IMAGE 2.0 Model. pp1-10. Bilthoven: RIVM.

Ali, A. 1996. Vulnerability of Bangladesh to climate change and seal level rise through tropical cyclones and storm surges: Water, Air and Pollution, 92, 171-179.

Anand, S. and A. Sen, 1994. *"Human Development Index: Methodology and Measurement"*, Occasional Paper 12, United Nations Development Programme, Human Development Report Office, New York.

Asian Development Bank 1994. *Climate Change in Asia: Sri Lanka Country Report*, Regional study on Global Environment issues, Manila.

Atlas of Sri Lanka 1997. Arjuna's, Dehiwala, Sri Lanka.

Beg, N., Morlot J.C., Davidson O., Afrane-Okesse Y., Tyani L., Deton F., Sokona Y., Thomas J. P., Rovere E., Parisk J., Parikh K., and Rahman A. 2002. Linkages between climate change and sustainable development, Climate Policy, Vol. 2: 129-144.

Blaikie, P., Cannon, T., Davis, I., and Wisner, B. 1994. At risk, natural hazards, Peoples' Vulnerability and Disasters (London, New York: Routledge).

CCD/DANIDA 1986. *Sri Lanka Mater Plan: Coast Erosion Management,* Coast Conservation Department, Colombo.

Central Bank of Sri Lanka 1990. Industrial Trend and Income, Colombo.

Chambers R. 2005. Responsible well-being: a personal agenda for development: in Ideas for Development, pp. 184-198, Earthscan, London.

Chen, M. A. 1991. *Coping with Seasonality and Drought*, Sage, New Delhi.

Datt, G. and M. Ravillion, 1998. *"Growth and Poverty in Rural India"*, http://econ.worldbank.org/files/500_wps1405.pdf

Domroes, M. and Schaefer, D. 2000. "Trends of recent temperature and rainfall changes in Sri Lanka" *Proceedings of the International Conference on Climate Change and Variability*, pp.197-204.

Downing, T.E. 1991. "Vulnerability to hunger in Africa: a climate change perspective", Global *Environmental Change* 1, 365-380.

EC 1998. European Commission: Sustainable urban development in the EU: A framework for action, communication from the commission to the council, the European Parliament, the Economic and Social Committee and the Committee of the Regions, Brussles: DG XVI, EC.

Fischer, G., M. Shah, and H. V. Valthuizen 2002. *"Climate Change and Agricultural Vulnerability"*, World Summit on Sustainable Development, Johannesburg, www.iiasa.ac.at/Research/LUC/JB- Report.pdf

Gibbs D. 2002. "Approaching local sustainability": in: *Local economic development and environment*, page 1-25, Routledge Press.

Gornitz V. 1995. Seal-level rise: A review of recent past and near-future trends: Earth Surface Process and LAndreforms, No. 1, pp. 7-20.

Hoozemans F.M.J., Marchand M., Pennekamp H.A. 1993. A global vulnerability analysis: vulnerability assessment for population: Coastal wetlands and rice production and a global scale, second revised version, Delft: Delft Hydraulics.

htpp://www.jmg.gov.my. Impact of the 26th December 2004 Tsunami on Groundwater Systems and Groundwater Based Water Supplies in Malaysia Report prepared by: Minerals and Geoscience Department Malaysia Tingkat 19-22, Bangunan Tabung Haji Jalan Tun Razak 50658 Kuala Lumpur, MALAYSIA.

htpp://www.surveyofindia.gov.in/tsunami.htm.

htpp://www.who.int/hac/crises/international/asia_tsunami/3months/report/en/print.html.

Indian Meteorological Department (IMD), 1996. *Tracks of Storms and Depressions in the Bay of Bengal and the Arabian Sea*, Available online at: http://www.imdmumbai.gov.in/cycdisasters.htm

Intergovernmental Panel on Climate Change (IPCC), 2001. *IPCC Third Assessment Report. Climate Change 2001,* in five volumes: Synthesis Report, Working Group I: The Scientific Basis; Working Group II: Impacts, Adaptation and Vulnerability; Working Group III: Mitigation, Geneva: Intergovernmental Panel on Climate Change and Cambridge: Cambridge University Press, Available online at www.ipcc.ch

IPCC. Climate Change 2001. Impacts, adaptation, and vulnerability. IPCC, Cambridge University Press, UK. 2001.

ISDR, 2002. Living with Risk, A Global Review of Disaster Reduction Initiatives (Geneva).

Kanji, N. and Greenwood L. 2001. Participatory approaches to research and development in IIED: Learning from experiences, IIED, London, October.

Kayane Y. 1995. Hydrological consequences of Global Warming revealed by the Century-long Climatological data in Sri Lanka an South West India. *In the Interim Report of IGBP Activities in Japan 1990-1994,* Science Council of Japan.

Ligon, E. and L. Schechter 2002. *"Measuring Vulnerability"*, Washington, D.C., USA, Available online at: www.ifpri.org/events/conferences/2002/092302.htm

Madduma Bandara, C.M and Wickramagamage, P. 2004. "Climate change and its impact on upper watersheds of the Hill country of Sri Lanka", in Srikantha Herath, Assela Pathirana and Weerakoon S.B., *Proceedings of the International conference on sustainable water resources management: In the changing environment of the monsoon region:* Volume 1, National Water Resources Secretariat, Colombo.

Madduma Bandara, C.M. 1989. Coastal change of Crow Island and Its Enviorns, *Coast Conservation Department,* Colombo.

Mansuri, G. and A Healy 2002. *"Vulnerability Prediction in Rural Pakistan"*, Washington, D.C., USA, Available online at: www.ifpri.org/events/conferences/2002/092302/mansuri.pdf

Munasinghe, M. 1998. Climate change decision-making: science, policy and economics: International Journal of Environment and Pollution, Vol. 10. No. 2, pp. 23-36.

Munasinghe, M. 2001. Sustainable development and climate change: applying the sustainomics trans-disciplinary framework: International Journal of Global Environmental Issues, Vol. 1, No. 1, 15-25.

Muralideran, D., Rolland Andradge, G., Laximnarayan, T., Swathi, T. and Swetha, T 2000) The imprint of 26 December 2004 Sumatra earthquake in aquifers of Hyderabad granite Pluton, Current Science 89(7).

NARESA 1991. *Natural Resources of Sri Lanka,* Condition and Trends, NARESA, Colombo.

Natcom 2000. Democratic Socialist Republic of Sri Lanka, *Initial National Communication under the United Nations Framework Conventionon Climate Change,* State printing corporation, Padukka.

Nathan, F, 2007. Natural Disasters, Vulnerability and Human Security, in Brauch, H. G. *et al.* (eds) *Facing Global Change: Environmental, Human, Energy, Food, Health and Water Security Concepts.* (Springer-Verlag).

National Remote Sensing Agency, 2005. *NRSA Newsl.*, January.

Nicholls R. J., Hoozemans M.J., Marchand M. 1999. Increasing flood risk and wetland losses due to global sea-level rise: regional and global analysis: Global Environmental Change, 9, S69-S87

Oliver-Smith, A. 2004. Theorizing vulnerability in a globalized world: A political ecological perspective, In Bankoff, G., Ferks, G and Hilorst, D (Eds.) Mapping Vulnerability, Disasters, Development and People (London-Sterling, Va.:Earthscan) 10-24.

Palanivelu, K., Nisha priya, M., Muthamil selvan, A. and Usha natesan. 2006) Water quality assessment in the tsunami affcected areas of Chennai, Current Science 91(5), 584p.

Patwardhan, A. and K. Narayanan 2003. "*Assessment of Vulnerability of Indian Coastal Zones to Climate Change*", Conference on Water Resources, Coastal Zones and Human Health.

Patwardhan, A., K. Narayanan, D. Parthasarathy and U. Sharma 2003. "Assessment of Vulnerability and Adaptive Capacity in Coastal Zones", in Shukla *et al.* (eds) "*Climate Change and India Vulnerability Assessment and Adaptation*", Hyderabad: Universities Press.

Peduzzi, P., Dao, H., Herold,C., Rochette,D., and Sanahuja, H. 2001 Feasibility report on Global Risk and Vulnerability Index- Trends per year for UNDP/ERD (Geneva: UNEP/DEWA/GRID, June.

Rao, C. H. H. 1994. *Agricultural Growth, Rural Poverty and Environmental Degradation in India*, Dehli, Oxford University Press.

Rao, D. G. and S. K. Sinha 1994. "*Impacts of climate change on simulated wheat production in India. Implications of Climate Change for International Agriculture: Crop Modelling Study*", US Climate Change Division Report EPA 230-B-94-003, India, 1-10.

Rekha Nianthi, K.W.G. 2003a. *Global Warming and Climate Change* (Sinhala language version), Dharshana printers, Kandy, Sri Lanka.

Rekha Nianthi, K.W.G. 2003b. "Recent trends of climate change in Sri Lanka", *The Journal of Geo-Environment*, 3: 37-49, Rajshahi University, Bangladesh.

Rekha Nianthi, K.W.G. 2005. "Climate change and its potential influences on agricultural economy of Sri Lanka". Unpublished Ph.D. Thesis, Dept. of Geography, North- Eastern Hill University, Shillong (India).

Riebsane, W. E., S. A. Changnon Jr. and T. R. Karl 1991. *Drought and natural resources management in the United States: Impacts and implications of the 1987-1989 drought*, West view, Boulder, CO, Available online at: ams.allenpress.com/pdfserv/i1520-0477-084-09-1231.pdf.

Sen, A. 1981. *Poverty and famines: an essay on entitlement and deprivation.* Oxford University Press, Oxford.

WHO 2004. Health action in crisis: Southeast Asia earthquake and tsunami update 27 Dec 2004. Geneva: World Health Organization WHO 2005a. Three months after the Indian Ocean earthquake-tsunami: health consequences and WHO's response. The Drinking Water Response to the Indian Ocean Tsunami, Including the Role of Household Water Treatment © World Health Organization 2005.

Chapter 8

Assessing the Economic Value of Coastal Ecosystem for Coastal Management

M. Suparmoko
Faculty of Economics, Jenderal Soedirman University, Purwokerto, Indonesia
E-mail: suparmoko@yahoo.com

Introduction

This chapter is mostly based on experiences during the membership of the author in the Task Force on Economic Evaluation of the UNEP/GEF Project entitled "Reversing the Degradation Trends of South China Sea and the Gulf of Thailand" 2002 -2007. In addition some information and comments are drawn from other reading materials and contacts with several authors of the reading materials.

Basically the objectives of economic evaluation may be grouped into 4 services:

1. Determining the value of the total flow of benefits from coastal ecosystem
2. Determining the net benefits of interventions that alter ecosystem conditions
3. Examining how costs and benefits of ecosystem are distributed
4. Identifying potential financing source for conservation

Hence, first of all management of coastal ecosystem should be understood about what are the objectives of the management, whether it is directed for the coastal ecosystem conservation or even preservation, or it is for the human intervention to alter the uses of the ecosystem to achieve the development objectives such as to increase national or the regional per capita income of the people, or to increase employment, or to increase the recreation uses, etc.

Many have understood the role and uses of the ecosystem, but that is not enough because the understanding must be followed with action either to protect or to preserve the environments. During the last two decades, the coastal ecosystems in Indonesia and Southeast China Sea are under the trends of degradation. UNEP and GEF are conducting projects to reverse the degradation trends of the five different ecosystems that are spread over the 7 countries in the regions. The seven countries are: Campuchia, China, Malaysia, Indonesia, Philippines, Thailand and Vietnam. The ecosystems covered with this project are mangrove, coral reefs, sea grass, wet lands, and one additional aspect is land based pollution. Covering the ecosystems different projects have been initiated in the seven member countries. Indonesia conducts rehabilitation projects for mangrove ecosystem, in Batu Ampar and West Kalimantan Province, sea grass in Bintan island of Kepulauan Riau Province, and coral reefs in Trikora island.

The following discussion is then adopted from paper presented earlier for the South China sea Project and it is hoped that the content is still appropriate for the present discussion.In fact the South China Sea is surrounded by 9 nations; the People's Republic of China to the north; the Republic of the Philippines to the east; Malaysia, the Republic of Singapore, the Republic of Indonesia, and the Kingdom of Brunei to the south; the Kingdom of Thailand, the Kingdom of Cambodia and the Socialist Republic of Vietnam to the west. It is understood that there has been an increasing environmental degradation in the South China Sea and the Gulf of Thailand. The sustainability of the natural environment of the South China Sea is threatened by high rate of population growth, pollution, excessive harvesting and habitat modification, resulting in rapid loss of habitat and impairment of the regenerative capacities of living systems. On the other hand, the food consumption patterns of the people in the region rely heavily on cheap protein derived from fisheries. The food consumption is increasing due to the increase in the populations.

The increases in both the populations and the demands for the marine environment services together with increases volume of pollutants dumped to the sea have caused the degradation of the marine habitats and marine environments in the South China Sea. As a result, there can be imbalance between the demand for and the supply of foods and other environment and biological services. In turn it may depress the standard of living of the people in the region; at least to the 270 million people (or 5 per cent of the world's population in 1999) who live in the coastal sub regions of those nations (UNEP SCS/SAP Ver.3, Strategic program for the South China Sea, 24 Feb. 1999).

Recognizing the important role of the South China Sea to those nations, the existing environmental degradation trends of the South China Sea must be stopped and reversed towards better environmental quality. Then actions to halt and reverse the going on degradation trend of the environment of the South China Sea are urgently needed; otherwise the natural resources and the environmental and biological services reserved by the environment of the South China sea will not be able to support the growing population in the region. UNEP and GEF, supported project "Reversing Environmental Degradation Trends in the South China Sea and the Gulf of Thailand" is a good step in this direction.

The project is expected to contribute to the development of the South China Sea and the Gulf of Thailand regions to be able to improve the present condition of the coastal natural resources and the environments contained in the region. In other words the project must be able to improve the environmental functions of the coastal and of the sea of the region. To derive the real benefits of the project, one of the index is to measure the net social benefits of the project. The analysis has to compare the net social benefits of having the project versus those of not having the project. The methodology is then can be "with and without project" approach or "before and after the projects" approach which involve the measurement of the direct and the indirect or the environmental costs and the direct and the indirect or environmental benefits.

The Task Force on Economic Valuation (RTF-E)

The Task Force on Economic Valuation (RTF-E) for the South China Sea Project has been formed formally during the first meeting which was held in Phuket, Thailand, 11th–13th September 2003. As mentioned in the draft terms of reference for the RTF-E the primary objective of the RTF-E is to provide advices and supports to the national committees, sub-committees and regional working groups, by providing the appropriate expertise and assistance in completing the envisaged economic valuations and cost-benefit analysis. The secondary objective is the development of a regional valuation framework for use in the evaluation of the cost effectiveness of alternative interventions for implementation under the Strategic Action Programme. The meeting also agreed to produce conceptual economic valuation framework for four habitats of the South China Sea: Seagrass, coral reefs, mangroves, and wetlands. Land based pollution has not been considered during the two-day meeting at Pukhet, only framework for economic valuation on mangrove was discussed intensively, whereas for the other habitats would be finalized by the assigned members after they have returned to their respective offices. At this time the conceptual frameworks for the four mentioned habitats have been completed as can be seen at the attachment of this paper.

The Objectives of the Economic Valuation

As already mentioned above the South China Sea and The Gulf of Thailand Project must be able to improve the environmental quality and therefore the environmental functions can also be maintained and even be developed. Every project should be evaluated in terms of its net benefits in order to come to a decision whether the project is economically feasible. A project has to show the technical feasibility, financial feasibility, economic feasibility, social feasibility, and environmental feasibility. This paper will only discuss the economic and the environmental feasibilities. Benefit-cost analysis is more widely used form of methodology in the region for evaluating the feasibility of a certain project.

When the function of the environment gains its appreciation and is accepted and approved as a very important dimension to be considered in determining the success of the development projects, the benefit cost analysis should include not only the direct costs and direct benefits of the project, but it must also include the external costs and external benefits or the environmental cost and the environmental benefits of the project. Therefore the new cost-benefit analysis may be called as the extended benefit-cost analysis.

In order the extended benefit-cost analysis obtains a single value, a summation of the flows of benefits and costs, which accrue in different time periods, must be done. The values can then be stated in a common single value such as in term of present value or future value, but commonly is in a present value form. The present value is obtained by discounting all the future values of benefits and costs by using a discount factor. The discount factor is based on the discount rate or the rate of interest. The rate of discount should be the opportunity rate of discount which takes into account the business, the economic, and the environmental costs. This discount rate is called the social rate of discount.

So benefit-cost analysis requires the comparison of the total social benefits and the total social costs (B/C ratio) or requires the subtraction of the total social benefits by the total social costs (Net Present Values = NPV) or Internal Rate of Returns (IRR) that is the value of the discount rate which can make the NPV equals to zero. In order to have a single value, the benefits and the costs of the project must be stated in money term. Therefore the physical results and physical impacts of the project must be given values. Hence economic valuation is actually the core of the extended benefit- cost analysis. The problem is then how to assign the economic values on the impacts of the project, where the impacts consist of economic impacts, social impacts and environmental impacts.

In connection with the UNEP/GEF Project: "Reversing Environmental Degradation Trends in the South China Sea and the Gulf of Thailand" a proper economic valuation of coastal resources has been considered critically important to establishing policy goals, and to the completion of cost-benefit analyses. These all are required in assessing the comparative net social benefits of alternated demonstration sites. The Regional Scientific and Technical Committee has also noted that, the national reports received to date from the seven participating countries, are weak in the areas of economic analysis and resource valuation. (Report of the First Meeting of RTF-E 11-13 September)

Methodology

A general approach for economic valuation has been developed earlier by Pearce *et al.*, 1989). A natural environment has three main functions, *i.e.* to provide natural resources as inputs in production of outputs, to provide environmental services, and to provide environmental or natural assimilator functions. The total economic value (TEV) of the environment therefore has to include the values of all the above functions. The above functions may be recognized as having use value (UV) and non-use value (NUV). The UV is constructed by the summation of *direct value* which comes from the extracted natural resource commodities such as woods and fish from mangrove forest, rock and fish from coral reefs, with the *indirect value* which mostly comes from the non-extracted natural resource commodities such as for tourism, research and education.

In addition to the direct and indirect values, an *option use value* can be added as the UV. Option value means a value given to not using the environment now but preserving the option of using the resource in the future. This option use value can be applicable either for the extractive commodities or the non-extractive commodities, or even to the environmental services and the biological diversity services.

The NUV is also disaggregated into three groups of value, namely the quasi option value, bequest value, and existence value. The services of the natural environment in terms of those three values can be derived from the services provided by the environment such as the environmental services and biological services. Examples of the environment services among others are shoreline protection, carbon sequestration, and water purification derived from seagrass; and also shoreline protection, carbon sink and nursery ground derived from mangrove forest. Examples of the biological services among others are the habitats for the migratory species and endangered species for mangrove forest and for the biodiversity for seagrass. Therefore economic values could be assigned to the services provided by the environment. This conceptual framework for economic valuation is depicted in Figure 8.1.

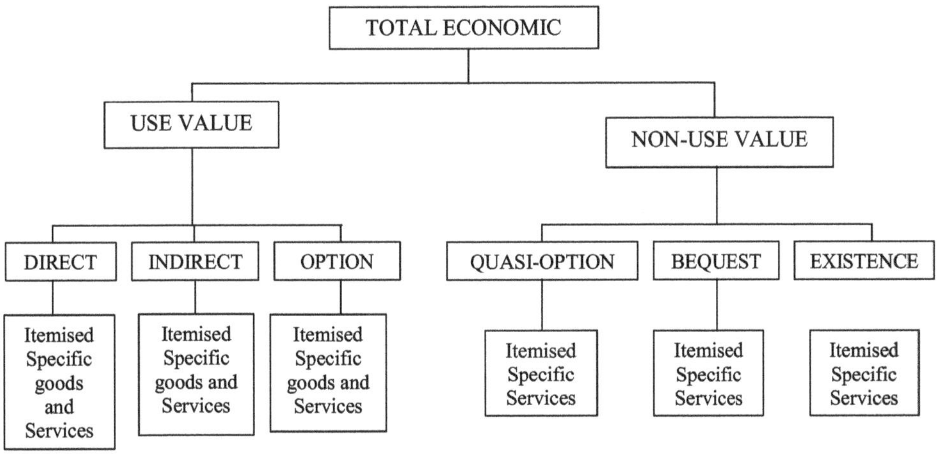

Figure 8.1: Economic Valuation Approach

Extractive Uses

When the functions and services of each habitat of the natural environment have been recognized and identified, the next step is to quantify the commodities and services related to the extractive commodities, the non-extractive commodities, the environmental services and the biological services. At this stage the size of the habitat must be measured, and then the quantity or the volume of the natural resource commodities must be computed. Since there are so many natural resource commodities could be extracted from the coastal and marine resources, it is impossible to describe all kinds of the coastal and marine natural resource commodities. The most common one is the fish resources, although it is difficult to measure the fish stock in a certain sea or ocean. But the quantity of fish harvested each year can be measured in tons per year. Also the volume of timber and the number of trees cut from the mangrove forest can be quantified in terms of m^3 per Ha of mangrove forest. The number of shrimp or crab can be measured in tons per Ha of mangrove forest or per Ha of coral reefs. Those kinds of data are needed because to economically value them relatively easy due to the availability of market price for them. When there is no market price, first try to find the substitutes of the products and find the market value for the substitutes and apply the price for the natural resource commodities.

In measuring the economic value of a natural resource, it is necessary to understand the purpose of the valuation; whether we want to value the *total stock of the natural resource* or do we want to value the resource commodity harvested from the stock. The former may be called the *wealth approach* and the latter is called the *production approach.* In the extended cost- benefit analysis with environmental dimension is taken into account, we are interested in measuring the net social benefits produced by a certain project. The value of the natural resource is represented by the *economic rent* of the resource. Economic rent is the different between the total revenue minus the total costs including the reasonable profit share for the enterprise. The economic rent divided by the total natural resource commodity harvested is called the *unit rent.*

Non-extractive Use

For the non-extracted use of resources such as mangrove forest and coral reefs area for recreation purposes, the data needed first of all is again the size of the area, and the number of visitors who come to the site for recreational purposes. Since the approach of valuation will be the *travel cost approach,* therefore data related to the travel such as the distance of the recreation site to the home origin of the tourist, the financial travel cost, the time travel cost, and the length of the stay, and the total expenses during the stay should be obtained mostly by survey method. The travel cost approach is more appropriate to use because there is no market price for such environmental services as recreation site.

Another example is the aesthetic value of the environment. The approach for valuing the environment services of this type, a *hedonic pricing method* may be used. In this case the price of an asset (house) located at an area with a nice environment quality will be different from the price of the same type and quality of asset which is located near the rubbish dumping area. The difference in prices between the two assets can be assumed to be the economic value of the difference in environments.

Another non-extractive use is the use of natural environment for education and researches. These kind of uses are grouped into the non extractive use and the economic valuation can be estimated by the real costs or expenses to finance the education and research either paid by the government or by the funding agencies. The more frequent the resources are used the more potential economic value can be assigned to the natural resources and the environment. Figure 8.2 illustrates these valuation techniques.

Environmental Services

The environmental services are considered providing services, which are indirectly used by the society. For the environmental services, various approaches are known but their uses are different from one case to the other. For mangrove forest with a shoreline protection function for example, the economic valuation for it may

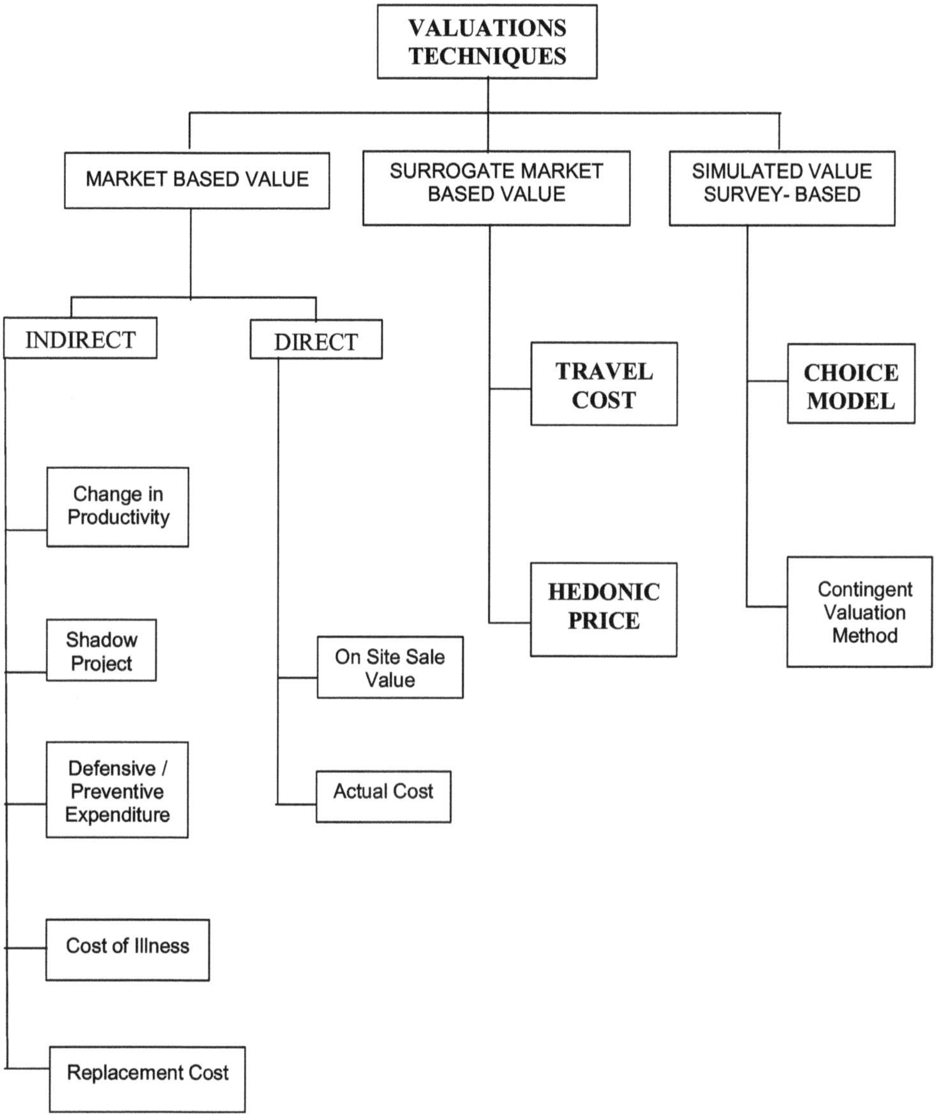

Figure 8.2: Valuation Techniques

***Source*: Report of the First Meeting of the Regional Task Force on Economic Valuation, Phuket, Thailand, UNEP/GEF, Bangkok, September 2003.**

be using a construction costs for the substitutes such as for building a wall or a dike to protect the seashore. To get the economic value, the length of the seashore must be measured and then multiplied by the cost of building the wall or the dike. The total costs of building the wall or the dike can be assumed to be the value of the mangrove forest as shoreline protector. Another approach is the use of *replacement cost* in estimating the cost of a damage mangrove forest. The economic value of a damage mangrove forest can be estimated by the replanting costs of the same size of the mangrove forest including the operation and maintenance costs until the forest becomes similar to the virgin mangrove forest.

The environmental services produced by seagrass and coral reefs such as carbon sequestration and water purification can be valued using costs of substitutes like the cost of water treatment plant or the cost of environmentally rice farming.

Biological Diversity Services

The biological diversity services such as the mangrove ecosystem and endangered species living in the mangrove forest mostly are assigned to have the non-use value meaning that the services are not in the form of extractive commodities and consumed directly by the users. Also the ecosystem of coral reefs as the habitat of fish, plankton, etc. is assigned to have non-use value. To value such kind of services of the natural environment, it is common to use willingness to pay approach (WTP) or willingness to accept approach (WTA). Both approaches are called contingent valuation method (CVM). The CVM involves direct surveys on individuals to obtain their preferences rather than analyzing the actual behavior revealed in the market. So the values is expressed as an individual's maximum willingness to pay (WTP) to safeguard the environment or certain natural resource, or it can be expressed as the minimum monetary compensation someone would accept WTA) to forego the good quality environment or useful natural resource. The quality of the valuation will depend on the description of the contingent market, the knowledge of the respondent about the environment, the respondent preferences, the respondent income constraints, and the availability of substitutes and complementary resources.

Recognizing that a certain habitat of the natural environment has many different functions or uses, therefore the total economic values (TEV) will be the sum of all values derived from those different uses. So the total economic value will be the sum of the direct use economic value, the indirect use economic value, the non-use economic value. From the discussion above, the basis for economic valuation can be market base value and non-market base value and the techniques used is market price mainly for the direct use value which is applied for the extractive commodities. Other techniques such as travel cost approach, hedonic pricing approach, surrogate market price approach, replacement cost approach are used to derive the non-use values which are applicable mostly for the environmental services and biological diversity services since most of those services do not have any transaction in the market (Figure 8.2).

The Need for Regional Values

The above discussion is focusing on the economic valuation of specific environments which disregards the interrelationship among the different environments. Each environment in fact has interdependency with other environments which form a certain ecosystem. An ecosystem then will have multi environmental functions. The multifunctional of an ecosystem must be economically valued and summed up to be the Total Economic Value. Also when an environment is utilized by many users, the value of the benefits is equal to the total values of individual benefits received by individual users. However it must kept in mind when the functions of the environment are mutually exclusive in uses, the individual value can not be added together; whereas if the functions are supplementary in use, their respective values must be added to each other to obtain the Total Economic Value of an ecosystem. Furthermore, there are inherent characteristics with the environment; among others is the externality character of the environment so that what happens to a certain natural environment may have impacts on other environments and on other users or nations.

For example, the South China Sea is being used as transportation media for many ships owned by different nations. The economic values derived from such services by each nation must be summed up together to form the total economic value of the sea as the transportation media. Another example, a poor management of mangrove forest in one country may reduce the total fish population and decrease the volume of fish catch by the fishermen in the region. The worst condition is when there is forest fire in one place or in Indonesia for example, not the area surrounding the fired forest were affected, but also the neighboring places or countries like Singapore, Malaysia, Thailand and the Philippines suffered the impacts of the forest fire. In this case it is clear that a regional value is important to show how crucial the functions of certain natural resource and the environment to the region.

Therefore as a basis for justification of the South China Sea project, a comparison between the total regional economic values for the South China Sea and the Gulf of Thailand with and without the project is imperative. This value will demonstrate how important will the project be for the whole nations of the world in general and in the South China Sea and the Gulf of Thailand in particular. In other words, it is imperative to make regional evaluation of the net social benefits of the UNEP/GEF Reversing Environmental Degradation Trends in the South China Sea and Gulf of Thailand project.

Although the South China Sea is surrounded by 9 countries, there are only 7 countries involved in the project because most of the coastal and sea resources are found only in those 7 countries: Cambodia, China, Indonesia, Malaysia, the Philippines, Thailand, and Vietnam. Specific natural resources such as mangrove forest or coral reefs besides their services are directly and indirectly consumed by the nations who own the resources, due to the nature of externalities and common property of the environment, the environmental services can be enjoyed by other nations. So, in measuring the regional economic values of the net benefits, each nation may conduct the economic valuation studies in their own country. Then to find the total regional economic values of certain resources or certain environment, the economic values found in each country for the same resources or environment may be added to the other economic values for the same resources or the same environment found in the same region which accrue to other nations. However, it must be kept in mind again that the nature of the services must be complementary in nature and not mutually exclusive. For the mutually exclusive use of natural resources such as the use of mangrove forest as source of charcoal and as source of timber. The value of the mangrove forest must only be represented by highest opportunity costs. If the price of charcoal is higher than the price of timber, then the values derived from charcoal will be the right value. Based on the conceptual frame work (Figures 8.1 and 8.2), Tables 8.1 to 8.4 detail the methodology for valuation in cases of mangrove, coral reef, sea grass and wet land habitats.

Table 8.1: Framework for Mangrove Valuation

Uses	*Use Values*			*Non-Use Value*			*Suggested Techniques*
	Direct	*Indirect*	*Option*	*Quasi-Option*	*Bequest*	*Exis-tence*	
			Extractive Use				
Timber	X						Market Price
Firewood	X						Market Price
Poles	X						Market Price
Leaves/Thatch	X						Market Price
Charcoal	X						Market Price
Poles	X						Market Price
Fish capture	X						Market Price
Crab	X						Market Price
Prawn	X						Market Price
Wildlife	X						Market Price
Molluscs	X						Market Price
Bark (Tanning and dyes)	X						Market Price
Fruits/propaguls	X						Market Price
Worms	X						Market Price
Medicine	X						Market Price
Sap (sugar, alcohol, acetic acid)	X						Market Price
Wood tar	X						Market Price
Shell fish collection	X						Market Price
Insect and larvae collection	X						Market Price
Zooplankton	X						Market Price
Jellyfish	X						Market Price
Honey and wax	X						Market Price
Algae	X						Market Price

Contd...

Table 8.1: Framework for Mangrove Valuation

Uses	*Use Values*			*Non-Use Value*			*Suggested Techniques*
	Direct	*Indirect*	*Option*	*Quasi-Option*	*Bequest*	*Existence*	
Non-Extractive Use							
Tourism/Recreation	X						Travel cost
Research/ Education	X						Actual costs (Gov. NGO, Academic)
Aesthetic	X						Hedonic Price method
Fish culture							Market Price
Crab culture							Market Price
Prawn culture							Market Price
Shelfish culture							Market Price
Other Aquaculture							Market Price
Environmental Services							
Shoreline protection		X					Replacement Cost (Replanting and Reestablishing Mangrove including operations and maintenance for a specified time) Or Single function replacement costs such as construction of coastal defenses
Windbreak		X					
Flood Protection							
Flood Protection							
Carbon sequestration		X					
Water purification		X					
Oxygen release		X					
Sediment removal							
Nutrient removal							
Aquaculture (pearl)		X					Market Price
Nursery area		X					Market Price
Shoreline accretion/ land increase		X					Market based value of land
Biological Diversity Services							
Biodiversity			X				Contingent valuation method
Migratory species		X				X	Contingent valuation method
Endangered species					X	X	Contingent valuation method
Mangrove ecosystem						X	Contingent valuation method
Wilderness value					X	X	Simulated value survey based

Table 8.2: Framework for Coral Reef

Uses	*Use Values*			*Non-Use Value*			*Suggested Techniques*
	Direct	*Indirect*	*Option*	*Quasi-Option*	*Bequest*	*Existence*	
Extractive Use							
Lime stone	x						Market price
Building materials	x						Market price
Echinoderm	x						Market price
Crustacean	x						Market price
Algae	x						Market price
Mollusks	x						Market price
Coral trade	x						Market price
Bioactive substance	x						Market price
Non-Extractive Use							
Tourism/recreation		x					Travel cost
Research and Education		x					Actual costs
Mari culture		x					Hedonic pricing
Environmental Services							
Nursery ground		x					
Beach protection		x					
Carbon sink		x					
Climate change record		x					
Biological Diversity Services							
Biodiversity Storage					x		
Coral reef ecosystem					x	x	
Food storage for other biota						x	
Secondary producers						x	

The Problems Faced

This chapter has presented the frameworks of approaches for the economic valuation of the coastal environment as proposed by the RTF-E which include mangrove forest, coral reefs, wetlands and seagrass. However, the frameworks have not been discussed any further to be accepted as the approved framework by the RTF-E members due to lacks of communication and unclear funding position for undertaking further actions at the country level. Some projects in each country are supposed to be completed by the end of the year 2003 and the budgets for the projects did not clearly allocate funding for the services of the member of RTF-E.

Table 8.2a: Framework of the Economic Valuation for Coral Reefs According to Stakeholders

Beneficiaries	*Use Values*						*Non-Use*
	Direct Use Values				*Indirect Use Values*	*Option Value*	*Non-Use Values*
	Recreation	*Fisheries*	*Marine Products*	*Research and Education*			
(i) Local stakeholders							
Communities/general public	H	H	H	H	H	H	H
Fishermen	H	H	H	?	H	H	H
Recreational Users	H	H	?	H	?	H	?
Conservation groups	H	?	H	H	?	H	H
Schools/Universities	H	?	?	H	?	H	H
Government	H	H	H	H	H	H	H
Businesses	H	H	H	H	H	H	?
(ii) Regional Stakeholders							
Communities/general public	L	H	H	H	L	H	H
Fishermen	L	H	H	?	L	H	L
Recreational Users	H	L	?	H	?	H	?
Conservation groups	H	?	H	H	?	H	H
Schools/Universities	H	L	L	H	?	H	H
Government	L	L	H	H	H	H	H
Businesses	L	L	L	H	L	H	L
(iii) National Stakeholders							
Communities/general public	L	L	H	H	L	H	H
Fishermen	?	L	L	?	L	?	?
Recreational Users	H	L	?	H	L	H	?
Conservation groups	L	L	H	H	?	H	H
Schools/Universities	H	?	?	H	?	H	H
Government	L	L	H	H	H	H	H
Businesses	L	L	L	H	L	H	?

Contd...

Table 8.2a–Contd...

Beneficiaries	*Use Values*						*Non-Use*
	Direct Use Values				*Indirect Use Values*	*Option Value*	*Non-Use Values*
	Recreation	*Fisheries*	*Marine Products*	*Research and Education*			
(iv) International Stakeholders							
Communities/general public	L	L	H	H	L	H	H
Fishermen	?	L	L	?	L	?	?
Recreational Users	H	H	?	H	?	H	H
Conservation groups	L	L	H	H	?	H	H
Schools/Universities	H	?	?	H	?	H	H
Government	L	L	H	H	L	H	H
Businesses	H	H	L	H	L	H	L

H: Potentially significant benefit; L: Potentially some benefit; ?: Unsure benefits.

So to activate the RTF-E, a certain amount of funding should be allocated in the executing agencies' budget or any other ways which are most feasible and convenient for the whole project. Closer coordination among the executing agencies, the RTF-E and the focal point needs to be developed further.

Conclusion

The paper presents the role of economic valuation for the impacts of UNEP-GEF project on Reversing the Environmental Degradation Trends in the South China Sea and Gulf of Thailand, which is indeed very crucial as a basis for decision-making. The instrument for judging whether a project is feasible is the extended benefit-cost analysis. The extended benefit-cost analysis requires the transformation of the physical impacts into the monetary impacts which involve the economic valuation of the total impacts. In measuring the impacts, attention must be given to the extractive use, non-extractive use, environmental services, and biological diversity services. Those different services involve the use values and the non-use values. The use values includes the direct use vales of the extractive resources, the indirect use values of the non extractive resources, and the environment resources, plus the non-use values of the environmental and biological diversity services.

The valuation techniques can be either market base or non-market base using either market price or non-market price such as the travel cost approach, hedonic pricing, surrogate pricing approach, replacement cost and prevention costs approaches.

For the transboundary resources, the valuation of it must be done carefully by keeping in mind whether the environmental functions of the resources are mutually exclusive or complementary in nature. For the mutually exclusive functions, the values

Table 8.3: Frameworks for Seagrass Economic Valuation

Uses	*Use Values*			*Non-Use Value*			*Value Basis*		*Suggested Techniques*
	Direct	*Indirect*	*Option*	*Quasi-Option*	*Bequest*	*Existence*	*Market*	*Non-Market*	
Extractive Use									
Fish	X						X		Market Price
Medicine	X						X		Market Price
Food	X						X		Market Price
Fertilizer	X						X		Market Price
Handicraft	X						X		Market Price
Non-Extractive Use									
Research and Education	X							X	Actual costs (Gov. NGO, Academic)
Aesthetic/culture	X							X	Hedonic Pricing method
Environmental Services									
Shoreline protection		X						X	Replacement Cost
Carbon sequestration		X						X	
Water purification		X						X	
Oxygen release		X						X	
Sediment and nutrient retention		X						X	
Nursery area		X						X	Market Price Value of off-shore catch (shrimps/fish)
Waste catchment		X						X	Replacement Cost
Erosion prevention		X						X	
Biological Diversity Services									
Biodiversity			X					X	Contingent valuation method

Table 8.4: Draft Frameworks for Wetland Economic Valuation

Uses	*Use Values*			*Non-Use Value*			*Value Basis*		*Suggested Techniques*
	Direct	*Indirect*	*Option*	*Quasi-Option*	*Bequest*	*Existence*	*Market*	*Non-Market*	
				Extractive Use					
Timber	X						X		Market Price
Firewood	X						X		Market Price
Leaves/Thatch	X						X		Market Price
Charcoal	X						X		Market Price
Poles	X						X		Market Price
Fish	X						X		Market Price
Crab	X						X		Market Price
Prawn	X						X		Market Price
Wildlife	X						X		Market Price
Mollusks	X						X		Market Price
Tanning Bark	X						X		Market Price
Fruits	X						X		Market Price
Worms	X						X		Market Price
Medicine	X						X		Market Price
Aquaculture (pearl) and Agriculture and Transportation	X						X		Market Price
Peat/energy	X							X	Market Price

Contd...

Table 8.4–Contd...

Uses	*Use Values*			*Non-Use Value*			*Value Basis*		*Suggested Techniques*
	Direct	*Indirect*	*Option*	*Quasi-Option*	*Bequest*	*Existence*	*Market*	*Non-Market*	
Non-Extractive Use									
Tourism/Recreation	X						X		Travel cost
Research and Education	X							X	Actual costs (Gov. NGO, Academic)
Aesthetic/culture	X							X	Hedonic Pricing method
Environmental Services									
Shoreline protection		X						X	Replacement Cost
Windbreak		X						X	
Carbon sequestration		X						X	
Water purification & ground water recharge		X						X	
Oxygen release		X						X	
Nursery area		X						X	Market Price Value of off-shore catch (shrimps/fish)
Flood Control		X						X	Replacement Cost
Sediment and Nutrient Retention and Export		X						X	
Climate Change Mitigation		X						X	
Salt water intrusion		X						X	
Water supply (subsistence value)	X							X	

Contd...

Table 8.4–Contd...

Uses	*Use Values*			*Non-Use Value*			*Value Basis*		*Suggested Techniques*
	Direct	*Indirect*	*Option*	*Quasi-Option*	*Bequest*	*Existence*	*Market*	*Non-Market*	
Biological Diversity Services									
Biodiversity			X					X	Contingent valuation method
Migratory species		X						X	Contingent valuation method
Endangered Species					X			X	Contingent valuation method
Mangrove Ecosystem						X		X	Contingent valuation method

must not be added up to each other, but for the complementary functions of natural resources, the values can be the total values of all functions together.

References

Anonym, *Reversing Environmental Degradation Trends in the South China Sea and the Gulf of Thailand, Regional Task Force on Economic Valuation, Report of the First Meeting, 11-13 September 2003*, UNEP/GEF, Bangkok, September 2003.

Anonym, Total Economic Valuation: Coastal and Marine Resources in the Strait of Malacca, UNEP/UNDP/IMO, September 1999.

Angelos, Marian Delos, Herminie Fransisco and Ansii Niskanen, "Evaluation of the Profitability of Forest Plantations in the Philippines, in Jennifer Rietbergen-Mcracken and Husein Abaza, editors, *Environmental Valuation; A Worldwide Compedium of Case Studies*, UNEP, Earthscan Publications Ltd, London, 2000, pp. 92-98.

Dixon, John A and Maynard Hufschmidt, "Valuation of Losses of marine Product Resources Caused by Coastal Development of Tokyo Bay", adopted from Materials prepared by Yuzuru and Hanayama and Ikuo Sano, in Dixon, John A and Maynard Hufschmidt, editors, *Economic Valuation Techniques for the Environment; A Case Study Workbook*, The John Hopkins University Press, London, 1986.

East Asian Seas Regional Coordinating Unit, UNEP, *Strategic Action Programme for the South China Sea*, Draft Version 3. 24 February 1999

Edgar Furst, David D. Burton and Gerardo Jeminez, Total Economic Value of Forest in Mexico, in Jennifer Rietbergen-Mcracken and Husein Abaza, editors, *Environmental Valuation: A Worldwide Compedium of Case Studies*, UNEP, Earthscan Publications Ltd, London, 2000, pp. 1982-1995

Edgar Furst, David D. Burton and Gerardo Jeminez, "The Costs and Benefits of Reefs Conservation in the Bonaire Marine Park, in the Netherlands Antilles," in Jennifer Rietbergen-Mcracken and Husein Abaza, editors, *Environmental Valuation; A Worldwide Compedium of Case Studies*, UNEP, Earthscan Publications Ltd, London, 2000, pp. 162-171

Edgar Furst, David D. Burton and Gerardo Jeminez, "Partial Economic Valuation on Mangroves in Nicaragua," in Jennifer Rietbergen-Mcracken and Husein Abaza, editors, *Environmental Valuation; A Worldwide Compedium of Case Studies*, UNEP, Earthscan Publications Ltd, London, 2000, pp. 196-206

Hufschmidt, Maynard M., David James, Anton D. Meister, Blair T. Bower, John Dixon, Environment, Natural systems, and Development: An Economic Evaluation Guide, The John Hopkins, University Press, Baltimore, 1983.

Panayotou, Theodore, *Instrument of Change: Motivating and Financing Sustainable*, UNEP, Earthscan Publicztions Ltd., London 1998.

Pearce, David W. and Jeremy J. Warford, World *Without End: Economics, Environment, and Sustainable Development*, Oxford Universituy Press, 1993.

Chapter 9

The Role of Mangrove Ecosystem as Natural Barrier to Environmental Hazards

Rudhi Pribadi

Department of Marine Sciences, Diponegoro University, Semarang, Indonesia

E-mail: rudhi_pribadi@hotmail.com

Introduction: Mangrove Ecosystems

The term "mangrove" may be used in two senses. First it describes the constituent plants of tropical intertidal forest communities of which the major plant components belong to only nine genera from five families, with minor components from 11 genera each from a different family (Tomlinson 1994). In a second sense, the term refers to the complex of tree-dominated plant communities fringing sheltered shores. Such mangrove ecosystems exist in tropical areas on all continents and frequently extend into subtropical areas. The most luxuriant and diverse mangrove forests occur in southeast Asia (Waisel 1972).

Mangroves have long attracted the attention of scientists because of their habitat (siltrich, saline, coastal waters), and the curious structures of many of their species (flying buttresses, stilt roots, pneumatophores, viviparous reproduction) (Whitmore 1984). The ecosystems are highly productive with a high litter production (Snedaker and Brown 1981, Proctor 1984) and are believed to enhance near-shore primary and secondary production (Macnae 1974; Mann 1982). The mangrove intertidal habitats are the major ecosystems along sheltered subtropical and tropical coasts and are reported to have a total area of 43.7 million ha (Rodin *et al.*, 1975).

Odum (1969) described mangroves as interface or open systems because of their flowthrough pathways for transporting matter. These pathways are driven by physical (tides, terrestrial runoff, and rainfall) and biological (litter production, decomposition, mineral uptake and faunal activities) factors that control the rate of matter import, export and storage (Pool *et al.*, 1975). The result of these flows and transformations is the import of inorganic matter from terrestrial ecosystems and the export of both particulate and dissolved organic matter to adjacent marine ecosystems.

Mangrove Component: Vegetation

A distinctive character of mangrove forest is its low species richness (Tomlinson 1994). However, since the architecture of the forests is influenced by the magnitudes and periodicities of tides, nutrients, and stresses such as hurricanes, drought and salt accumulation, and because the action of these factors varies widely over geographic regions, mangrove stands exhibit wide regional and local variations in structural characteristics. High structural variation is associated with high species richness. Where species diversity is high, structural variation is greater (Cintron and Novelli 1984).

Tomlinson (1994) recognised three elements of mangrove vascular plant species: (a) major, (b) minor and (c) mangrove associates. Major elements are recognized have all or most of the following features:

1. Fidelity to the mangrove environment and not extending into other terrestrial communities.
2. A major role in the mangrove structure and the ability to form a pure stand.
3. Morphological specialisation such as aerial roots and vivipary of the embryo.
4. Some physiological mechanism for salt tolerance. They frequently visibly excrete salt.
5. Taxonomic separation from their relatives at least at the generic level and often at the subfamily or family level.

The minor element species are not conspicuous but occupy peripheral habitats and only rarely form pure communities. Their taxonomic isolation is mostly at the generic level. The mangrove associates are never inhabitants of strict mangrove communities and occur only in transitional vegetation.

Zonation, often into monospecific stands, is typical in mangroves. This phenomenon is sometimes used to explain a succession process since the zonation in space is often accepted as matching a zonation in time (Lugo 1980).

The Ecological Functions

Two important functional roles of mangroves have become the subject of considerable research and debate; namely, to what extent do they export nutrients and detritus to the coastal waters and how important are they as nursery areas for fish and crustaceans of commercial importance (Chong *et al.*, 1990)?

Litterfall occurs throughout the year, and through the feeding activities of microbial decomposers and the larger detritivore animals the litter is processed into fine particles of organic detritus (Odum and Heald 1975). This organic detritus forms a significant fraction of the available food particles in mangrove-lined estuaries (Mann 1972; Odum and Heald 1975). The organic particles that become colonized by bacteria, fungi, and other micro organisms are a main food source for a range of detritus-consuming omnivorous organisms (Odum and Heald 1975), including prawns, crabs, and a number of fishes. This food web, with the principal flow of energy along the route: mangrove leaf detritus □□→bacteria and fungi → detritus consumers → lower carnivores → higher carnivores, is characteristic of shallow muddy estuaries with extensive plant communities such as mangroves, seagrasses, or microalgae.

Mangroves provide important feeding and nursery areas for many fish species, especially mullets (*Mugyl* sp.) (Knox and Miyabara 1984). Their most important function for fisheries however is their role as nursery areas. The freshwater prawn *Macrobrachium* spp., with more than 100 species in tropical areas, is dependent on the brackish-water environment for the completion of its life cycle. Pregnant females migrate from freshwater rivers and lakes to the mangrove swamps along the coasts, where the eggs hatch into free-swimming larvae. After completion of the pelagic larval stage, the larvae metamorphose to a benthic stage and crawl back upstream to freshwater. The marine species, on the other hand, spawn in near-shore waters, and the newly hatched larvae are transported to the mangroves along the coasts, where they stay until they reach the juvenile stage and migrate offshore (Knox and Miyabara 1984).

Attempts have been made to determine the relationship between mangrove areas and offshore shrimp production. Martosubroto and Naamin (1977) suggested that there is a significant linear relationship between the two, and that any reduction in mangrove area will cause a reduction in shrimp production. A regression analysis of annual prawn production with mangrove area and latitude, using a world-wide data set (Turner 1977), showed that in Indonesia one could expect a production of about 100 kg of prawns/ha/year of mangrove. Nair (1977) has attempted to quantify the economic tradeoffs between mangrove forestry and offshore prawn fisheries in Sabah.

In Australia, Staples *et al.* (1985) found a high positive correlation between the length of the mangrove-lined section of rivers and the mean annual commercial catch (1973 to 1982) of the banana prawn (*Penaeus merguiensis* de Man) in different areas of the Gulf of Carpentaria. Adults of the commercially important penaeid prawns usually spawn at sea and after a short larval life (2–3 weeks), the post larval stage settles in the near inshore areas and estuaries (Dall *et al.*, 1990). Many species spend their juvenile phase in or associated with vegetation such as mangroves or seagrass beds. In an analysis combining many penaeid species from several countries, Turner (1977) demonstrated a positive correlation between the offshore yield of prawns and the amount of mangrove present in the nursery area.

Ecologically mangrove also to be believed protect the surrounding seagrass and coral reef ecosystems by absorbing or trapping 'pollutant' entering an coastal area either from land, transporting by river water, or from nearby fishpond. Many research have been conducted in this area and mostly indicating that into certain level mangrove can be indeed trapping the pollutant, however further investigation should be conducted to determine the fate of trapped pollutant once they released to the ecosystem through decomposition processes.

The Physical Functions

Physically mangrove can reduce erosion using their complex root system. The smaller, pencil-like root mats of *Avicennia* and the slightly bigger stick of *Sonneratia* will effective on trapping sediment transported by sea water, while the giant stilt roots of *Rhizophora* will disperse and reduce the wave energy into smaller units. The effectively of this process of course depend on the mangrove size and thickness and the altitude of the waves itself. Similarly in good condition mangrove are also able to reduce wind energy to a significant level.

No such comprehensive research has been conducted to determine the effect of mangrove vegetation on salt intrusion, however some field reports indicate that many area have been severely damaged (especially for agriculture purposes) and at one time the coastal greenbelt consisting of mangrove forest was destroyed.

The Resources Functions: Mangrove Utilization in Indonesia

For centuries Indonesians have traditionally utilized mangroves, mostly for firewood (almost all mangrove woody plants), charcoal (*Bruguiera* and *Rhizophora*), tannin (*Bruguiera* and *Rhizophora*), medicine (*Acanthus* and *Sonneratia*), timber (*Xylocarpus, Bruguiera, Rhizophora* and *Ceriops*) and even constructing boats (*Heritiera*) (Soegiarto 1984).

The Riau Archipelago Province (Sumatra) has long been known as a centre for mangrove charcoal, and the Forestry Service there (unpublished) reported in 1984 that in Riau there were about 836 charcoal kilns with a total production of 22,207 tons per year^{-1}. The products are exported mostly to Singapore and Hong Kong.

The large-scale commercial exploitation of mangroves in Indonesia has begun relatively recently with the production of charcoal, logs, poles and chip-wood for paper pulp. In early 1982 when the total mangrove area in Indonesia was about 4.25 million ha, mangrove areas that had been surveyed for logging amounted to 674,200 ha of which 4,01,000 ha were already exploited (Darsidi 1984). In the same year, for instance, 1,63,000 ha of the mangrove forests in Irian Jaya were already on lease (Darsidi 1984) and no less than 10 private corporations were ready to get a share in the remaining 205,000 ha of the allocated 368,000 ha (Direktorat Bina Program Kehutanan 1981).

Mangrove Destructions

During the past decades, the demand for mangrove resources has been steadily increasing, not only for the products but also for the land itself. Of these two demands, the latter is potentially more destructive, since the end result is a completely different environment such as human settlements, industrial sites, agricultural lands and brackish water fish ponds (Soemodihardjo and Soerianegara 1989). The most extensive form of mangrove land conversion is undoubtedly the "tambak" or brackish-water fish ponds. There is no doubt that fish are a very important source of animal protein for the people of tropical Asia, therefore to improve nutrition and to cope with projected population increases must involve substantial increases in fish production. Among the South-east Asian Nations, Indonesia has the second largest total catch, both for marine and freshwater fishes. The greatest potential for expansion of fish production in Indonesia would appear to lie in brackish-water culture and the harvesting of natural marine living resources. Both of these have their greatest impact in the coastal zone (Knox and Miyabara 1984). There were 155,081 ha of tambak in Indonesia in 1980, mainly in Java, Sulawesi and Sumatra (Soewito 1982). In the Fourth Five-Year-Development Programme of 1984 -1989, where the effort to increase foreign exchange earning from fishery sectors was encouraged, the government sponsored the opening of thousands of hectares of mangrove and their conversion to brackish-water fishponds, particularly for the culture of prawns. It was to be carried out mostly outside Java (Soemodihardjo and Soerianegara 1989).

The mangrove wood chip industry is quite new in Indonesia. The first chip mill was established in Tarakan, East Kalimantan and produced around 70,000 m^3/year of chips (Burbridge and Koesoebiono 1980). In 1984 the Bina Lestari Timber Company completed the construction of a chipmill in Riau with a maximum capacity of about 176,000 m^3/year (Soegiarto and Polunin 1981), and in 1988 another chipmill industry was already operating in Bintuni Bay in Irian Jaya with a production capacity of about 300,000 m^3/year (Erftemeijer *et al.*, 1989). Fortunately, most mangroves regenerate naturally, their species composition is simple, and their dynamics relatively easy to study (Tamai and Iampa 1988), therefore effort should be conducted to make sure that once the forest cleared it will regenerate naturally, or if in some cases failed it should be assisted in proper ways.

Even though commonly mentioned that mangrove ecosystem can reduce wave and wind energy in the coastal line, in some extent extreme natural processes can be tolerated and destroyed the ecosystem. The huge tsunami strike on the 26 December 2004, for example, left tremendous damage in the coastal area including the mangrove of parts of Sumatra. A powerful earthquake (8.9 in RS) which hit Nias Island in the western coast of Sumatra also destroyed hectares of mangrove due to the shifting and subsidence effect in the island.

Mangrove Conservation and Rehabilitations

The management of mangrove ecosystems requires knowledge of their regeneration and species characteristics. There are two main concepts on mangrove conservation: reservation and rehabilitation. Reservation means protecting of an area with mangrove in good condition such as including them into a reserved area, national park, wetland sanctuary etc., while rehabilitation will cover efforts to resurrect the functions of the ecosystem in the degraded mangrove area.

Mangroves have little capacity for vegetative reproduction and are dependent on seedling recruitment. Consequently, survival of seedlings directly affects species distribution and abundance. All mangrove tree species are dispersed by sea water, and the propagule (fruit, seed or seedling) has some ability to float, even if for a limited time (Tomlinson 1994). Establishment of the seedling is often difficult because of the unstable, variable substrates and the tidal influence (Tomlinson 1994). Ball (1988) and McKee (1993a,b) discussed how tidal inundation influences soil redox potential, salinity, pH, and concentrations of nutrients and phytotoxins. Biotic factors such as seed predation may vary across the intertidal zone and influence mangrove species distribution patterns (Smith 1987a, Smith *et al.*, 1989, Mc Kee 1995a). Macnae (1968) and Aksornkoae (1975) had earlier pointed out that light conditions are the main physical factors in the establishment of mangrove trees.

References

Aksornkoae, S. 1975. Structure of mangrove forest at Amphoe Khlung, Changwat Chantaburi, Thailand. *Forest Research Bulletin*, 38, 42-47

Ball, M. C. 1988. Ecophysiology of mangroves. *Trees*, 2, 129–142.

Chong, V. C., Sasekumar, A., Leh, M. U. C. and D'Cruz, R. 1990. The fish and prawn communities of a Malaysian coastal mangrove system with comparison to adjacent mud flats and inshore waters. *Journal of Estuarine, Coastal and Shelf Sciences*, 31, 703-722.

Cintron, G. and Novelli, Y. S. 1984. Methods for studying mangrove structure. *The Mangrove Ecosystem: Research Methods* (eds S.C. Snedaker and J.G. Snedaker), pp 91–113. UNESCO, Paris.

Dall, W., Hill, B. J., Rothlisberg, P. C. and Staples, D. J. 1990. Biology of the Penaeidae. *Advances in Marine Biology* (eds J.H.S. Blaxter and A.J. Southward). Academic Press, London.

Darsidi, A. 1984. Pengelolaan hutan mangrove di Indonesia. *Prosiding Seminar II Ekosistem Mangrove* (ed S. Soemodihardjo), pp. 19–28. Man and Biosphere–Lembaga Ilmu Pengetahuan Indonesia, Jakarta.

Direktorat Bina Program Kehutanan 1981. *Potensi dan Penyebaran Hutan Payau di Indonesia, Edisi Khusus Direktorat Jendral Kehutanan no. 40.* Departemen Kehutanan Republik Indonesia, Jakarta.

Erftemeijer, P., Allen, G. and Zuwendra 1989. *Preliminary Resource Inventory of Bintuni Bay and Recommendations for Conservation and Management.* Direktorat Perlindungan Hutan dan Pelestarian Alam Departemen Kehutanan Republik Indonesia/Asian Wetland Bureau, Bogor.

Knox, G. A. and Miyabara, T. 1984. *Coastal Zone Resource Development and Conservation in The South East Asia, with Special Reference to Indonesia.* UNESCO–ROTSEA–East West Centre, Jakarta–Honolulu.

Lugo, A. E. 1980. Mangrove Ecosystems: Successional or Steadystate? *Biotropica*, 2, 65–72.

Macnae, W. 1968. A general account of the fauna and flora of mangrove swamps and forests in the Indo-West-Pacific Region. *Advance Marine Biology*, 6, 73-270.

Mann, K. H. 1972). Macrophyte production and detritus food chains in coastal waters. Detritus and its role on aquatic ecosystems (eds U. Melchori-Santalini and J. Hopton). *Memoir Institute Italiano Hydrobiologie* (suplement), 29, 13–16.

Mann, K. H. 1982. *Ecology of Coastal Waters: a system approach.* University of Califonia Press, Berkeley.

Martosubroto, P. and Naamin, N. 1977. Shrimp production in Indonesia. *Marine Research Indonesia*, 18, 81–86.

McKee, K. L. 1993a. Soil physicochemical patterns and mangrove species distribution–reciprocal effects? *Journal of Ecology*, 81, 477–487.

McKee, K. L. 1993b. *Determinants of mangrove species distribution in neotropical forest: biotic and abiotic factors affecting seedlings survival and growth* (Rhizophora mangle, Avicennia germinans, Laguncularia racemosa). Ph.D. dissertation, Louisiana University, Baton Rouge, Louisiana.

McKee, K. L. 1995a. Seedling recruitment patterns in a Belizean mangrove forest: effects of establishment ability and physico–chemical factors. *Oecologia*, 101, 448-460

Nair, M. Y. 1977. *An Appraisal of The Economic Potential of Mangrove Swamps*. Msc. thesis, University Pertanian Malaysia.

Odum, E. P. 1969. The strategy of ecosystem development. *Science*, 164, 262–270.

Odum, E. P. and Heald, E. J. 1975. Mangrove forest and aquatic productivity. *Introduction to Land–Water Interaction (Ecological Study Series)*, pp. 129-136. Springer-Verlag, Berlin.

Pool, D. J., Snedaker, S. C. and Lugo, A. E. 1975. Litter production in mangrove forest of South Florida and Puerto Rico. *Proceedings of International Symposium on Biology and Management of Mangroves Vol 1* (eds G. Walsh, S.C. Snedaker and H. Teas), pp. 213–237. Institute of Food and Agriculture Sciences, University of Florida, Gainesville.

Proctor, J. 1984. Tropical forest litterfall: 2. The data set. *Tropical Rain Forest: Ecology and Management, Supplementary Volume.* (eds S. L.Sutton, T. C.Whitmore and A. C. Chadwick), pp 83–113. Blackwell Scientific Publications, Oxford.

Rodin, L. E., Bazillevich, N. I and Rozov, N. N. 1975. Productivity of the world's main ecosystems. *Productivity of The World Ecosystem*, pp. 13–26. National Academy of Science, Washington D.C.

Smith, T. J. III 1987a. Seed Predation in relation to the dominance and distribution in mangrove forests. *Ecology*, 68, 266–273.

Smith, T. J. III, Chan, H. T., McIvor, C. C. and Robblee, M. B. 1989. Comparisons of seed predation in tropical todal forests from three continents. *Ecology*, 70, 146–151.

Snedaker, S. C. and Brown, M. S. 1981. Primary productivity of mangroves. *CRC Handbook Series of Biosolar Resources, Vol. 1: Basic Principles* (eds C.C. Black and A. Mitsui), pp. 477-486. CRC Press, West Palm Beach, Florida.

Soegiarto, A. and Polunin, N. 1981. *The Marine Environment of Indonesia: A Report to The Goverment of Indonesia–Sponsored by IUCN/WWF*. University of Cambridge, Cambridge.

Soemodihardjo, S. and Soerianegara, I. 1989. The status of mangrove forest in Indonesia: country report, Indonesia. *Proceedings of Symposium on Mangrove Management: Its*

Ecological and Economic Considerations, Biotrop Special Publication no. 37 (ed I. Soerianegara), pp. 73–114. Biotrop, Bogor.

Soewito 1982. Status ekosistem hutan bakau bagi perikanan di Indonesia dan langkah pembinaannya. *Pertemuan Teknis Evaluasi Hasil Survai Hutan Bakau*, pp. 19–32. Direktorat Jenderal Perikanan, Departemen Pertanian Republik Indonesia, Jakarta.

Staples, D. J., Vance, D. J. and Heales, D. S. 1985. Habitat requirement of juvenile penaeid prawns and their relationship to offshore fisheries. *Second Australian National Prawn Seminar* (eds P.C. Rothlisberg, B.J. Hill and D.C. Staples), pp. 47–54. Queensland, Australia.

Tamai, S. and Iampa, P. 1988. Establishment and growth of mangrove seedlings in mangrove forests of south Thailand. *Ecological Ressearch*, 1, 227–238.

Tomlinson, P. B. 1994. *The botany of mangroves*. Cambridge University Press, Cambridge–New York–Melbourne.

Turner, E. R. 1977. Intertidal vegetation and commercial yields of penaeid shrimp. *Trans American Fishery Society*, 106, 411–416.

Waisel, Y. 1972. *Biology of hallophyte*. Academic Press, New York.

Whitmore, T. C. 1984. *Tropical Rainforest of the Far East*. Clarendron Press, Oxford.

Part II

Impact Studies Related to Tsunami, Tropical Cyclones and Toxic Contamination

Chapter 10

Some Aspects of Coastal Zone Management in Sri Lanka Including Impact of Tsunami: A Review

***M.S. Rupasinghe and M.S.J. Perera**

Department of Natural Resources, Sabaragamuwa University of Sri Lanka

**E-mail: mrmahinda@yahoo.com*

ABSTRACT

2 km wide band of ocean and the adjoining strip of land extending 300 m inland along the 1,760 km coastline is described as the coastal zone of Sri Lanka. It is rich in both living and non living resources including a diversity of coastal ecosystems; fringing coral reefs, sea grass beds, estuaries, lagoons, mangroves, salt marshes, sandstone reefs, sand dunes, sandy and rocky beaches, mud flats, gentle seashore vegetation and maritime grasslands. These habitats are important in restraining coastal erosion and sustaining coastal industries.

Over exploitation of resources, coastal constructions and pollution are major issues within the zone. Resources are being subjected to extractive (fisheries, sand mining), and non-extractive (tourism) pressures. Even though coastal capture fishery (edible and ornamental) resources are almost fully exploited, offshore fishery could still be expanded. Most of coastal constructions are related to fisheries and tourism industry.

Coast Conservation Department manages the zone, under the Coast Conservation Act (1981). The first Coastal Zone Management Plan developed in 1990 was revised with the assistance of Coastal Resources Management Project in 1997. Key management areas are coastal permit system and Special Area Management.

The Indian Ocean Tsunami took Sri Lanka by surprise, as the island is not subjected to many coastal hazards except for localized coastal erosion and tropical cyclones. In addition to the huge humanitarian loss, it also affected coastal ecosystems. With understanding of the need for preparedness and disaster management, the Sri Lanka Disaster Management Act (2005) was enacted and several other measures were taken. It is evident that if the Coast Conservation Act and the CZMP were effectively implemented the tsunami damages could have been greatly reduced. It is also proven that the presence of natural ecosystems in coastal zone assist in minimizing impacts of coastal hazards, hence the importance of conserving them is emphasized.

Keywords: *Coastal zone management, Resources, Conservation, Hazards, Tsunami, Sri Lanka.*

Introduction

Sri Lanka is a tropical island in the Indian Ocean with a history of continuing human occupation for over 25 centuries. The country's total population is around 19 million and with a population density of 280 persons per sq km is one of the highest in the world. Over 40 per cent of the people are engaged in activities directly dependent on the environment and about 25 per cent people live in urban or semi urban areas (UNEP, 2001). Almost a third (32 per cent) of the country's population, two-thirds (65 per cent) of the total urban population, two-thirds (67 per cent) of the industrial facilities and over 80 per cent of the tourist infrastructure are accommodated within, only one fourth (24 per cent) of the island's land area having a coastal boundary (MoFE, 1999; Olsen *et al.*, 1992). Projections estimate that coastal migration will continue to increase coastal population densities to even over 1,000 persons per sq km (De Silva, 1997).

The Coastal Zone of Sri Lanka, in the context of the Coast Conservation Act (CCA) of 1981 and amendments (MoFARD–Ministry of Fisheries and Aquatic Resources Development), is defined as the 2 km wide band of ocean and an adjoining strip of land extending 300 m inland. In the event of a water body connecting to the sea, the zone extends two kilometers inland from the mouth of the water body. The coastal zone thus, could be considered to include an area of land or water where ecological processes of both land and marine environments are inter-linked, with the influence from human activities in the region.

The term "coastal resources" refers to anything that is useful for human, both living and non-living, found within the coastal zone. Economically, the coastal region is considered a distinctive area, where development is influenced to a large extent by access to the sea (Savundranayagam *et al.*, 1994). The increase of the coastal sector's contribution to the GDP from 35 per cent in 1983 to 40 per cent by 1990s is indicative of increasing economic activity in the coastal area (Savundranayagam *et al.*, 1994, Brown, 1997, UNEP, 2001).

The length of the Sri Lankan coastline is about 1,760 km from which the Exclusive Economic Zone (EEZ) extends 200 nautical miles and covers an area of over 230,000 437,400 sq km. Extending outwards from the coastline is the continental shelf, with an average width of about 40 nautical miles and a total area of about 26,000 sq km (nearly half the total land area of the island). The seas overlying the shelf are referred to as coastal. Beyond the edge of the continental shelf is the continental slope, which descends steeply to the floor of the deep ocean. The territorial limits of the island extend to 19.2 km offshore, beyond which lies the contiguous zone, under the provisions of the UN Law of the Sea (Figure 10.1) (Arulpragasam, 2000).

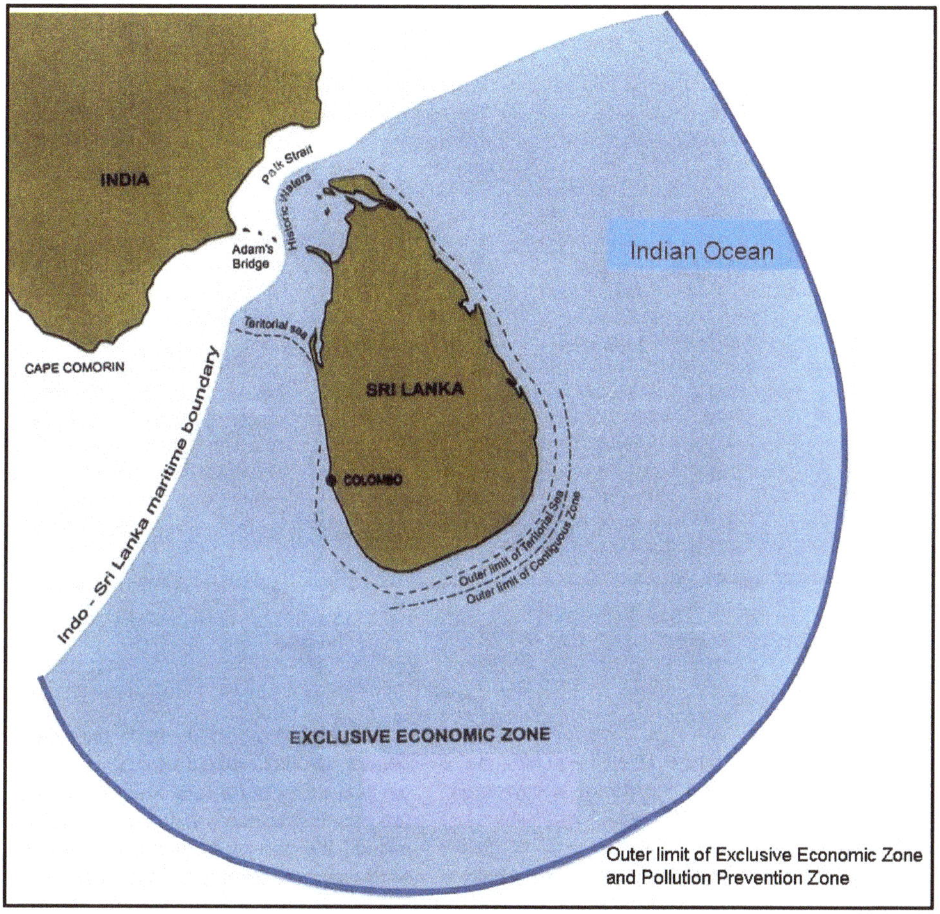

Figure 10.1: Coastal and Maritime Jurisdiction of Sri Lanka
(*Source*: Natural resources of Sri Lanka 2000, National Science Foundation)

The coastline and adjacent waters support highly productive marine ecosystems such as fringing coral reefs, shallow beds of coastal and estuarine sea grasses and an extensive system of 45 estuaries and 40 lagoons (UNEP, 2001). In addition, coastal ecosystems such as mangroves, salt marshes, sandstone reefs, sand dunes, sandy and rocky beaches, mud flats, gentle seashore vegetation and maritime grasslands occurs within sea-land interface.

The present paper reviews about the coastal resources, their depletion and other coastal issues in Sri Lanka with higher emphasis on coastal hazards. The paper also summarizes attempts and trends in Coastal Zone and Hazard Management in Sri Lanka.

Coastal Resources in Sri Lanka

Living renewable coastal resources in the form of marketable "goods" in Sri Lanka include fish, shark fin, shrimp, prawn, lobster, crab, bivalve, squid, cuttlefish,

Figure 10.2: Some Coastal Habitats Found in Sri Lanka: (*a*) Mangrove, (*b*) Salt Marsh, (*c*) Sand Dune and (*d*) Sandy Beach

(*a*) © W.P.N. Perera, IUCN – Sri Lanka Photo Collection

(*b*) © M.S.J. Perera, IUCN – Sri Lanka Photo Collection

Contd...

Figure 10.2–Contd...

(*c*) © IUCN – Sri Lanka Photo Collection

(*d*) © M.S.J. Perera, IUCN – Sri Lanka Photo Collection

brine shrimp (*Artemia*) sea weed (*Gracillaria, Sargassum*), chank (*Xancus pyrum*), sea cucumber (beche-de-mer) and other invertebrates for the export aquarium trade (UNEP, 2001). In addition to those marketable living resources Sri Lankan coast is rich in an intrinsic biodiversity value; nationally threatened estuarine crocodiles, 5 species of globally threatened sea turtles, wading birds, sea birds and marine mammal species including dolphins, whales and dugongs.

Non-living marketable coastal resources include mineral resources such as energy minerals, metallic minerals and non-metallic minerals. Among energy minerals, peat deposits that cannot as yet be economically used are estimated to be about 50 million tons. Radioactive minerals such as monazite, metallic ores such as iron ore (2.2 million tons), copper- magnetite deposits (4 million tons), mineral sands (4 million tons at Pulmoddai), ilmenite (3 million tons), rutile (6 million tons), zircon (4 million tons), high purity silica sands (6 million tons) as well as limestone, clays, deposits of inland coral and seashells, granite, granitic gneisses, marbles, quartzites and chamockites are also found in the coastal area (Ranasinghe, 1997). The mineral resources found in beach sand as well as terrestrial areas of coastal zone are utilized and marketed as salt, graphite, limestone, clay, sand, feldspar, lime, kabook, brick, gem and mineral sands.

The "services" provided by coastal resources include a wide range, from flood protection, nutrient flows, nursery and breeding grounds of species harvested for commercial and ornamental purposes, and tourism, including recreational activities such as turtle, whale and dolphin watching. It is also important to note that many more coastal resources may still remain unknown since cataloguing and inventorying of resources, particularly those in the sea, have not been carried out comprehensively (UNEP, 2001).

Resource Exploitation and Depletion

Diverse coastal resources are being subjected to extractive pressures such as fish capture or sand mining, as well as non-extractive pressures like aesthetic enjoyment and tourism (UNEP, 2001). Resource depletion, which occurs when, pressure of exploitation, exceeds generation rates are observed in some coastal resources, especially in the near-shore fishery industry. Excessive exploitation causes pressure directly on the resource itself by its physical overuse and/or by indirectly exerting pressure on the resource-related ecosystem, which disrupts resource generation rates.

The coastal capture fishery of edible species is the major economic activity using biological resources of the coastal area. The coastal sea is the site of activity for the capture of finfish and shellfish (mostly for domestic consumption) as well as fish and invertebrates (for the marine ornamental export trade). Bays and other near shore marine waters support the fishery of over 500 edible fish species. It has been recorded that about 70 per cent of coastal living resources are composed of small pelagic fish like sardines, herrings, anchovies and mackerels and small demersal fish such as sciaenids and carangids (UNEP, 2001). The coastal fisheries provide nearly 97 per cent of the marine fish production and 90 per cent of the employment in fisheries (Arulpragasam, 2000).

Capture fishery, which was 174,500 mt in 1994 rose to 152,750 mt in 1997 and to 171,950 mt in 1999 (MoFARD, 1999). Estimates of the maximum annual sustainable yield range from 250,000 mt to about 300,000 mt of fish (UNEP, 2001). Therefore, the exploitation pressure is gradually reaching the maximum sustainable capacity of the coastal ecosystem.

Possible over-collection and the use of inappropriate techniques, for aquarium/ ornamental fish capture, as well as inappropriate modes of transport in the export aquarium trade have also caused severe threat to reef fish populations in near shore habitats. According to UNEP (2001) although the freshwater fish export industry cultures most of its exported fish, no marine species are cultured in Sri Lanka. They are collected mostly from inshore areas where corals occur, often using the "moxy net" that damages the reef structure, while individuals are being caught from the sea by snorkelers and divers.

In the capture aquarium fishery industry, 530 mt of organisms (both collected from marine and freshwater) comprising approximately 200 species with a value of about Rs. 300 million was exported in 1996, which almost doubled to 1043 mt (with a value of Rs. 531 million) in 1998 (MoFARD, 1999; Ekaratne, 1997).

Offshore fishery resources in Sri Lanka is under-exploited and further expansion of this sector could yield much more economic gains. Offshore fishery which contributed 4,259 mt or 2.2 per cent to the total marine fishery in 1987, and 7.1 per cent in 1990, has been expanded to around 25 per cent in 1995 and 1996 (NARA, 1998) and to 29 per cent in 1997 (UNEP, 2001).

Some invertebrate species in the edible marine capture fishery have been overexploited resulting in dwindling numbers. Sea cucumber (beche-de-mer) fishery is a good example as, although 13 of the 70 species recorded in Sri Lanka are consumed in various parts of the world, only the dominant species (*Holothuria scabra*) is being selectively harvested. Unsubstantiated reports indicate that several species are now being over harvested. Lobster and crab are also exported with 1998 export quantities being 164 mt and 486 mt, respectively. The harvesting of lobster could have harmful effects on wild stocks, as gravid lobsters as well as small-sized lobsters are regularly collected in spite of bans on such collection (UNEP, 2001).

The collection of coastal non-living resources has also expanded gradually, increasing pressure on the resource and stability of the habitat. For example, garnet sand mining creates instability of the shores at Hambantota in the south east of Sri Lanka, inland coral mining and seashell mining result in ground pits that alter the habitat profile in Kalametiya and may other areas in the south east. Sand mining carried out extensively has affected the beach nourishment and the integrity of the shoreline (UNEP, 2001). The volume of sand extracted has increased from 523,780 cubes in 1984 to 625,662 cubes in 1991(CCD, 1997). Sand mining from the offshore seabed has attracted discussion in recent times for purposes of infrastructure development such as highways.

The impacts of over exploitation can have a direct effect on the related coastal resource and/or affect the resource generating process through effecting on habitat and ecosystem quality. Coastal resources may have a resilient capacity to withstand, or to rebound from, impacts by natural phenomena. This resilience may, however, be adversely affected by impacts arising from man-induced pressures as discussed above.

Development in the Coastal Ecosystems and Pollution

Development in the coastal zone of Sri Lanka is primarily related to economic activities conducted in order to utilize living and non-living coastal resources. In return most of these unplanned and unsustainable developments have adverse impacts on coastal ecosystems such as mangroves, salt marshes, coral reef as well as sand dunes. Major industries that sustain on the coastal resources in Sri Lanka, which leads to coastal constructions are fisheries and tourism. In addition, housing, roads, railways and other infrastructure have excreted a pressure on the coastal ecosystems in urban areas.

Infrastructure development for the fishing industry such as the construction of harbours creates pressure on coastal ecosystems and coastal resources. Additional pressure is created by pollution from boats, anchor damage, unsustainable fisheries such as catching small sized fish, use of dynamite and other destructive practices (UNEP, 2001).

Coastal aquaculture progressed massively during past three decades having severe impacts on mangroves and coastal wetlands, due to the high profits in its initial stages. Prawn farms (Tiger prawn export industry) were established in cleared ecologically sensitive mangrove areas in the northwest exerting tremendous pressure on mangroves and degrading associated lagoons and estuaries. It is estimated that 400 to 500 ha of mangroves or mangrove associates (NARA, 1995) have been cleared for prawn farms in North West.

Increasing tourist arrivals will continue to be a source of pressure on coastal resources, particularly because coastal tourism is showing rapid growth worldwide (Brown, 1997). The pressure on coastal habitats through hotel construction, which was low before 1980's, is now seeing a gradual increase. Tourism in Sri Lanka has primarily focused on its scenic sandy beaches and coastal estuaries and lagoons. About 85 per cent of tourism revenue comes from facilities in coastal areas. Today over 75 per cent of graded hotels and over 80 per cent of the hotel rooms are located along the coast (NARESA/USAID, 1991). Land filling for infrastructure facilities such as roads and housing, which occur in coastal wetlands, especially in urban and sub-urban areas has also become a threat to coastal habitats.

Considerable pressure is exerted on coastal water quality by ships operating on the major international shipping route off the southern and south-western coast with an estimated annual traffic of over 5,000 tankers (NARESA/USAID, 1991). The routine dumping of engine oil into the water by fishing craft further pressurizes water quality as well as the food web functions. Occasional grounding of oil tankers, oil spills and leaks has also occurred in the recent past (UNEP, 2001).

Industrialization in the coastal zone also creates pressures on the natural ecosystem. The discharge of tannery industry waste is one of major issues discussed during recent past. Pressure on coastal ecosystems from upland or upstream sources are common, such as the seepage of chemicals, including agrochemicals, into the sediment load that eventually reaches the coastal habitats through river run-off. Pollution can therefore arise from point and non-point sources (UNEP, 2001).

Coastal Zone Management in Sri Lanka

In resource management, the coastal region can be considered to represent an area of transition where terrestrial and marine environments interact to form unique environmental conditions. The coastal region therefore includes inshore waters, intertidal areas and extensive tracts of contiguous land (Brown, 1997).

Population growth has increased pressure on the resources of Sri Lanka's coastal zone, creating new stresses in the coastal environment. Industrial development since the late 1950s, beach based tourism development since the late 1960s and the open economic policies of the late 1970s have all placed strains on the coastal environment. By the early 1970s problems associated with coastal development had begun to arise. The need for a more management-oriented approach was first recommended in 1971 by the coast protection unit of the Colombo Port Commission. In 1978 the subject of coast conservation was transferred to the Ministry of Fisheries, by the creation of Coast Conservation Division. This division was upgraded to the status of a Government Department in 1984 (Samaranayake, 2000).

Coast Conservation Act No. 57 of 1981

Amended by Act No 64 of 1988 the Coast Conservation Act was primarily intended to conduct a survey and prepare a Coastal Zone Management Plan (CZMP) in addition to regulating development activities within the coastal zone. This Act prohibits any development activity within the coastal zone without a permit from the Director of CCD, which will be issued only if the proposed activity is consistent with the CZMP and such activity would not have any adverse impact on the stability, productivity and environment quality of the coastal zone (Jayawardane, 2005).

Coastal Zone Management Plan (CZMP)

As required by the coast conservation act, the first CZMP was prepared and approved by the cabinet of ministers in 1990. The 1990 plan deals with the coastal problem of erosion, loss and degradation of natural coastal habitats, as well as loss and degradation of archeological, cultural and historical sites, as well as recreational and scenic areas. Objectives and policies for the management of each problem were identified along with specific management techniques. In particular, the rationale and procedures for continuing the coastal permit system are detailed. Implementing actions in this plan are regulation, direct development, research, co-ordination, education, and planning and policy development (Samaranayake, 2000).

The CZMP permitted construction activities within the coastal zone of 300m subject to the setbacks defined in it. But, law enforcement was not very effective for several reasons resulting in ad-hoc construction within the coastal zone. Difficulty in effective implementation of the Act has been compounded due to provisions in other government Acts such as Mines and Mineral Act. No 33 of 1992 and Sri Lanka Land Reclamation and Development Corporation Amending Act No. 52 of 1982 facilitating development bypassing CCA, even though Town and Country Planning

Amending Act No. 57 of 1981 restrict any development within Coastal Zone except after consulting the Minister in charge of coast conservation (Jayawardane, 2005).

The CCA requires the CZMP to be revised every 4 years. The CCD with the assistance of the Coastal Resources Management Project (CRMP) reviewed the main objectives, policies and activities of the CZMP-1990. The process included several national level workshops to review set-back regulations and other local level meetings and workshops. The revised plan–CZMP-1997 was effective from 1st November 1997. New focuses are the reduction of coastal pollution and Special Area Management (Samaranayake, 2000). The CZMP was further revised in 2004 following a similar process in order to strengthen the capacity of CCD in managing the coastal zone. The Special Area Management (SAM) can be defined as a collaborative, adaptive and flexible approach to planning resource management within a defined geographic area. Initially two SAM sites were chosen in the south coast–Rekawa and Hikkaduwa, and ecological profiles as well as SAM plans have been prepared. SAM process is now being implemented by CRMP in seven sites (Kalpitiya, Negombo, Lunawa, Madu Ganga, Hikkaduwa, Unavatuna and Tangalle), among 23 sites identified as eligible for SAM sites within the island (pers. com., CCD, 2006).

Major Milestones in Sri Lankan Coastal Zone Management

Sri Lanka have passed several milestones in the coast conservation arena during past 25 years, starting from the CCA in 1981. Few of main achievements could be listed as:

1. Promulgation of laws (such as the CCA, Environment Acts, and regulations promulgated under the provisions of the Fisheries and Aquatic Resources Act)
2. Preparation and implementation of National CZMP of 1990 and 1997
3. Requirement for Environmental Impact Assessments and issue of permits by the CCD for major developmental activities in the coastal zone
4. Outlawing unsustainable methods of fishing, such as "light course" and dynamite fishing, as well as the capture of gravid lobsters and undersized lobsters, and coral mining.
5. Increasing inputs into deep water oceanic fisheries so as not to increase pressure on coastal near-shore fishery resources
6. Declaring protected areas for in-situ habitat conservation such as the Hikkaduwa, and Pigeon Island National Parks and the Bar Reef Sanctuary
7. SAM Programme and the Wetlands Conservation Project

In addition, Sri Lanka is a signatory in several International Conventions dealing on the management of coastal resources. They include the UN Convention on the Law of the Sea (UNCLOS), Conventions within the UNEP Regional Seas Programme, Convention on Biological Diversity (CBD), Convention on International Trade in Endangered Species of Wild Fauna and Flora (CITES), Convention on wetlands of international importance especially as waterfowl habitat (RAMSAR Convention), International Convention for the Prevention of Pollution from Ships (MARPOL

Convention, Convention for the Prevention of Marine Pollution from Land-based Sources (Paris Convention) (UNEP, 2001).

Coastal Hazards

Natural disasters in Sri Lankan coasts are commonly caused by localized coastal erosion and tropical cyclones. The devastation caused by Indian Ocean tsunami in 2004, however, took Sri Lanka by surprise warning that Sri Lanka is also vulnerable to low-frequency high impact events with extensive damage. Earthquakes have also been recorded over the past 400 years with relatively small magnitudes with no significant damage. But, any future seismic damage to Sri Lanka cannot be excluded specially due to increased seismic activity in the region with a new tectonic boundary formed below Sri Lanka (Jayawardane, 2005)

Coastal Erosion

Coastal erosion has been a problem over the years especially in the south, west and northwestern coasts. At certain locations, net erosion has been recorded up to 1 m per year. Accretion rates, on the other hand, have not exceeded 0.1 m per year, except in the northeast where the rate is 0.3 m per year. The average for the entire country is a net mean rate of erosion of 0.20 to 0.35 m per year (CCD, 1997). Mainly Sri Lanka's south-western and southern coastal belt is affected by coastal erosion damaging infrastructure such as coastal railway, road system, communication and other economic activities along the coast such as fishing and recreational activities. Damages to residential buildings, hotels and degradation of valuable land also occur on regular basis (Jayawardane, 2005). Two major factors make the western and southwestern coasts of the island particularly vulnerable; the impact of the southwest monsoon and the geology of the coastline. The latter has produced a bay and headland topography in those coastlines (Arulpragasam, 2000).

The expenditure incurred annually to mitigate shore erosion is at great cost to the national economy (Berg *et al.*, 1998). Shoreward wave force in some areas in the southern coast has shown a statistically significant increase after post bleaching death of corals in 1998 and the subsequent collapse of the reef structure (UNEP, 2001).

Causes of coastal erosion include damming of rivers, sand mining in beaches and rivers, destruction of protective reefs through coral mining, collection of coral rubble, removal of coastal vegetation, improperly sited shore protection structures and buildings and loss in river discharge due to upland water use (Jinendradasa and Ekaratne, 2000).

Tropical Cyclones

Sri Lanka's eastern and north-eastern parts are vulnerable to cyclones especially during November–December with occasional cyclone impacts in some parts of north-central and north-western provinces. Most of tropical storms developed in the south-west or south-east Bay of Bengal may become a cyclone storm (67-117 km/hr) or cyclone (above 118 km/hr) just after passing Sri Lanka's latitudinal region and hence

Sri Lanka is less vulnerable to direct impact of a cyclone (Jayawardane, 2005). Major cyclones have occurred in the years 1907, 1922, 1978 and 2000 (DMC, 2005).

Coral Bleaching

Coral reefs are an important coastal resource that constrain coastal erosion and sustain coastal fisheries. The mass coral bleaching caused by El Nino Southern Oscillation (ENSO), that occurred in April 1998, resulted in extensive (around 80 per cent) coral death followed by serious impacts on the reef ecosystem and in changes in fish populations, species composition, reef structure, biodiversity, succession and ecosystem functions. Most of the branching and tabulate corals were completely destroyed within the depth range of 1 to 8 m. Even after two years of the bleaching, existing coral species have not recovered their reproductive capacities and there is little likelihood that reef ecosystems would regenerate and provide their normal services (Abeysirigunawardana and Ekaratne, 2000a, 2000b; Arulpragasam, 2000; Jinendradasa and Ekaratne, 2000).

Tsunamis

Sri Lanka's coastal belt covering the districts of Jaffna, Mullaitivu, Trincomalee, Batticaloa, Amparai, Hambantota, Matara, Galle, Kalutara, Colombo, Gampaha and Puttalam (Figure 10.3) was severely affected by the Indian Ocean (Boxing Day) Tsunami on December 26, 2004. The affected region in Sri Lanka harbors many key marine and coastal ecosystems such as coral reefs, mangroves, seagrass beds, coastal sand dunes, mudflats, salt marshes, and lagoons, all of which plays a vital role in sustaining living natural resources on which many people in the affected region depend. The effect of the tsunami on these ecosystems has the potential to be very severe ecologically, which could also reduce the provision of environmental goods and services to the society and economy.

The Indian Ocean Tsunami took Sri Lanka by surprise making it the worst natural disaster ever experienced in Sri Lanka killing about 35,000 people. At least 16,000 were injured. Almost one million people suddenly became homeless, and an estimated 350,000 lost their jobs, including fishermen, other informal traders, and microenterprise employees representing tourism as well as other small businesses, infrastructure, schools, tourist hotels and other commercial buildings were damaged (Jayawardane, 2005, Preuss, 2005). But, in the present paper we concentrate more on the impact of the tsunami on coastal environment, despite the damage to property and human society.

The North East coastline appeared to have borne the brunt, with the affected areas reaching 2-3 km inland. The tsunami has affected a much narrower strip in the Southern and Western coastlines, with the affected areas limited to approximately 500 meters or less and at elevations below about 2.5 to 3 meters. The damage caused by the tsunami in the Ampara and Batticaloa districts in the eastern Sri Lanka, both to human infrastructure and property as well as to natural ecosystems, was far more severe than that observed in the southern coast of Sri Lanka. However, the tsunami damage to natural and human-modified ecosystems has not been uniform everywhere.

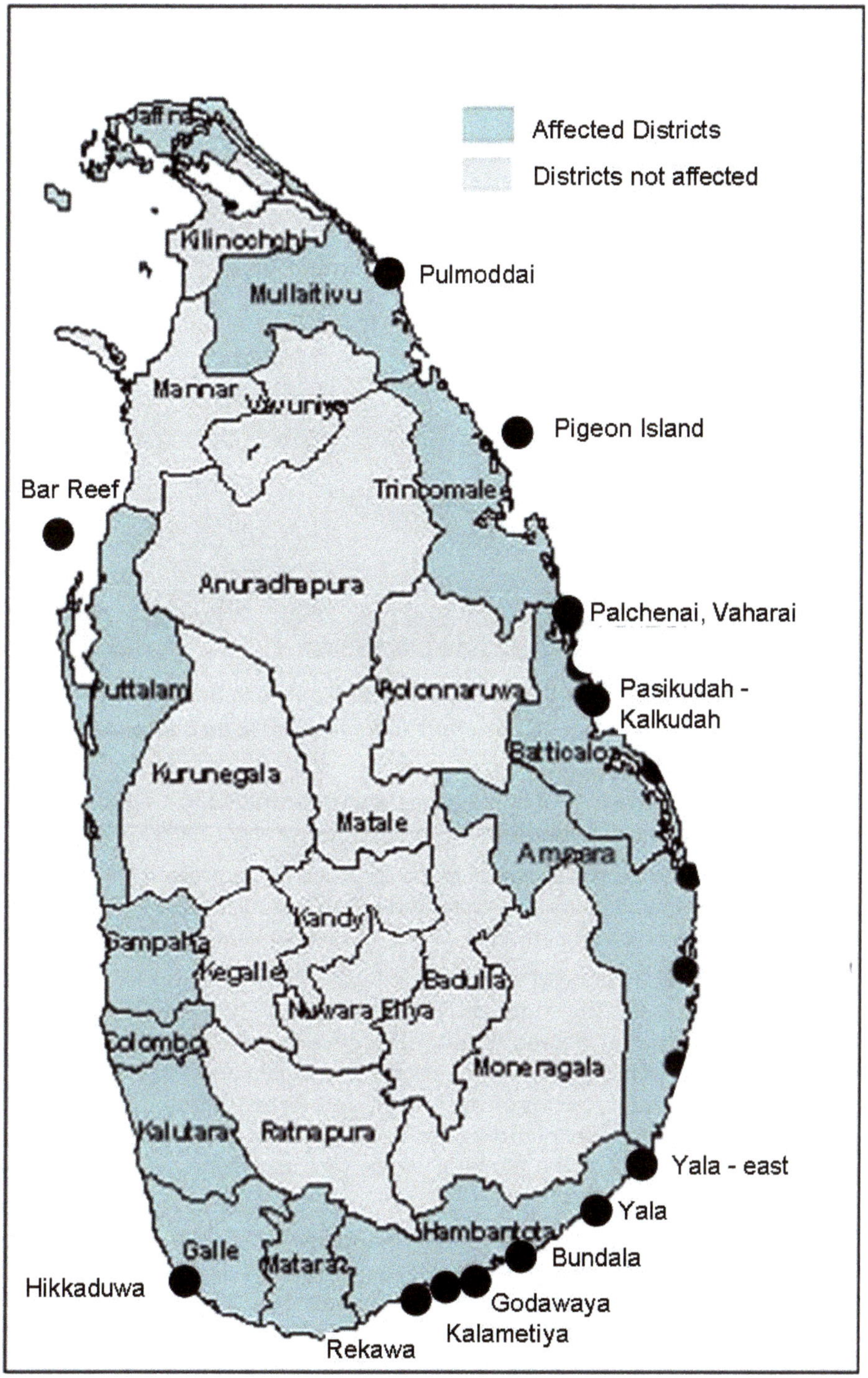

Figure 10.3: A Map of Sri Lanka Showing Districts Affected by the Indian Ocean Tsunami and Sites Mentioned in the Text

Figure 10.4: Damage Caused by Tropical Cyclone in Sri Lanka, 1978
***Source*: http://datum.gn.apc.org/resources/slide-10jpg & slide-4.jpg**

Both the extent and intensity of damage has been determined to a significant extent by site-specific factors and conditions (Bambaradeniya *et al.*, 2005c).

There has been a very little loss of fauna recorded in the protected areas affected by the tsunami. Few wild boars (*Sus scrofa*) and a peafowl (*Pavo cristatus*) were recorded dead in Yala (pers. obs. and pers. com., Park officials of Yala National Park, 2005).

Apart from fish, other dead vertebrate animals observed in Rekawa, Kalametiya and Godawaya areas in the south eastern coast include two specimens of Mouse Deer (*Tragulus meminna*), a Land monitor (*Varanus bengalenisis*) and a soft-shelled Terrapin (*Lissemys punctata*). Dead invertebrate animals observed in the affected habitats include several species of freshwater and terrestrial mollusks, as well as hermit crabs. A large number of turtle nests in Rekawa area and the Bundala National Park had been destroyed by the Tsunami waves. (Bambaradeniya *et al.*, 2005a, 2005c; Bambaradeniya, 2005)

In contrast there has been a significant impacts on coastal habitat and floral diversity in and outside protected areas such as Yala East, Yala, Bundala, Kalametiya, Rekawa and Pigeon Island (Figure 10.3) (Jayawardane, 2005, Perera *et al.*, 2005; Bambaradeniya *et al.*, 2005b; Bambaradeniya, 2005). The gentle sea-shore vegetation consisting of creeping plants and *Pandanus* stands have been severely affected by facing the tidal surge being in the beach front. Sandy beaches, seaward slopes of mature sand dunes, and immature sand dunes have been eroded. Salt marshes, coastal grasslands and mangrove ecosystems have been affected due to large volumes of sand and marine sludge being transported and deposited by the tsunami waters. In mangrove stands, the tree line facing the waves has been subjected to moderate to severe damage. The invasive alien Prickly Pear Cactus (*Opuntia dillennii*) that occurred in patches in the beachfront habitats have been removed and transported to distant inland areas by the tsunami waves, and these propagules are established in new areas. The spread of another invasive alien plant–Mesquite (*Prosopis juliflora*), may increase rapidly over time, as it exhibits a competitive dominance under increased saline conditions (Bambaradeniya *et al.*, 2005a).

It was also expected that the damage to the inter-tidal and sub-tidal area could have been extensive. Many coral reefs might have lost their structure and biota and

Figure 10.5: Tsunami Impact on Coastal Ecosystems in Sri Lanka

(a) Funneling of tsunami waves in Kalametiya lagoon; (b) Beach erosion caused by tsunami in Kahanda-modara (c) Piles of dead coral rubble washed up on Pasikudah beach (d) Uprooted mangrove trees, sand and other debris deposited on salt marshes in Ambalantota (e) Impact on gentle sea shore vegetation dominated by Pandanus in Oruwella, Rekawa (f) Destroyed frontline mangroves in Kahanda-modara.

(a) © M.S. J. Perera, IUCN – Sri Lanka Photo Collection

Contd...

Figure 10.5–Contd...

(b) © IUCN – Sri Lanka Photo Collection

(c) © IUCN – Sri Lanka Photo Collection

Figure 10.5–Contd...

(d) © M.S. J. Perera, IUCN – Sri Lanka Photo Collection

(e) © IUCN – Sri Lanka Photo Collection

(f) © IUCN – Sri Lanka Photo Collection

might even be reduced to rubble in certain areas due to mechanical damage (Jayawardane, 2005). Large piles of coral rubble that have been accumulating on the beach were observed in Pasikudah-Kalkudah in the eastern province of Sri Lanka. These piles consist of a variety of coral types, the majority of which are branching corals in small broken pieces. Similarly in Palchenai (Vaharai) in the east coast many pieces of branching coral have been broken and are being abraded amongst the rubble. There were also occasional observations of table corals and massive coral colonies having been overturned (Bambaradeniya *et al.*, 2005c). There could also be significant contamination by runoff from land, with large quantities of wastes and pollutants, debris, soil and organic matter. Many marine organisms that may have survived the wave itself may later be adversely affected or killed due to deposition of debris and sand on reefs and seagrass beds (Jayawardane, 2005).

Greater ecological impacts are also on freshwater bodies and fishery breeding grounds in the protected areas, which have now been contaminated with saline water. Groundwater contamination and the disposal of debris was another significant environmental issue brought up due to the tsunami.

Coastal Hazard Management

Hazards or disasters were previously perceived as isolated events, which need to be responded with 'emergency relief' as the standard norm. It is now recognized that disaster management is more effective if it is preceded by preparedness and the relief is followed by rehabilitation and reconstruction (Jayawardane, 2005). Consequent to two workshops conducted in July and December 1994 to mark the International Decade for Natural Disaster Reduction and to exchange national experiences in disaster management, the government established the National Disaster Management Centre (DMC) in July 1996 under the Ministry of Social Services (Karunaratne, 2002).

Although several initiatives were taken by Sri Lankan government to mitigate the damages caused by the Indian Ocean Tsunami, they were initially reactive emphasizing relief and recovery rather than proactive with damage prevention or minimization strategies. Later the need for damage prevention strategies was emphasized and it was also discussed to develop an improved communications system to alert communities of impending dangers. The government by newspaper notices in February 2005, declared two zones of coast conservation buffer zones. Zone 1 includes western and southern coastal belt with 100 m buffer zone and Zone 2 includes eastern and northern coastal belt with 200 m buffer zone from the mean high water line (Jayawardane, 2005). Due to a considerable delay in settling tsunami victims lived within these buffer zones in other safer areas and protests by public and opposition political parties these limits have further been amended. It is evident that if the CCA and the guidelines in the CZMP were effectively implemented the damages caused by 2004 tsunami could have been greatly reduced. (Jayawardane, 2005).

However, 2004 tsunami has made responsible parties to act collectively for a comprehensive, long term and holistic disaster risk management framework. In May 2005, the Sri Lanka Disaster Management Act No 13 of 2005 was enacted providing a solid legislative and institutional arrangement for disaster risk management establishing a powerful National Council for Disaster Management under the President and the DMC as the lead agency for disaster risk management. In November 2005, the Ministry of Disaster Management was established to provide undiluted leadership. The Ministry declared its Road Map in December 2005 focusing on seven thematic components. They are policy, institutional mandates, and institutional development, hazard vulnerability and risk assessment, multi hazard early warning systems, preparedness and response plan, mitigation and integration of disaster risk reduction into the development process, community based disaster risk management, and public awareness, education and training. It is expected that proper implementation of this Road Map will go a long way towards safer Sri Lanka from natural disasters (Jayawardane, 2005).

In addition to the humanitarian needs, several short-term environmental recovery needs and priorities were identified during the aftermath of tsunami. They are rubble and debris disposal, study of the role of natural resources in minimizing tsunami damage, clean up of dug wells and tube wells in affected areas, study the impacts on Protected Areas and other environmentally sensitive areas affected by the tsunami as well as a Strategic Environmental Assessment of the rehabilitation and reconstruction Programs. Most of these short term needs were fulfilled in varying degrees during the first year after the disaster, by the government with the fullest involvement of a large number of international and local non-governmental organizations, as well as community based organizations in those affected areas.

Studying long term environmental impacts on the marine ecosystems, development of sustainable waste management systems for affected areas, actions to rehabilitate the coastal ecosystems damaged by the tsunami, as well as reconstruction of damaged and destroyed buildings in National Parks were identified as medium to long term tsunami response priorities concerning on the environment. The requirement to undertake environmental impact assessments for all rehabilitation and reconstruction activities was also emphasized to ensure adverse environmental impacts are minimized and appropriate mitigation measures are included when designing reconstruction and rehabilitation projects.

Protective Function of Natural Landscapes and their Protection Importance

It has been clearly proven by post tsunami environmental assessments in Sri Lanka (Ekanayake *et al.*, 2006; Bambaradeniya *et al.*, 2005a; Bambaradeniya *et al.*, 2005c; Dahdouh-Guebas *et al.*, 2005) as well as else where (Danielsen *et al.*, 2005; Hiraishi and Harada, 2003; Stokstad, 2005) that, the presence of healthy natural ecosystem components in coastal belt have assisted in minimizing tsunami impacts considerably. Intact mature sand dunes (*i.e.*, old and broad dunes covered with gentle sea shore vegetation) have functioned as an effective barrier against the tsunami waves, thereby protecting inland ecosystems and human settlements. Intact *Pandanus* vegetation, as well as *Cassuarina* plantations have also served as a frontline defense by absorbing the wave energy of tsunami. Broader stands of mangroves, especially served as a protective barrier against the wave at places where, coastal wetlands like lagoons and estuaries are connected to the sea. As widely reported, extensive areas of mangroves can reduce the loss of life and damage caused by tsunamis, but narrow mangrove strips can have limited positive effects. Coastal wetlands, including mangrove swamps, salt marshes, broad estuaries and lagoons also have diffused the destructive power of sea-water and sediments brought in by the tsunami waves, thereby protecting managed landscapes such as paddy fields and settlements. In addition, it was noted that coral reefs, rocky beaches and sandstone reefs have reduced the energy of incoming waves. Moreover, it has been observed that tsunami waves have penetrated inland with a greater force in areas where natural sand dunes have been exploited and/or converted into managed landscapes such as Coconut plantations and home gardens. In areas, where near shore coral reefs have subjected to prolonged mining or destroyed by bottom set netting, etc. were also considerably damaged by tsunami waves (Ekanayake *et al.*, 2006; Kruse, M. 2005).

In light of above evidence it was clearly proven that, sites shielded by tall mature sand dunes, and *Casuarina* plantations on at the beach front as well as broad intact mangrove stands, bordering coastal wetlands have functioned as effective barriers against the tsunami protecting inland landscapes. Therefore, while it may be a good investment to establish early warning systems for the next tsunami, it could be far more effective to restore and protect coastal landscapes as a natural defence in parallel.

Concluding Remarks

It has been observed that unplanned development and land use, ad-hoc human settlement and lack of implementation of existing regulations have caused severe encroachments in the coastal zone and they should be managed in order to avoid future losses of human life as well as property during coastal hazards. In the view of that the implementation of the CZMP should be strengthened with adequate enforcement of law. It is hoped that proper implementation of SAM plans and establishment of new SAM areas will improve the ecological as well as socio-economic status in sensitive coastal landscapes in the island.

It is also extremely important to identify the importance of protecting natural coastal landscapes and habitats like sand dunes as well as mangroves taking lessons learnt during the 2004 Indian Ocean tsunami experience. An early warning system as well as an effective system of communication should be placed within the coastal zone, in order to facilitate quick action in a crisis like a tsunami or a tropical cyclone.

The hazard or disaster management has not been considered as a priority issue in social security in Sri Lanka till the devastating tsunami made it the worst disaster ever experienced by Sri Lankans. Therefore, even though disaster management initiatives were taken to recover from localized environmental hazards, they were not sustaining, and the focus was mainly on relief. But with the recent initiatives taken through enacting the Disaster Management Act as well as the new Ministry and the National Council for Disaster Management, with the assistance from international and local non-governmental organizations as well as the academia, Sri Lankan coastline is hoped to be well protected from future environmental hazards.

Acknowledgements

Authors wish to acknowledge the assistance and information provided by Dr. Mrs. Sunethra Gunatilake, Sbaragamuwa University of Sri Lanka; Dr. Channa N.B. Bambaradeniya, Asia Regional Office, IUCN-The World Conservation Union, and Mr. Gamini Hewage, Deputy Director, Coast Conservation Department of Sri Lanka, during the compilation of the present article.

References

Abeysirigunawardana, M.D. and Ekaratne, S.U.K. 2000a. Responses by corallivorous drupellids to bleach-induced changes in coral species composition. International Coral Reef Symposium. Oct., 2000.

Abeysirigunawardana, M.D. and Ekaratne, S.U.K., 2000b. Changes in corallivorous and herbivorous fish assemblages associated with coral bleaching at a marine reserve in Sri Lanka. International Coral Reef Symposium. Oct., 2000.

Arulpragasam, K.D., 2000. Coastal and Marine Resources. In: Natural Resources of Sri Lanka 2000. National Science Foundation, Sri Lanka. pp. 167-193.

Bambaradeniya, C.N.B., 2005. Impacts of the recent tsunami on the Bundala National Park–the first Ramsar wetland in Sri Lanka. Available at http://www.iucn.org/tsunami/docs/tsunami-impact-bundala-s.pdf. (accessed 14 July 2006).

Bambaradeniya, C.N.B., Ekanayake, S.P., Perera, M.S.J., Rodrigo, R.K., Samarawickrama, V.A.M.P.K. and C. Asela. 2005a., A report on the terrestrial assessment of tsunami impacts on the coastal environment in Rekawa, Ussangoda and Kalametiya (RUK) area of Southern Sri Lanka. Available at http://www.iucn.org/tsunami/docs/ruk-rapid-assessment-report.pdf. (accessed 14 July 2006)

Bambaradeniya, C.N.B., Perera, M.S.J., Rodrigo, R.K., Samarawickrama, V.A.M.P.K. and M.D.C. Asela. 2005b., Impacts of the recent tsunami on the Lunama-Kalametiya Wetland Sanctuary. Available at http://www.iucn.org/tsunami/docs/tsunami-impact-lunama-kalametiya-s.pdf. (accessed 14 July 2006)

Bambaradeniya, C., Sengupta, S., Perera, S., Tamelander, J., Meynell, M., Rust, M., Vidanage, S.and S. Perera. 2005c., Rapid environmental and socio-economic assessment of tsunami damage in terrestrial and marine coastal ecosystems of Ampara and Batticaloa districts of Eastern Sri Lanka. Available at http://www.iucn.org/tsunami/docs/rapid-ass-easte-sri-lanka.pdf. (accessed 14 July 2006)

Berg, H., Ohman, M.C., Troeng, S., and Linden, O., 1998. Environmental economics of coral reef destruction in Sri Lanka. Ambio., 27 (8), pp. 627-634.

Brown, B.E., 1997. Integrated Coastal Management: South Asia. Department of Marine Science and Coastal Management, University of Newcastle, Newcastle-upon-Tyne, U.K.

CCD., 1997. Revised Coastal Zone Management Plan. Coast Conservation Department, Colombo, Sri Lanka.

Dahdouh-Guebas, F., Jayatissa, L.P., Di Nitto, D., Bosire, J.O., Lo Seen, D. and Koedam, N., 2005. How effective were mangroves as a defense against the recent tsunami? Current Biology, 15(12), pp. 443–447.

Danielsen, F., Sørensen, M.K., Olwig, M.F. Selvam, V. Parish, F. Burgess, N.D., Hiraishi, T., Karunagaran, V.M., Rasmussen, M.S., Hansen, L.B., Quarto, A., and Suryadiputra, N., 2005. The asian tsunami: A protective role for coastal vegetation. Science, 310, pp. 643.

De Silva, S., 1997. Population. In: Arjuna's Atlas of Sri Lanka. (Eds.) T. Somasekeram, M.P. Perera, M.B.G. de Silva and H. Godellawatta, Arjuna Consulting Co. Ltd., Dehiwala, Sri Lanka. pp. 47-52.

DMC, 2005. Towards a Safer Sri Lanka–Road Map for Disaster Risk Management. Disaster Management Centre, Ministry of Disaster Management, Government of Sri Lanka. Dec., 2005.

Ekanayake, S.P., Perera, M.S.J., Bambaradeniya, C.N.B., Rodrigo, R.K., Samarawickrama, V.A.M.P.K. and Asela, M.D.C., 2006. Protective function of natural coastal landscapes against the impact of tsunami: A case study from Medilla to Godawaya in South-eastern Sri Lanka. Proceedings of the 10th Anniversary International Symposium on 'The Issues and Challenges of the 21st Century, Sabaragamuwa University of Sri Lanka. July, 2006. pp.10.

Ekaratne, S U K., 1997. Research and Training for Conservation and Sustainable Management of Coral Reef Ecosystems in Sri Lanka: Present Status and Future Directions. Proc. Regional Workshop on Coral Reefs, Chennai, India. Dec. 1997, C, 1-11.

Hiraishi, T. and Harada, K., 2003. Greenbelt Tsunami Prevention in South-Pacific Region, available at http://eqtap.edm.bosai.go.jp/useful_outputs/report/hiraishi/data/papers/greenbelt.pdf. (accessed 14 July 2006).

Jayawardane, A.K.W., 2005. Disaster mitigation initiatives in Sri Lanka. Available at http://www.iiirr.ucalgary.ca/archives/Kochifullpaper-final.pdf. (accessed 15 July 2006).

Jinendradasa, S.S. and Ekaratne, S.U.K., 2000. Post-bleaching changes in coral settlement at the Hikkaduwa marine reserve in Sri Lanka. International Coral Reef Symposium. Oct., 2000.

Karunaratne, G., 2002. Natural Disasters and Sustainable Development: Sri Lanka's Contribution Towards Agenda 21. 3rd International Seminar on The Engineer in Sustainable Social and Economic Development–Regional Contribution towards Agenda 21, Institution of Engineers, Colombo, Sri Lanka.

Kruse, M., 2005. Rehabilitation of tsunami affected mangroves needed: Should be part of integrated coastal area management. FAO newsroom. Available at http://www.fao.org/newsroom/en/news/2005/89119/(accessed 15 July 2006).

MoFARD, 1999. Six Year Fisheries Development Programme 1999-2004. Ministry of Fisheries and Aquatic Resources Development, Colombo, Sri Lanka.

MoFE, 1999. Biodiversity Conservation in Sri Lanka: A Framework for Action. Ministry of Forestry and Environment, Colombo, Sri Lanka.

NARA, 1995. Report on the Preparation of a Zonal Plan and Identification of Land for Further Development. National Aquatic Resources Research and Development Agency, Colombo, Sri Lanka.

NARA, 1998. Sri Lanka Fisheries Year Book: Report on Offshore Large Pelagic Fish Resource Survey. National Aquatic Resources and Research and Development Agency, Colombo, Sri Lanka.

NARESA/USAID, 1991. Natural Resources of Sri Lanka: Conditions and Trends. Ed: Baldwin, M.F. Natural Resources, Energy and Science Authority of Sri Lanka, and United States Agency for International Development. XI + 279 pp.

Olsen, S., Sadacharan, D., Samarakoon, J.I., White, A.T., Wickremeratne, H.J.M. and Wijeratne, M.S. (Eds.). 1992. Coastal 2000: Recommendations for A Resource Management Strategy for Sri Lanka's Coastal Region. Volumes I and II. CRC Technical Report No. 2033, Coast Conservation Department, Coastal Resources Management Project, Sri Lanka and Coastal Resources Centre, The University of Rhode Island. 102 pp.

Perera, M.S.J., Bambaradeniya, C.N.B., Ekanayake, S.P., Rodrigo, R.K., Samarawickrama, V.A.M.P.K. and M.D.C. Asela., 2005. Impacts of the recent tsunami on the Rekawa proposed sanctuary and turtle refuge. Available at <http://www.iucn.org/tsunami/docs/tsunami-impact-rekawa-ps.pdf. (accessed 14 July 2006).

Preuss, J., 2005. Learning from Sri Lanka: A Wake Up Call for Coastal Hazards Mitigation. Natural Hazards Observer, XXIX(5). Available at http://www.colorado.edu/hazards/o/may05/may05c.html. (accessed 15 July 2006).

Ranasinghe, N.S., 1997. Mineral Resources. In: Arjuna's Atlas of Sri Lanka. (Eds.) T. Somasekeram, M.P. Perera, M.B.G. de Silva and H. Godellawatta, Arjuna Consulting Co. Ltd., Dehiwala, Sri Lanka. pp. 44-46.

Samaranayake, R.A.D.B., 2000. Sri Lanka's Agenda for Coastal Zone Management. EEZ Technology. Available at www.sustdev.org/journals/others/iczm/05.d.pdf. (accessed 15 July 2006).

Savundranayagam. T., de Alwis, L., Joseph, L. and Siripala, N., 1993. The Economic Significance of the Coastal Region of Sri Lanka. Coastal Resources Management Project, Sri Lanka.

Stokstad, E., 2005. Taking the pulse of Earth's life-support systems. Science, 308, pp. 41-43.

UNEP, 2001., Sri Lanka: State of the Environment, 2001. United Nations Environment Programme. pp. 86-99. Available at http://www.rrcap.unep.org/reports/soe/srilanka_coastal.pdf. (accessed 14 July 2006).

Abbreviations

UNEP: United Nations Environment Programme

MoFE: Ministry of Forestry and Environment

CCD: Coast Conservation Department

CCA: Coast Conservation Act

MoFARD: Ministry of Fisheries and Aquatic Resources Development

EEZ: Exclusive Economic Zone

NARA: National Aquatic Resources Research and Development Agency

USAID: United States Agency for International Development

NARESA: Natural Resources, Energy and Science Authority of Sri Lanka

CZMP: Coastal Zone Management Plan for Sri Lanka

CRMP: Coastal Resources Management Project, Sri Lanka.

SAM: Special Area Management

DMC: National Disaster Management Centre.

Chapter 11

Mapping of Aquaculture Pond Damaged After Tsunami Along East Coast: Assessment of Damage Levels

Z.A. Muchlisin
Department of Marine Sciences Faculty of Sciences,
Syiah Kuala Unversity, Banda Aceh, NAD
E-mail: muchlisinza@yahoo.com

ABSTRACT

The December 2004 tsunami has caused severe damage to live, properties and infrastructure in Aceh, Indonesia. In this chapter, we describe a still on-going project to map the tsunami-affected coastal aquaculture areas in the districts east of Pirdie, Bireuen and, North Aceh. The damage to aquaculture ponds has serious repercussions on the livelihoods of fishing communities in Aceh. The objective of the present study is to validate the previously prepared maps with high resolution satellite image generated maps. A preliminary damage map of aquaculture ponds in the eastern coastal areas of Aceh has been produced by visual interpretation of SPOT-5 multispectral images at 10-m spatial resolution and it has validated. According to direct observation and comparison with pre-map imagery show that most of pre-prediction are correct. Most of the coastal ponds along east coast were partly damaged, while highly damaged ponds were found around river mouth and unprotected coast by coastal forest.

Keywords: Tsunami, damage assessment, pond and aquaculture

Background

The economy of the people of Aceh is largely dependent on agriculture, plantations and fishing all at a small scale using low levels of investment and technology. Agriculture (including fishery) is the main source of income for about 50 per cent of the population in Aceh. Tsunami disaster has caused agriculture and fishery land to be hyper-saline, and in most cases beyond use. This has adversely affected the livelihood of Acehness people. The losses of income in this sector make up over one-third of total losses due to the disaster. In order to minimize losses in the flow of income, it is important to resuscitate these sectors as quickly as possible.

The most severe destruction occurred in western coast (Aceh Barat, Aceh Jaya, Aceh Besar) and Banda Aceh district in the northwestern tip. The more extensive coastal pond aquaculture activities that are located in the districts east of Aceh Besar (in the districts of Pidie, Bireuen, North Aceh and East Aceh) were also affected to varying degrees. This has serious repercussions on the livelihoods of fishing communities in Aceh. Table 11.1 describes the workforce by sector. Majority (47.6 per cent) of workforce is engaged in agriculture, plantation, forestry and fishing. Table 11.2 gives the estimated losses due to Tsunamis. One to two years later, the comprehensiveness of damage assessments, especially with regard to specific fishery locations, is still far from complete. Therefore, there is an urgent need for planning and prioritization of rehabilitation efforts. Mapping of the tsunami affected coastal areas is needed to assess the current condition of aquaculture sites for better-targeted initiatives for rehabilitating affected aquaculture activities.

Table 11.1: Aceh–Work Force by Sector

Primary Work	*Workforce*	
	Number	*%*
Agriculture, Forestry, Plantations and Fishing	1,073,454	47.6
Industry	87,636	3.9
Construction	93,705	4.2
Retail, Wholesale, Restaurants and Hotels	468,057	20.8
Transport, Storage and Communication	101.292	4.5
Finance, Insurance, Rentals and Building Services	12,756	0.6
Services	407,130	18.0
Mining, Quarrying, Electricity, Gas and Water	10,125	0.5
Total	2,254,155	100

Source: World Bank (2005).

Objectives

The main objective of the whole project is to produce base maps of coastal areas of Aceh province using high resolution satellite images such as those acquired by SPOT5 and IKONOS, to be used for village-level targeting and planning of rehabilitation of aquaculture activities as part of reconstructing the livelihoods of fishing communities affected by the tsunami. The objective of the present study is to validate the previously produced maps.

Table 11.2: Aceh–Estimated Losses of Agricultural Land, Trees and Livestock

District	*Rice Field (ha)*	*Tree Crops (Trees)*	*Dry Land (ha)*	*Livestock (Head)*
Sabang	–	4,147	–	32,061
Banda Aceh	75	50	–	332,505
Aceh Besar	5,611	7,048	9,465	500,000
Pidie	1,859	11,304	3,072	238,301
Bireun	2,118	9,575	567	153,961
Aceh Utara	1,224	–	612	74,460
Kota Lhokseumawe	–	–	–	27,292
Aceh Timur	2,119	–	–	–
Aceh Barat	1,432	14,950	1,114	251,962
Nagan Raya	757	14,895	1,560	137,765
Aceh Jaya	1,645	12,240	3,068	156,280
Simelue	3,410	14,937	79	–
Aceh Selatan	–	9,636	–	–
Aceh Barat Daya	3,080	3,729	4,758	
Estimated Total Losses	23,330	102,461	24,345	1,904,587
% of Total Losses	48 per cent	59 per cent	75 per cent	67 per cent

Source: World Bank (2005).

Scientific Objectives

1. Develop the methodology for detecting mapping status of aquaculture ponds in tsunami-affected areas.
2. Devise a systematic strategy for ground trusting the interpreted maps.
3. Use the mapped outputs to assess nature and extent of damage to aquaculture ponds.

Methodology

1. Acquire and process or enhance satellite imagery for available dates of acquisition before and after the tsunami event. Attempt to get the highest resolution imagery available.

   SPOTS–multispectral 10 m resolution for most of the coast.

   SPOTS–combination of panchromatic (2.5 m) and multispectral imagery for certain portion of coast.

   IKONOS–1 m resolution for a limited stretch of the coast.
2. Visual interpretation: to distinguish

   (*a*) pond aquaculture from other use–non water features

   (*b*) "Intact" vs "damaged" ponds.

If substantial length of bunds are broken

If ponds are sited close to damaged coastline and river mouth.

3. Online digitizing of areas/polygons
   - intact aquaculture ponds
   - damaged
   - salt pans
   - uncertain

The above three steps were discussed in our first study, the present study pertains to following the step 4, discussed below.

4. Field Assessment: What information content is relevant for planning and restoration?

The pre-map was used as a guide-line during assessment and direct observation were done at every coordinate point which were chosen and marked in the map. Physical parameters used to exam the damages level of aquaculture pond and all of the parameter converted to score value.

(*a*) Nature and Extent of damage
   – Physical damage to bunds for restoration: Physical condition of pounds–dykes, water inlet and outlet

(*b*) Access to water source
   – river/channel source
   – land level

(*c*) Scale operation

(*d*) Physical accessibility

(*e*) Others important information

Results and Discussion

The assessment were done by two phases ; the firstly on 17 to 24 December 2005 were focused around Aceh Utara and Lhoseumawe districts and secondly on 12 to 17 January 2006 focused around Pidie, Bireun and some part of Aceh Utara.

In Pidie, ground checking was done in 27 sampling points and involved ten sub districts namely; Batee, Pidie, Kota Sigli, Simpang Tiga, Kembang Tanjung, Bandar Baru, Pante Radja, Trieng Gadeng, Meureudu and Ulim. Ground checking was also conducted in 10 sub districts with 31 sampling points in Bireun. While, in Aceh Utara, ground checking was done in 4 sub districts with 19 sampling points and three sub districts in Lhokseumawe (Table 11.3).

The Damage levels in every sub districts show that Seuneudon were highly damaged (66.6 per cent) compared with other locations. The survey was done after one year tsunami attacked and field condition has changed in some locations. The results show that 52.87 per cent of dyke were partly damaged and only 24.14 per cent totally broken. Water inlet and outlet were also affected by tsunami and around 29.89 per cent of 87 locations were totally damaged. Our observation show that out of river mouths affected by tsunami, around 39.08 per cent of river mouths were totally damaged, 47.13 per cent were partly damaged and only 13.79 per cent were not damaged (Table 11.4).

Table 11.3: Damage Level in Each Sub Districts

Sl.No.	*District/Sub District*	*Damage Levels*			*Number of Sampling Points*
		Totally (%)	*Partly (%)*	*No Dam. (%)*	
(A)	**District: Pidie**				**27**
1.	Batee	33.33	0	66.66	3
2.	Pidie	0	100	0	1
3.	Kota Sigli	0	50.00	50.00	4
4.	Simpang Tiga	33.33	66.66	0	3
5.	Kembang Tanjung	20.00	80.00	0	5
6.	Bandar Baru	0	100	0	2
7.	Pante Raja	0	100	0	2
8.	Trienggadeng	0	66.66	33.33	3
9.	Meureudu	0	100	0	1
10.	Ulim	33.33	66.66	0	3
(B)	**District: Bireun**				**31**
11.	Peulimbang	33.33	33.33	33.33	3
12.	Samalanga/Meuliek	50.00	50.00	0	4
13.	Jeunib	0	100	0	1
14.	Pandrah	33.33	66.66	0	3
15.	Peudada	0	100	0	4
16.	Jeumpa	60.00	40.00	0	5
17.	Kuala	0	100	0	2
18.	Jangka	40.00	60.00	0	5
19.	Gandapura	25.00	50.00	25.00	4
(C)	**District: Lhokseumawe**				**11**
20.	Muara Batu	0	0	100	2
21.	Muara Dua	28.56%	42.88	28.56	7
22.	Banda Sakti	0	0	100	2
(D)	**District: Aceh Utara**				**19**
23.	Dewantara	0	0	100	6
24.	Samudera	57.14	42.86	0	7
25.	Tanah Pasir	33.33	0	66.66	3
26.	Seuneddon	66.66	33.33	0	3

Table 11.4: Summary of Damages Level of 87 Sampling Points

Sl.No.	Parameters	Answers (refers to questioner chat)		
		A	B	C
1.	Dyke	24.14%	52.87%	22.99%
		Totally	Partly	Not
2.	Water inlet-out let	29.89%	47.13%	22.99%
		Totally	Partly	Not
3.	Sand bar	29.89%	44.83%	25.29%
		Totally	Partly	Not
4.	Coastal forest	9.2%	58.62%	32.18%
		Totally	Partly	Not
5.	River mouth	39.08%	47.13%	13.79%
		Totally	Partly	Not
6.	Debris	3.49%	72.41%	24.14%
		Many	Not so	Not present
7.	Tsunami affect on	5.75%	80.46%	13.79%
	access to sea water	Severe	Partly	Not
8.	Tsunami affect on access	5.75%	78.16%	16.09%
	to fresh water	Severe	Partly	Not
9.	Accessibility to location/pond	17.24%	50.57%	32.19%
		Not easy	Moderate	Very
10.	Reliability of supply (feed/fry)	3.49%	91.95%	4.6%
		Not easy	Moderate	Very
11.	Dependence on aquaculture	34.48%	63.22%	2.3%
	Conclusion score in each locations	Totally damaged 10 (11.49%)	Partly damaged 46 (52.87%)	No damage 31 (35.63%)

In general, most of the aquaculture ponds in east coast were partly damaged (52.87 per cent) and only 11.49 per cent of ponds were broken totally. Severity of damages pond usually happened around a river mouth and unprotected pond. The ponds located far away from coastal line and river mouth area are not affected and availability of coastal forest is also protecting the pond from tsunami. The facts show that ponds were protected by coastal forest and sand bars are not broken. However, most of the sand bars were removed and some of them were lost, fortunately most of coastal forests around east coast are still present and only 9.2 per cent was broken by tsunami.

Our observation and comparison with pre-map imagery show that most of our prediction are correct; a long coastal line (around 10-50 meters) were broken totally, while broken area of 500 meters around river mouth. In generally, coastal ponds in Aceh are located more than 500 meters from coastal line and not faced directly to open sea; it gives an advantage for protecting of the ponds from tsunami.

According to our observation during survey that most of the locations are moderate easy to access (50.57 per cent) and only 17.24 per cent of locations are difficult to reach by vehicle. Our interviews with farmer show that reliability of feed and fry are moderate easy due to availability of mini hatchery near to location (Table 11.5). Beside from hatchery, fry are also supplied from natural resources (wild fry). Interesting result show that 97.70 per cent of ponds were tenanted by farmer or fisherman and most of fisherman admitted that they have secondary job beside as a fisherman such as husbandry, home trading and home industry.

There is no different animal cultured before and after tsunami, most of the fishermen have been rearing shrimp and fish (milk fish, tilapia and grouper) in their pond. Our observation and discussion with fisherman show that most of the aquaculture efforts are extensive (67.82 per cent) and only 5.75 per cent of intensive cultured (Table 11.5). The main problem is lack of capital for their business, beside that skill and technology cultures are also still low and have to improve both skill and technology.

Table 11.5: Summary of Ground Checking

Code	*Questions*	*Answer*			*Total*
		A	*B*	*C*	
A.1	What was cultured before the tsunami?	8 (9.205)	3 (3.49%)	76 (87.36%)	87
A.2	What is cultured after the tsunami?	73 (83.91%)	1 (1.15%)	13 (14.94%)	87
A.3	Scale of operation	59 (67.82%)	23 (26.44%)	5 (5.75%)	87
A.4	Tenancy	85 (97.70%)	0 (0%)	2 (2.30%)	87
A.5	Source of sea water	59 (67.82%)	4 (4.60%)	24 (27.59%)	87
A.6	Source of fresh water	4 (4.60%)	46 (52.87%)	37 (42.53%)	87
A.7	Source of capital for operating ponds	85 (97.70%)	0 (0%)	2 (2.30%)	87
A.8	Type of soil	4 (4.60%)	40 (45.98%)	43 (49.43%)	87

Conclusions

A preliminary damage map of aquaculture ponds in the eastern coastal areas of Aceh has validated and according to direct observation and comparison with pre-map imagery show that most of prediction are correct. Most of the coastal ponds along east coast were moderately damaged, while highly damaged ponds were found around river mouth and unprotected coast by coastal forest.

References

Liew, S.C., Kam, S.P., Chen, P and Muchlisin, Z.A. 2005. Mapping Tsunami-Affected Coastal Aquaculture Areas in Northern Sumatra Using High Resolution Satellite Imagery. National University of Singapore, Singapore.

Muchlisin, Z.A., Edirudi., Nasir, M and Sutekad, D. 2006. Kondisi dan Potensi Lahan Basah di Kota Lhokseumawe dan Kabupaten Aceh Utara Pasca Tsunami. Wetland International-Universitas Syiah Kuala, Banda Aceh.

World Bank. 2005. Master Plan for Reconstruction. Banda Aceh.

Chapter 12

The Impact of Cyclone "2A" on the Tidal Link Drain and Adjacent Areas Based on Historical Data

Syed Moazzam Ali
Ocean Technology and Coastal Zone Management,
National Institute of Oceanography, Karachi, Pakistan
E-mail: moazzam_nio@yahoo.com

Introduction

A Tropical Cyclone landed on the Southern and Southeastern parts of Sindh Coast of Pakistan during 20-21 May 1999 and caused huge losses to life, property and to some development infrastructure along southeastern part of the coast. The Indus River Delta constitutes most of the coastal Sindh, which is among the least developed areas of the country with the exception of Karachi City. This report focuses on the description and characteristics of Southeastern Coast of Sindh with a view to assess the impact of Cyclone "2A" on this part of the coast and on the Tidal Link Drain, LBOD located therein. A number of small fishing villages, small human settlements, dhand (brackish water) fishery, coastal fishery, limited coastal forestry and some important infrastructure for agriculture drainage and exploratory work for oil and gas exist along Southeast Coast of Sindh. There were healthy mangrove forests in the creek islands and in inter-tidal mud flats about 70-90 years ago which provided a dependable and natural coastal defense to the area from onslaught of cyclones and severe cyclonic storms in the area. The area has been acting as a natural passage for storm water drains and also as a temporary reservoir of drain waters and as watersheds for land-run-off and agriculture drain waters from croplands in the North.

Previous Relevant Studies

A number of reports have been prepared on the impacts of cyclone and storms on the coastal areas in the world and particularly on the coast of Bangladesh and India dealing mostly with the impacts in terms of loss of life and property and other socio-economic aspects. In case of Cyclone "2A" there are some un-published reports dealing with the socio-economic impacts of the cyclone. However, there is hardly any study dealing specifically on the impacts of Cyclone 2A on the morphology and hydraulic regime in the Indus Delta. Some of the relevant reports on the cyclone "2A" and Indus Delta from Pakistan include NIO 1999, NIO 2000, Kella, 1999 and Keerio and Bhatti 1999. Similarly limited information is available on the man made agriculture drainage systems in the deltaic areas except for the Delft Hydraulics (1988) who carried out a Pre-design Study for tidal link outfall at the same site and have submitted three reports to WAPDA, Ministry of Water and Power, Government of Pakistan covering the following aspects of Tidal Link Drain: 1) Field surveys, 2) Mathematical Modeling, and 3) Desk study and Physical modeling and by NIO on the monitoring of Tidal Link Drain. The patterns of deterioration in the Nile Delta due to reduced discharge have also been reported (U.A.R.,1976, 1977; Coleman 1981). The importance of acute measurements of land subsidence was emphasized for low lying deltaic areas such as Yangtze River Delta has been highlighted by Chen and Stanely (1995). Smith and Fitzgerald (1994) have described a seaward sediment transport witnessed during increasingly dominated ebb tidal currents in Essex River Inlet, Massachusetts, USA.

Scope and Objectives of the Study

In accordance with the TOR (Terms of Reference) agreed upon by NIO and UTP the contract was to conduct a short term Desk Study and to prepare an Assessment Report to cover the following components:

1. Review, assessment and synthesis of collected data/information on the impact of Cyclone "2A" on the Tidal Link and adjacent areas based on historical data;
2. Survey area map (updating of old map with the major morphological and physical changes after the cyclone);
3. Preparation of an Assessment of impact of Cyclone "2A"on the Tidal Link and adjacent areas based on the available data/information and historical data.

Methodology and Approach

General Approach

The Desk Study while focusing on the Terms of Reference encompassed all site specific information of the area covering also the major morphological changes and changes in the hydraulic regime and extent of tidal flooding in the area of interest. The following points were given major consideration in the preparation of the Assessment Report:

1. The present study includes the Tidal Link area, adjacent dhands, the tidal mud flats and creeks (Shah Samando Creek), and part of Rann of Kutch.

2. Impact assessment of the cyclone on the marine environmental features based on the previous reports and development infrastructure in the area. The following parameters were included in the study:
   (*i*) Major morphological features: Sediment load and discharges in Tidal Link, Erosion and Sedimentation patterns.
   (*iii*) Meteorological condition: such as wind speed and direction, air temperatures, etc. Water Salinity distribution at Shah Samando Creek, Tidal Link Drain and adjacent dhands in the area.
   (*iii*) Tidal levels, Seawater currents, water drainage and water discharges in dhands, Tidal Link Drain and at Shah Samando Creek.
   (*iv*) Updating the area map prepared during 1991 taking into account the existing situation (after the cyclone).
   (*v*) The demarcation of new High Water line and H.A.T. around Shah Samando Creek and Tidal Link Drain area and dhands in the area using maps of the area based on satellite imagery 1990, 27th October 1998, 23rd May 1999 and 25th April 2000.
   (*vi*) Detecting recent major morphological and physical changes after the cyclone in the area interpreting the map of the area prepared during 1991 and the satellite imageries of 27th October 1998, 23rd May 1999 and 25th April 2000.

Assessment of the impacts of Cyclone "2A" on the coastal Geo-morphological parameters such as Tidal inundation and H.A.T. in the area. Impacts on Hydraulic Regime (tidal heights, water circulation patterns, water discharge) and on siltation/sedimentation–erosion patterns. Impacts on the Socio-Economic aspects including life and property and on development infrastructure in the area.

General Characteristics and Morphology of the Area before the Landfall of Cyclone "2A"

Meteorology

Data on Air Temperature, Wind Speed and Direction was recorded at Shah Samando Creek and Jati area as well during the monthly field trips to the Tidal Link Drain site. The Meteorological Data was also collected from Meteorological Department for their Meteorology Station at Badin. The data was collected for the period June 1992 to June 1994 and averaged data for 1995 to 1997.

Air Temperatures

The air temperature data from Badin indicate an annual range between 10.1 to 41.1°C. The maximum temperatures were recorded during the SW monsoon period in May-June period while the minimum temperatures were recorded NE monsoon (December-January) period. The data indicate 34-35 °C during June to 14-15 °C during December through January.

Wind Speed and Direction

The wind speed ranged from 1.8 to 9.1 Knots during 1993, 2.2 to 8.2 Knots during 1994, 2.6 to 9.4 Knots during 1995, 2.3 to 8.7 Knots during 1996, at Badin Site. The maximum wind speed was recorded during May to September (Southwest Monsoon) while minimum wind speed was recorded October-November and February NE monsoon period. The wind direction during December-January was from Northeast while it was from Southwest during May to September. The direction of wind was variable during March, April (on set of SW monsoon period) and October-November period (post SW monsoon period).

Rain Fall

The monthly maximum rainfall for 1992-97 was reported during January-February and July through December. However, the monthly maximum data on rainfall for the last 5 years indicates 0 to 14.3 mm (5 mm Average) rainfall during January. In February the rainfall record was between 0 to 28.5 mm (12 mm Average) rain, 12 to 196.2 mm rainfall during July, 0 to 358.6 mm (90 mm Average) rain fall during August, 0 to 358.5 (89.6 mm Average) mm rain fall during September, 0 to 40.6 mm (10.15 mm Average) rain fall during October and 0 to 36.5 mm (9.1 mm Average) rain fall during December. The maximum rainfall of 350-360 mm has been reported from Badin during last five years. Higher rainfalls are generally associated with SW monsoon period.

General Morphology

The area around Tidal Link Drain of LBOD, in general and Shah Samando Creek in particular (Figure 12.1) is the part of Indus Deltaic region. The area falls into semi-arid zone with and average rainfall of about 200 mm per year. The area consists of a large intertidal delta of Indus River and adjacent belt of low lying areas (dhands) and shallow depressions which receive run-off from land directly or through storm water drains and drain waters from crop lands in the lower Sindh. The agriculture drains from lower Sindh area are drained into area through LBOD leading to the sea via Tidal Link Drain.

The area below Jati and Badin has agriculture lands upto a level where the prevailing water logging and salinity problems allow some crops to grow. Below the limit of agriculture activity there are large low-lying areas with depressions and salt encrustation upto to the tidal limits that are considered as wasteland. The area close to the tidal limits has minimal development infrastructure and communication network minimum human population. The fishery is the main economic activity in the coastal settlement and they depend on the coastal vegetation and mangroves for fuel and fodder for their cattle and camels.

The site conditions of the Indus Deltaic region including Shah Samando Creek, dhands and the Rann of Kuchh are described below.

Indus Delta

The Indus Deltaic region covers an area of approximately 1,000 square miles. The delta is the product of the enormous amount of silt and sand brought down annually by the Indus River in to the Northern Arabian Sea. The present day deltaic

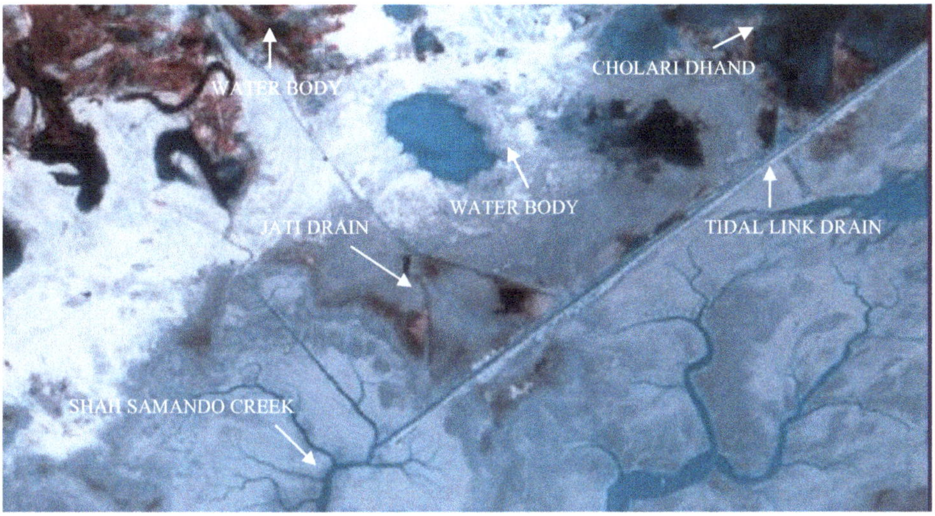

Figure 12.1: Location Map of Tidal Link Drain LBOD

complex is comprised of an upper alluvial part consisting of the deltaic flood plan and lower tidal Zone, which is the tidal delta. The lower margin of the deltaic flood plain is rimmed by tidal delta. The tidal data is comprised of three main parts, Bordering the deltaic flood plain is an arcuate zone of older tidal deposits, formed as a result of the drying and silting up of the distributary channels of the Indus. This zone is followed by the active tidal delta seaward. With the drying and silting up of the major distributions of the Indus the tidal delta has been isolated from the alluvial process and has assumed the form of a large tidal mud flat. The lower remnants of the distributaries have turned into tidal Creeks.

Dhands

There are a few very large prominent depressions/low-lying areas locally called Dhands are located between the agricultural lands and the High Tidal zone which are filled with brackish/saline water. The low lying areas covered about 400 sq km area and consist of four major Dhands namely, Pateji Dhand, Cholri Dhand, Sanhro Dhand, Mehro Dhand and all are connected to Cholri Dhand through narrow open inter-connections: *i.e.* between Cholri and Pateji Dhands, Mehro and Sanhro Dhand and between Sanhro and Cholri Dhands. The Pateji Dhand is a land locked shallow basin connected to the western end of Rann of Kuchh at an elevation of about 1.03 m. average mean sea level. The Cholri Dhand is directly connected to Rann of Kuchh through in the Southeast for the drainage of excess waters from the Dhand areas. The average depths in these Dhands vary between 0.37 m to 1.8 m. The maximum water level in Dhand areas has been reported to be 1.7 m during 1983 (M. McDonald and Partner, 1984). Dhands with brackish waters act as brackish/saline water lakes and support a diverse ecosystem, fisheries and as sanctuary for a variety of water fowls and migratory birds.

Rann of Kuchh

Rann of Kuchh is the large swampy area, which is situated partly in the Indian Territory adjacent to the Pakistan border at the southwestern part. It has been used as a natural drainage area for the most part of the lower Sindh area through Kadan Pateji Outfall Drain (KPOD). The waters from all the major dhands in the southwest coast of Sindh also have their natural drainage area in the Rann of Kuchh through Cholri and Pateji dhands. This is an ideal choice for using it as a drain area for the LBOD. (as it falls within the Indian Territory). However it could only be used through KPOD and Tidal Link Drain. Secondly the water quality of dhands might have adverse effect on its ecosystem and fisheries resources if the drainage of saline and agriculture waste water were allowed to drain in the dhand areas. Hence the option of conveying the drainage water from KPOD to the Arabian Sea through the Tidal link appear to be environmentally safe alternative. However there is an increase risk of seawater influence and flooding of seawater in the dhand areas from the Tidal Link.

Shah Samando Creek, Tidal Mud Flats and Adjacent Land Area

The area around Tidal Link Drain of LBOD, in general and Shah Samando Creek in particular is a part of Indus Deltaic region. The present Link Drain between KPOD and Shah Samando creek passes through intertidal area and is designed to discharge saline water in the tidal mud flat area. All the drainage from Tidal Link Drain is designed to drain in the Shah Samando Creek area creeks, leading to Sir Creek, Indus Delta. The delta is the product of the enormous amount of silt and sand brought down annually by the Indus River in to the Arabian Sea.

Tidal Link Drain

The area of the Tidal Link Drain is a very large tidal flat with a very low angle of beach slope resulting in a very wide inter-tidal area (25 -35 km wide). This has resulted because of the protection from intense wave action due to Indus Deltaic area and the deposition of silt and clay in the area. Little is known about the environmental characteristics of this large tidal mud flat. However, the area appears to be slightly different from other inter-tidal and inshore areas of the Indus Delta. The area is frequented with a number of low lying and swampy areas called "Dhands". The Pateji Dhand is a land locked shallow basin connected to the western end of Rann of Kuchh at an elevation of about 1.03 m average mean sea level. These shallow and large "Dhands" (brackish water lakes) support a small-scale fishery, dependent bird's population and a brackish water ecosystem. The area experiences subtropical, hot and dry weather conditions with salinised land. This area has long been used as a passage of the natural flow of surface drain waters, run-off and storm water drains. Although sparse mangrove vegetation is visible alongside the Sir Creek area but there are no plants seen in the Link Drain area. Sindh Forestry Depth has planted/ transplanted mangroves on the Link of Shah Samando Creek which is on the side of Tidal Link Drain but these have not survived to be visible.

Hydraulic Regime in the Area Before the Cyclone

There are four main water bodies in the area besides run-off from land which provide water flows in the area: (1) Shah Samando Creek, (2) Tidal Link Drain,

(3) Dhands, 4) Out-fall Drains in the area and swampy areas of Rann of Kuchh. The water flows by gravity gradient in the area through KPOD, Tidal Link Drain and dhands are by and large influenced by the Tidal Flows from Shah Samando Creek in to Tidal Link Drain and water Exchange between dhands and the Tidal Link Drain. The Tidal Levels recorded at Shah Samando Creek/RD -155, RD -125 and at -55 are summarized below.

Tidal Levels at Shah Samando Creek and at RD -155

Tides During 1993-94

The tides were typically semi-diurnal nature with two high waters and two low waters per 24 hours. The vertical Tide difference between Highest and lowest waters is about 6.0 meters (2.2 m to–4.0 m). The tides appear to be have similar values to those collected by Delft Hydraulics in December 1987 at the same site. However, a comparison has been made for the tidal data collected from the same site during November-December 1987 by Delft Hydraulics, November- December, 1994 data recorded by NIO and November- December, 1997 data recorded/predicted by NIO. No significant difference in the tidal behaviour at the Shah Samando Creek site during the pre-operational period of the Tidal Link/LBOD in 1993 to 1994 and the tidal data in 1995 after the operation of Tidal Link, LBOD.

Tidal Levels During 1995

The tides were typically semi-diurnal nature with two high waters and two low waters per 24 hours. The vertical Tide Range appears to be about 6 m (between 2.2 to -3.8). The range of maximum and minimum Spring Tidal Levels observed/predicted during 1995 was 2.3 m in January 1995 to -3.75 m in April 1995 respectively. The average range of Neap Tidal Levels during 1995 was 0.5 m and -1.16 m.

Tides 1996-97

The maximum and minimum Spring Tidal Levels observed/predicted during 1996 were 2.37 m in May/2.41 m in December 1996 to -3.76 m in January 1996 respectively. The average range of Neap Tidal Levels during 1996 was 0.65 m and –1.2 m. The maximum and minimum Spring Tidal Levels observed/predicted during 1997 was 2.42 m in June/July 1997 to -3.88 m in May 1997/–3.92 in November respectively.

Cyclones in the Area

The coastal Sindh is prone to storm surges associated with the severe cyclonic storms generated in the adjacent Arabian Sea. A significant number of the cyclonic storms produced in the Arabian Sea move towards North and Northeast and some of them hit Pakistan Coast. The Arabian Sea is known to generate frequent cyclonic storms and some these were among the worst severe cyclonic storms of the world resulting in huge losses to life and property in the coastal areas. However, most of these cyclones leading towards Southern parts of Pakistan Coast very often curve towards eastern coast of India. The southern and southeastern part of Pakistan Coast are comparatively more vulnerable than the west coast of Pakistan.

Cyclone "2A": The Cyclone 2A was formed over the Arabian Sea on May 16 and initially started moving Northward. Later, it changed its course to Northeastward and hit Sindh Coast. The cyclone remained stationary about 400 km south of Karachi on 18th May 1999. It was located at 21.5 N, 68.5 E. Its tidal surge at coastal waters was anticipated to range between 3.65 m–4.6 m. However, the observations reported from the Southeast coast of Sindh revealed that during high waters the water levels were above 7.1 m confirming the anticipated height of the storm surge in the area.

The cyclone wind speed varied between 100 knots to 150 knot. The internal sustained wind speed of above 110 to 115 knots. On 20th May 1999 its location was Latitude 23.9 N, Longitude 68.1 E which was 170 km South-east of Karachi. It then re-curved from NNE to NE. Wide spread heavy rain and thunderstorms were associated with gusty winds that ranged from 80 knots to 100 knots. Its landfall was recorded at 11 a.m. on May 20, 1999 on the coasts of Thatta and Badin covering Indus River Delta at the Southern coast of Sindh. The storm and gusty winds were accompanied by a tidal storm surge between 4.6 m to over 7.2 m. The seawater inundated low-lying coastal areas adjacent lands in this part of Indus Delta. The main hit areas include the Districts of Thatta, Badin and Tharparker. On land the South and southeastern parts of Thatta and Southern parts of Badin received the main thrust of the cyclone. The worst hit villages being Kadhan, Serani, Ali Bandar, Keti Bandar, Shah Bandar and the worst hit coastal towns include Jati, Mirpur Sakro and Sujawal.

Reported Cyclones in the Area

The available data on cyclones and storms for the last 109 years demonstrate that the coastal Pakistan is vulnerable to the severe cyclonic storms mostly during the period April to June (on set of SW monsoon) while no storms was observed during January to March (NE monsoon). The period of 100 years 4 cyclones landed at Balochistan Coast and about 15 Cyclones landed on Sindh Coast. In general, all the cyclonic storms originating from the Arabian Sea have been observed either to curve sharply into the Gulf of Kuchh or to cross the Arabian Sea from East to West and end up in the Arabian peninsula creating some storm surge at the coast. Some times depressions do cross the coast in June and generate winds of 30-35 knots. The maximum Storm surge of 2.6 feet have been recorded at Karachi due to the cyclonic storms. In addition, a storm surge of about 1.5 feet has also been observed at Karachi when cyclones cross over the Arabian Sea about 300-400 miles away from Pakistan Coast.

Cyclones Reported Along Pakistan Coast

Reported Cyclones along the Pakistan coast line are given in Table 12.1

Depressions and Cyclonic Storms

Frequency of Depressions

During the last 109 years about 106 Depressions have been reported from the Arabian Sea. The average annual number of depressions is about 1 while the annual number of depressions is about 47.75 per cent of the annual number of cyclonic disturbances. The frequency of Depressions ranged between 8 and 22 during September to December with highest in November, which is about 21 per cent of annual total. It ranged from 8 to 20 during May to July with 19 per cent of annual total in June. During rest of the months about 4 depressions per month were formed. In February no Depression was reported.

Table 12.1: Cyclones Reported Along the Pakistan Coast

Month	*Year*	*Coastal Area*
June	1897	Sindh Coast
April	1901	Makran Coast
April	1902	Sindh and Rann of Kutch
May	1902	Sindh Coast
June	1902	Sindh Coast
June	1907	Sindh Coast
October	1907	Sindh Coast
September	1925	Sindh Coast
September	1936	Sindh Coast
June	1945	Makran Coast
June	1985	Makran Coast
November	1994	Sindh Coast, Rann of Kutch
May	1999	Sindh Coast and Rann of Kutch

Cyclonic Storms

During the last 109 years about 51 severe cyclonic storms have been reported from the Arabian Sea. May–June and September–December are favourable periods for the formation of cyclonic storms in the Arabian Sea with highest frequency in October which is about 27 per cent of the annual total. There is secondary highest frequency of 7 in June, which is about 14.5 per cent of the annual total. There were 6 cyclonic storms reported in November and 5 cyclonic storms each during May, September and December and 2 each in January, April, July and August. There were no cyclones in February and March. The average annual number of cyclonic storms is about 0.5. The annual number of cyclonic storms is about 22.1 per cent of the annual number of cyclonic disturbances.

Severe Cyclonic Storms

The cyclonic and severe cyclonic storms combined together are termed as "cyclones". The average number of cyclones over the Arabian Sea is about 1.2 per year. The maximum frequency of 27 and 20 in November and May respectively while minimum frequency during January-March and July-August periods. The low number of cyclonic storms formed during these months is attributable to the strong vertical wind shear present during these months over the Arabian Sea. On the contrary, during the months of May-June and October-November, little or no vertical wind shear situation in the atmosphere offer favourable conditions for the formation of cyclonic and severe cyclonic storms in the Arabian Sea.

During the last 109 years about 69 Severe Cyclonic Storms have been reported from the Arabian Sea. The maximum number of severe cyclonic storms were noticed during May–June and October–November with highest frequency of 21 in November which is about 31.3 per cent of the annual total and second highest of 15 in May which is about 22.4 per cent of the annual total. There were 12 severe cyclonic storms formed in October 4 in April 3 in September and 2 in December. No severe cyclonic storm has been reported in January, February and March.

Vulnerability of Pakistan Coast to Landfall of Cyclones

The available data on tropical cyclones from (1891 to 1999) indicate that only 10 tropical cyclones crossed Pakistan Coast of which 4 cyclones crossed Makran Coast and 6 cyclones crossed Sindh Coast during 1891 to 1999. The Sindh Coast is more vulnerable to cyclones than Makran Coast. Monthly distribution of cyclones show that May and June are the most vulnerable months for Pakistan Coast with highest frequency of landfall for the cyclones while September also shows occurrence of landfall for the tropical cyclones along the coast.

Impacts of Cyclone "2A"

Morphological Changes and Impacts of Cyclone "2A"

The following morphological changes were detected after the cyclone. This study was evaluated by the available data and specific area information. The data collection was not specifically focused to evaluate on the impact assessment. For an in-depth assessment a detailed survey and focused field study is recommended.

Morphological changes in the High Water Mark: As most part of the coastal area of Sindh constituting Indus Delta mud flats, creeks, marshes/mangrove swamps are inundated with water during High Tides and are usually in-accessible. Therefore, the satellite images of the area provide a synoptic view of the general morphology and position of High water mark before and after the landfall of cyclone 2A on the coast. These are read and interpreted with reference to the area map of WAPDA based on satellite image of the area in 1991.

Cyclones/Storm Surge Prediction and Mitigation

The coastal areas of Indus deltaic region are flat and comprise natural weeks and swamps (Rann of Kutch), which are subjected to flooding from the sides as well as from the water channels/Creeks during the SW monsoon periods. Flooding coupled with incidents of Cyclones and storm surges inundates vast tracts of low-lying lands for several days causing irreversible damage. When the flooding waters recede. The silted water from the creeks, Tidal Drain make the topsoil too saline. As the ground water table is generally close to the surface in this area so intrusion of salt water through aquifers in the coastal areas present a persistent threat to soil degradation and loss in agriculture productivity. A study of the cyclones storms in the area through models of flooding/storm surge will improve our understanding of seawater penetration and persistence in this part of Indus Delta. Another emerging factor, which could have an influence on cyclones and storm surges, is climate change. Even through there is no universal agreement, the projected sea level rise and changes in cyclone frequency and storm surge intensity are correlated to global temperature variations (warming) coupled with increase in Globules Green House Gas emission, it could make things worse in the years to come. A rise in sea level predicted by one meter as a result of global warming of 2-3 °C over a century and melting/breaking off large polar ice packs into the oceans will have a serious repercussion in the South Asia as a whole. Although, it is difficult to estimate quantitatively the effectiveness of disaster preparedness and prevention measures, in terms of saving lives and protecting properties. There has been instance where millions of rupees of property were lost in a single storm. While a cyclone/storm prediction early warning centre will not avoid all damage but reduction of cyclone path and landfall will bring about a reduction in loss of life, property and damage to the infrastructure in the region. The accuracy of forecast of cyclonic storm and the lead-time on warning have to be improved and the uncertainties in the landfall timing and location need to be reduced. Cyclone and storm prediction demands accurate surface wind forecast for several hours before cyclone landfall hence there is a dire need for development/upgrading of a suitable cyclone prediction centre in the region.

References

Academy of Scientific Research and Technology, U.A.R., 1976. Proceedings of the Seminar on Nile Delta Sedimentology, Alexandria, April, 1975. UNDP/UNESCO/Dept. Geol., Alexandria Univ., Acad. Sci. Res. Technol. U.A.R., 257 pp.

Academy of Scientific Research and Technology, U.A.R., 1977. Proceedings of the Seminar on Nile Delta Coastal Processes with special emphasis on Hydrodynamical Aspects, Alexandria, 2-9 October 1976. UNDP/UNESCO/Dept. Civ. Engr., Cairo Univ., Acad. Sci. Res. Technol. U.A.R., 624 pp.

Bengston, R. L.; Carter C. E.; Fous, J. L.; Southwick, L. M.; and Willis, G. H., 1996. Agriculture drainage and water quality in Mississippi Delta. J. Irrig. Drainage Eng., Vol. 121, No. 4, pp. 292-295.

British Standard Institution, 1975. B.S.1377: Methods of test for soils for civil engineering purpose. British Standards Institution, 143pp.

Chen, Zhongyuan and Stanely, D. J., 1995. Quaternary Subsidence and river channel migration in the Yangtze Delta plain, Eastern China. J. Coast. Res., Vol. II, No.3, pp. 927-945.

Coleman, J. M., H. H. Roberts, S. P. Murray, and M. Salama, 1981. Morphology and dynamic Sedimentology of the Eastern Nile Delta shelf, Marine Geology, Vol. 42, pp.85-100.

Coakley, J.P., and J.P.M.Syvitski, 1991. Sedigraph techniques: in Principal, methods and application of Particle Size analysis, edited by J.P.M Syvitski, pp. 129-142, Cambridge University Press New York.

Delft Hydraulics, 1988. Left Bank Outfall Drain Stage-1 Project, Pre-design study for tidal link Drain and outfall–1. Report on Survey Activities. pp. 1-72 (including 27 Figures and 44 pages).

Hydrography Department, Pakistan Navy, 1994.Pakistan Tide Tables. Hydrography Department, Pakistan Navy.

Kazmi, A. H., 1984. Geology of the Indus Delta. In: Marine Geology and Oceanography of Arabian Sea and Coastal Pakistan, (Eds: Haq, B.U. and Milliman, J. D.). Von Nostrand Reinhold Company Inc. New York. pp. 71-84.

Kella Lekhraj, 1999. Protective Role of Indus Delta Mangroves with particular Reference to cyclone mitigation. pp. 8-90, In: Proc. National Seminar on Mangrove Ecosystem Dynamics of the Indus Delta. June 14-16, 1999, Karachi. Sindh Forest and Wild Life Department and the World Bank. 204 pp.

Keerio, G. R. and M.A. Bhatti, 1999. Major factors of degradation of Indus Delta Mangrove Ecosystem. Pp. 91-101. In: Proc. National Seminar on Mangrove Ecosystem Dynamics of the Indus Delta. June 14-16, 1999, Karachi. Sindh Forest and Wild Life Department and the World Bank. 204 pp.

M. MacDonald and Partners Limited, 1984. Left Bank Outfall Drain Stage-1 Project Preparation, ANNEX 12, Outfall System and Tidal Link. Sir M. MacDonald and Partners Limited. In association with Hunting Technical Services Limited, January, 1984.

Nair, R.R. and Hashimi, N.H., 1987. On the origin of the cohesive and non-cohesive sediment boundry on the western on the western shelf of India: in Contributions in marine sciences. Edited by Rao, T.S.S., *et al.*, p. 413-425. Published by NIO, India.

N.I.O. 1988. Hydraulic Monitoring of Tidal Link Canal (LBOD)- a Project Proposal to SMO South, WAPDA,Hyderabad. National Institute of Oceanography, Karachi. pp. 1-40.

N.I.O. 1988. Environmental Monitoring for impact of Tidal Link Canal (LBOD)- a Project Proposal for WAPDA,Hyderabad. National Institute of Oceanography, Karachi, pp. 1-40.

N.I.O. 1994. First Annual Report Hydraulic Monitoring of Tidal Link Drain (LBOD) for the period November 1992 to December 1994. Prepared for SCARP Monitoring South, WAPDA, Hyderabad. National Institute of Oceanography, Karachi. pp. 1-50.

N.I.O. 1995. Second Annual Report Hydraulic Monitoring of Tidal Link Drain (LBOD) for the period January to December 1995 (Revised in August 1997). Prepared for SCARP Monitoring South, WAPDA, Hyderabad. National Institute of Oceanography, Karachi. pp. 1-59.

N.I.O.,1998. Hydraulic Monitoring of Tidal Link Drain (LBOD): Final Technical Report, for the period November 1992 to December 1997. Prepared for SCARP Monitoring South, WAPDA, Hyderabad. National Institute of Oceanography, Karachi, Pakistan. pp. 1-147, I-VII Appendices.

Chapter 13

Effects of Hurricane Dennis on Jardines de la Reina

*Fabián Pina Amargós, Andrés Jiménez Castillo, Félix Martín Blanco, Abdel Zayas Fernández and Wilbert Acosta de la Red

Coastal Ecosystems Research Centre, Cuba

**E-mail: fabian@ciec.fica.inf.cu*

ABSTRACT

On the 8th of July 2005, the 135 mph winds of Hurricane Dennis hit Jardines de la Reina archipelago, the most pristine group of islands of Cuba. A team of researchers from the Coastal Ecosystems Research Centre, who had left the place before the hurricane, returned three days later to continue work, but first to assess the effects of this hurricane on Jardines de la Reina. Though information from other habitats was recorded, the quantitive assessment was focused on coral reef habitats. Taking advantage of the knowledge of the area, eight belt transect (200 m x 10 m) were surveyed on the reef crest (front part, average 2 m depth) and reef slope (average 10 m depth). On the reef crest, the white scars produced by the breaking of Elkhorn coral (*Acropora palmata*) branches were counted and on the reef slope, the number of coral heads loosed from the upper part and observed on the sand area contiguous to it. None of these effects were observed before the hurricane strike and were qualified as good indicators of the impact. An average of approximately 10 scars per 100 m^2 was recorded on the reef crest and an average of approximately 1 coral head loosed per 2000 m^2 was recorded on the reef slope. We worked on the same area until mid September and none of the effects of hurricane Dennis on Jardines de la Reina, either qualitative or quantitative, seemed to have brought about substantial changes on the abundance, species composition and behavior, thus ecological processes seem to go as normal as before Dennis. The scars on Elkhorn coral disappeared in less than two weeks and on the reef slope, only us could notice the coral heads loosed.

Keywords: *Effects, Hurricane, Reef crest, Reef slope, Seagrass, Mangrove, Cuba.*

Introduction

Coral reefs appear to be stable and persistent ecosystems. However, they can suffer extensive destruction particularly from the effects of tropical storms and hurricanes. Damage from these phenomena must be seen as part of natural processes since tropical ecosystems and tropical storms have co-evolved during thousands years.

Several studies show that storms often cause significant reductions in coral cover. In St. John, U.S. Virgin Islands following tropical storm Klaus (1984) a significant reductions in coral cover was detected (Rogers and Zullo, 1987). Heron Island reef was hit by two hurricanes in 1967 and 1972. The first caused a decrease in cover from about 58 per cent to 0 per cent and the second from about 28 per cent to 5 per cent (Connell, 1978). Nevertheless, Hurricane Andrew (1992), a category 5 hurricane, surprisingly caused little damage (primarily scouring of coral tissue) to Ball Buoy Reef (Florida), in spite of the fact that *Acropora palmata* (a branching form) contributes most of the approximately 30 per cent coral cover on this shallow reef (Rogers, 1993).

Coral damage from hurricanes depends on various factors. Delicate branching species are less resistant than massive corals. In coral reefs of Barbados (Porter *et al.*, 1981) and Jamaica (Woodley *et al.*, 1981 and Mah and Stearn, 1986), branching corals (specially Acroporidae) were the most severely damaged by Hurricane Allen (1980). Other important variables to predict damage caused by tropical storms are depth and the primary source of stress. Deep reef zones generally experience less disturbance than shallow zones (Woodley *et al.*, 1981; Fenner, 1991). In general, coral reefs can rapidly recover from hurricane effects if the primary effect of the storm is physical damage to corals caused by waves (Shinn, 1976; Woodley *et al.*, 1981) but situation worsens if extreme temperature, heavy rainfall and/or high sedimentation also act on site (Boaden and Seed, 1985).

The worst cyclones destroy the entire coral architecture and redistribute rubble and biogenic sediments (Dollar and Tribble, 1993; Scoffin, 1993; McClanahan, Polunin and Done, 2002). It is estimated that some 25-30 years may be required for reefs to recover completely from the effects of particularly destructive hurricanes (Boaden and Seed, 1985).

But effects of tropical storms are not limited to sessile biota. The resulting destruction of fish habitats due to the strike of Hurricane Allen (1980) in Jamaica was followed by changes in predator abundance, herbivore behavior and the distribution of territorial damselfishes. These phenomena gradually abated during the first month following the storm, but one year later, fish assemblages differed between patches of reef more or less severely damaged by the hurricane (Kaufman, 1983).

In Cuba, most evaluations of the effects of tropical storms had been carried out on human populations, settlements and economical activities until Hurricane Ivan (2004) caused significant damage on wilderness of western Cuba. Since then, a group of Cuban scientific and environmental institutions have been working to prepare a methodology to assess the effects of tropical storms on ecosystems and complement the well-deserved prestige of Cuba on managing natural hazards. This paper, focuses on a particular place and storm, as a part of the above effort on managing natural hazard.

Materials and Methods

On the 8th of July 2005, the 135 mph winds of Hurricane Dennis (Figure 13.1) hit Jardines de la Reina archipelago (Figure 13.2), the most pristine group of islands of Cuba and a Marine Reserve since late 1996. A team of researchers from the Coastal Ecosystems Research Centre, who had left the place before the hurricane, returned three days later to continue work, but first to assess the effects of this powerful hurricane on Jardines de la Reina.

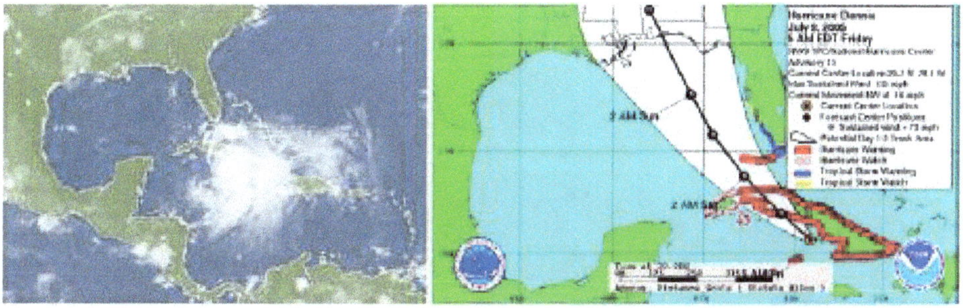

Figure 13.1: Hurricane Dennis around Jardines de la Reina

Though information from other habitat was recorded, the quantitive assessment was focused on coral reef habitats.

Taking advantage of the knowledge of the area, as a result of two weeks of work there immediately before the hurricane, eight belt transect (200 m x 10 m) were surveyed on the reef crest (front part, average 2 m depth) and reef slope (average 10 m depth) in

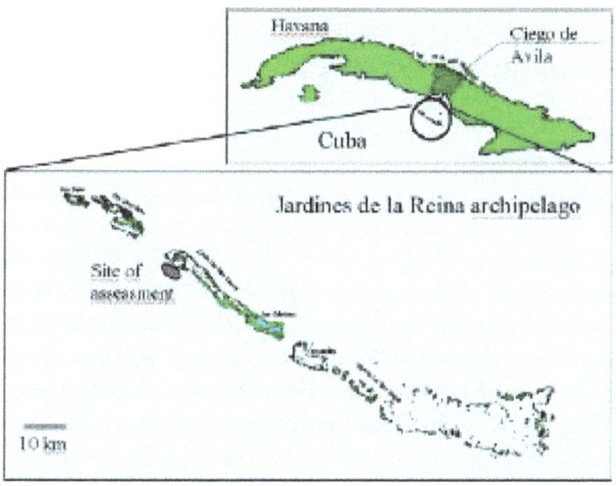

Figure 13.2: Location of Jardines de la Reina and the Site of Assessment

Figure 13.3: White Scars Produced by the Breaking of Elkhorn Coral (*Acropora palmata*) Branches (Upper photos) and Coral Heads Loosed (Lower photos)

a area known as Boca Grande. On the reef crest, the white scars produced by the breaking of Elkhorn coral (*Acropora palmata*) branches (Figure 13.3. Upper photos) were counted and on the reef slope, the number of coral heads loosed from the upper part and observed on the sand area contiguous to it (Figure 13.3. Lower photos). None of these effects were observed before the hurricane strike and were qualified as good indicators of the impact. Lack of detailed information about coral cover on the site impeded us to use the most common indicator of hurricane damage on coral reefs.

Results and Discussion

Qualitative Assessment

During a 48-mile trip to Jardines de la Reina from Jucaro, a small fishing community located south of Cuba, we observed drifting patches of seagrass in the Golfo de Ana Maria (Figure 13.4). Taking into account our observations on Jardines de la Reina seagrass beds, this effect did not seem to suggest a significant damage to the ecosystem. Winds and waves also detached small groups of mangroves (less than 10 m^2 each) (Figure 13.5). Eight of them were recorded in our trip Jucaro–Jardines de la Reina.

The assessment of few islands of Jardines de la Reina did not reveal dramatic impacts on flora and fauna. Only dead branches of trees and bushes were noted

(Figure 13.6), accelerating their input to food web through detritus way. The most important effect on vegetation was on leaves. They looked "burned" by strong winds and marine spray but the usual pattern of summer rain in Jardines de la Reina (almost everyday late in the afternoon) allowed their recovery. The conspicuous terrestrial fauna such as birds (Figure 13.7), jutias (endemic rodent, *Capromys pilorides*, Figure 13.8) and iguanas (endemic reptile, *Cyclura nubila*, Figure 13.9) was recorded on usual densities.

Figure 13.4: Drifting Patch of Seagrass

Figure 13.5: Effects on Mangroves

Figure 13.6: Dead Branches of Trees and Bushes

Figure 13.7: Pelican

Figure 13.8: Jutia

Figure 13.9: Iguana

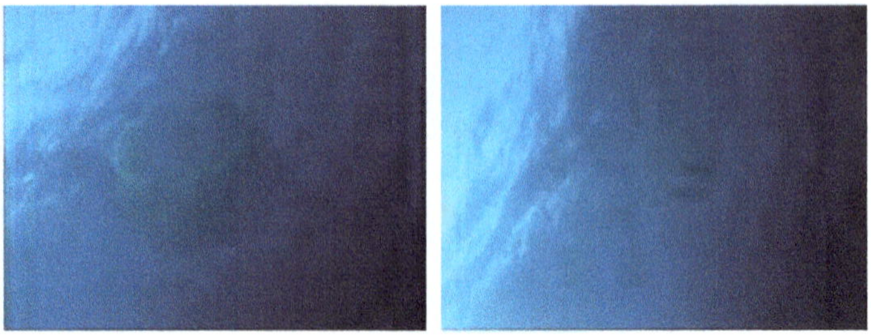

Figure 13.10: Jellyfish

Underwater qualitative assessment produced the following results. Thick schools of Jellyfish (*Aurelia aurita*) (Figure 13.10) drifting along reef crests for a few days after Dennis were also seen. The authors did not agree about changes on abundance of algae (same or decrease abundance).

In general, fish assemblages looked the same after and before Dennis and none of the long lasting effects seen by Kaufman (1983) in Jamaica were recorded in Jardines de la Reina. Unusual schools of Foureye butterflyfish (*Chaetodon capistratus*, Figure 13.11), Hogfish (juveniles, *Lachnolaimus maximus*, Figure 13.12), and Gray angelfish (*Pomacanthus arcuatus*, Figure 13.13) were remarkable but they did not last more than

Figure 13.11: Foureye Butterflyfish

Figure 13.12: Hogfish

Figure 13.13: Gray Angelfish

Figure 13.14. Whale Shark

a week. Small tuna fish schools and their usual companion, whale sharks (Figure 13.14), were observed during a couple of weeks after the hurricane, but their visits are unpredictable, thus may or may not be related to the event.

Quantitive Assessment

An average of approximately 10 scars per 100 m^2 was recorded on the reef crest and an average of approximately 1 coral head lost per 2 000 m^2 was recorded on the reef slope (Table 13.1).

Table 13.1: Data of the Effects of Hurricane Dennis on Boca Grande, Jardines de la Reina

Habitats	*Indicator*	*Transect*								*Average*
		1	*2*	*3*	*4*	*5*	*6*	*7*	*8*	
Reef crest	Scars/ 100 m^2	11.15	7.90	12.15	4.10	15.30	6.90	10.45	10.80	9.75
Reef slope	Heads/ 2000 m^2	2.00	1.00	1.00	0.00	2.00	0.00	1.00	2.00	1.12

As reported by other authors (Porter *et al.*, 1981; Woodley *et al.*, 1981 and Mah and Stearn, 1986) branching corals, in this case *Acropora palmata,* suffered the most damage and was used in this research as indicator of impact. This species shows a very high mortality in Jardines de la Reina (about 80 per cent of all colonies are dead presumably since early 90's of the XX century) and most of the scars recorded belong to dead corals. Thus, we assume Hurricane Dennis did not impact significantly coral cover in Jardines de la Reina although we did not measure it (no data in the specific area) but also because it did not obliterate large reef areas. This is another example of different effects of hurricane strikes: in some cases they provoke devastation (Connell, 1978; Woodley *et al.*, 1981, Rogers and Zullo, 1987) but in other little damage (Rogers, 1993). Woodley *et al.* (1981) and Fenner (1991) observed that deep reef zones (about 10 m in our case) generally experience less disturbance than shallow zones (about 2 m in our case).

We worked on the same area until mid September and returned in December and none of the effects of Hurricane Dennis on Jardines de la Reina, either qualitative or quantitative, seemed to have brought about substantial changes on the abundance, species composition and behavior, thus ecological processes seemed to go as normal as before Dennis. The scars on Elkhorn coral, very white the first day of assessment, disappeared in less than two weeks and on the reef slope, only we could notice the coral heads lost. The resilience of nature has played its role again as remarked by Shinn (1976) and Woodley *et al.* (1981) when the primary effect of the storm is physical damage to corals caused by waves.

Conclusions

None of the effects of Hurricane Dennis on Jardines de la Reina, either qualitative or quantitative, seemed to have brought about substantial changes on the abundance, species composition and behavior, thus ecological processes seemed to go as normal as before Dennis.

Acknowledgements

The authors gratefully acknowledge the collaborative assitance of Azulmar staff, our main supporter on Jardines de la Reina researches. Special thanks are owed to Giuseppe Omegna, Azulmar's manager.

References

Boaden, P.J.S. and R. Seed. 1985. An introduction to coastal ecology. Blackie, Glasgow, 219 p.

Connell, J.H. 1978. Diversity in tropical rain forests and coral reefs. Science 199: 1302-1310.

Dollar, S.J. and G.W. Tribble. 1993. Recurrent storm disturbance and recovery: a long-term study of coral communities in Hawaii. Coral Reefs 12: 223-233.

Fenner, D.P. 1991. Effects of hurricane Gilbert on coral reefs, fishes and sponges at Cozumel, Mexico. Bull. Mar. Sci. 48: 719-730.

Kaufman, L.S. 1983. Effects of Hurricane Allen on Reef Fish Assemblages Near Dicovery Bay, Jamaica. Coral Reefs (2): 43-47.

McClanahan, T. N. Polunin and T. Done. 2002. Ecological states and the resilience of coral reefs. Conservation Ecology 6(2).

Mah, A.J. and C.W. Stearn. 1986. The effect of Hurricane Allen on the Bellairs fringing reef, Barbados. Coral reefs 4: 169-176.

Porter, J.W., Woodley, J.D., G.J. Smith, J. Neigel E.A., J.F. Battey, M.D. Dallmeyer. 1981. Population trends among Jamaican reef corals. Nature 294: 249-250.

Rogers, C.S. 1993. Hurricane and coral reefs: the intermediate disturbance hypothesis revisited. Coral Reefs 12: 127-137.

Rogers, C.S. and E. Zullo. 1987. Initiation of a long term monitoring program for coral reefs in Virgin Islands National Park. Biosphere reserve research report no. 17. Viegin Islands Resource Management Cooperative/National Park Service, pp 1-33.

Scoffin, T.P. 1993. The geological effects of hurricane on coral reefs and the interpretation of storm deposits. Coral Reefs 12: 203-221.

Shinn, E.A. 1976. Coral reef recovery in Florida and Persian Gulf. Environ. Geol. 1, 241-254.

Woodley, J.D., E.A. Chornesky, P.A. Clifford, J.B.C. Jackson, L.S. Kaufman, N. Knowlton, J.C. Lang, M.P. Pearson, J.W. Porter, M.C. Rooney, K.W. Rylaarsdam, Tunnicliffe V.J., C.M. Wahle, J.L. Wulff, A.S.G. Curtis, M.D. Dallmeyer, B.P. Jupp, M.A.R. Koehl, J. Neigel, E.M. Sides. 1981. Hurricane Allen's impact on Jamaican reefs. Science. 214: 749- 755.

Chapter 14

Petroleum Hydrocarbons and Organochlorine Contaminants in Sediment and Fish Samples of the Western Coast of Alexandria, Egypt

Tarek Othman Said
National Institute of Oceanography and Fisheries, Kayet Bay, Alexandria, Egypt
E-mail: tareksaideg@yahoo.co.uk

ABSTRACT

The coastal area of Alexandria, which comprises the biggest and main harbours in Egypt and is considered as one of the most industrialized zones, receives considerable amounts of petroleum and its products. Another important source of contamination in the area are due to the discharge of huge amounts of land drainage water from the highly polluted Lake Mariut and the sewage water, they may include the wastes of lubricants and other mineral oils. Untreated domestic waste-water with agricultural and industrial wastes are still released through a number of drainages and outfalls along the coastal area of study. In order to understand more clearly the behaviour of these pollutants on a global scale and to prepare the future environmental policies, a baseline study is aimed to determine the occurrence of POPs in sediment and fish samples collected from the western coast of Alexandria. The range of total hydrocarbons (aliphatic + PAHs) in sediment was 683.8–34670.1 ng/g with an average of 9286.88 ng/g. The present study declared that the Sum of C16–C34 of Aliphatic fractions was < 4000 ng/g indicating the presence of fresh petroleum source. Also, the ratio of Pristane/Phytane (Pr/Ph) was >1 indicating fresh petroleum source. The concentration of total n-alkanes in various fish species was varied in the range of 253 to 11132 ng/g of wet weight and the concentration of total PAHs was ranged from 3862 to 35746 ng/g with an average of 4472 ng/g. For all sediments of our study, Anthracene/Phenanthrene (ANT/PHE) ratio was > 0.1 indicates the dominance of combustion source via pyrolytic process. Polychlorinated biphenyls (PCBs) concentrations were ranged from 0.79–64.9 ng/g with an average 12.14 ng/g of wet weight. The concentrations of organochlorines in fish species (*Euthynnus alleferatus, Scomberomorus commerson, Sphyraena Sphyraena, Diplodus vulgaris* and *Alepes djedaba*) decreased in the order: PCBs>DDTs>HCHs>total cyclodienes (aldrin, dieldrin and endrin). Concentrations of DDTs in fish tissues were ranged from 4.89 to 36.37 ng/g with an average of 16.4 ng/g of wet weight. The concentrations of total HCHs (alpha-, beta- and gama-HCH) were ranged from 0.3 to 65.7 ng/g with an average of 16.35 ng/g. Our data indicates: 1) new petroleum input where Pr/Ph > 1. (2) PAHs in sediment < 4000 ng/g, so there is no adverse effect on the biota in the area. (3) PAHs in tissue fishes << acceptable limit. (4) DDTs in sediment < effective range low level. (5) PCBs < effective range medium. And (6) pesticides and PCBs in all fishes < acceptable limit.

Keywords: *Petroleum hydrocarbons, Organochlorine, Sediment, Fish, Alexandria coast, Egypt.*

Introduction

The Mediterranean Sea is polluted from huge amounts of discharges resulting from: industrial and sewage wastes; loading/unloading operations in harbours; rivers and lakes outfalls and accidental oil spills. The inputs of these different sources amount 750 thousands tones discharged into the Mediterranean Sea (UNEP, 1984). Estuaries and coastal areas have traditionally been sites of industrial activity because of ease of transport and because the sea offers a convenient place for the disposal of waste substances (UNEP, 1989). It has been estimated that at least 40 per cent of all marine contaminants are coming from land-based sources (GESAMP, 1993). The coastal sea area of Alexandria, which comprises the biggest and main harbour in the country and is considered one of the most industrialized zones, receives considerable amounts of petroleum and its products. A part from this is SUMED pipeline which transports more than 100 million tons per year from crude petroleum of the Arabian Gulf countries. It was opened in December 1976 with its terminal at about 35 km west of Alexandria. Another important sources of oil contamination in the area are due to the discharge of huge amounts of land drainage water from the highly polluted Lake Mariut and the sewage water; they may include the wastes of lubricants and other mineral oils. Limited attention was given to follow up and evaluate the petroleum-contamination problem in the coastal region of Alexandria (Abu El-Dahab, 1980; Wahby and El-Deeb, 1980; Emara and Shriadah, 1995. The present investigation was carried out along the western coast of Alexandria at two sites; El-Max Bay and tourist villages distributed along the western part of Alexandria. The area includes the newly constructed El-Dekhela Harbour. The southwestern part of the region is El-Max Bay, which represents semi-closed shallow basin with an average depth of about 9.0 m bordered from its northeastern side by the Western Harbour outlet and by El-Dekhela Harbour at the northwest. EL Max Bay, which extends about 15 km between El-Agamy headland to the west and from the Western Harbor to the east and from the coast to a depth line of about 15 meter, has a mean depth of 10 m (Figure 14.1). It receives a heavy load of wastewater both directly from industrial outfalls and indirectly from Lake Maryut via El- Max pumping station (Shriadah and Tayel 1992). The main outfalls are Misr Chemical Industries effluent, the combined wastewater effluent from the tanneries and slaughter House, the Noubariya canal and El-

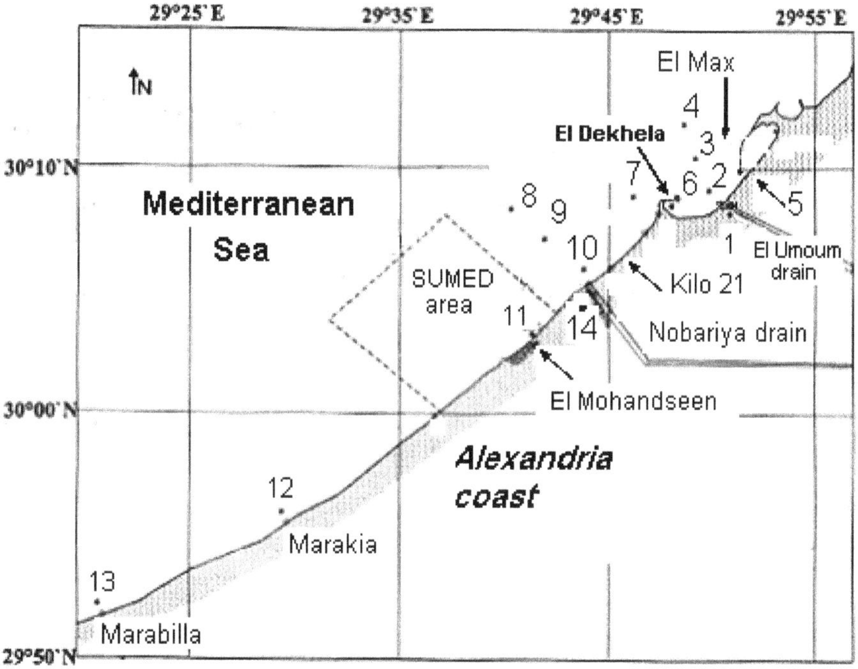

Figure 14.1. Area of Study and Sampling Stations

Mohmoudiya canal. El-Max pumping station, lies about one Kilometer upstream on El-Umum Drain Canal and pumps about 2.8×10^9 m^3/year. This is not only agricultural drainage water collected by El Umum Drain, but also comprises the overflow from Lake Maryut. Lake Maryut receives wastewater from the three Sources in its eastern section, consisting of domestic, industrial and agricultural wastes. These wastes fill the Lake and overflow west to El Umum drain to be discharged to the sea via El-Max pumping station. El Umum canal drains the south-western part of EL Beheira Governorate of about 45.8 Km long with a bottom depth of 20 m and an average depth of 3-4 m. It receives water from several pumping stations located at El Delengate, El Sher-shera, Farough, El Qualaa, El Deshoudy, Hares and Maryut. The level of drain is maintained in between 2.6 and 2.9 meter below the sea level through the action of El Max pumping (Mahmoud, 1985).

Untreated domestic waste-water with agricultural and industrial wastes are still released through a number of drainages and outfalls along the coastal area of study. Persistent organic pollutants (POPs) have a wide range of industrial, anthropogenic and agricultural applications. They include pesticides such as DDT (dichlorodiphenyltrichloroethane) and lindane (γ-HCH), in addition to petroleum hydrocarbons, which are organic chemicals, composed of fused benzene rings formed during incomplete combustion of coal, oil, petrol and wood (Laflamme and Hites, 1988; NRC, 1989). Svobodova *et al.* (1993) stated that organochlorine pesticides act as nerve poisons and are highly toxic to fish (48-hour LC_{50} <1.0 mg/L). Because of their chemical structure and their persistence, their use is now strictly controlled or banned (Nelson and Eisenreich, 2001). On the other hand, several polycyclic aromatic hydrocarbons (PAHs) are known to be potential human carcinogens these include; benzo[a]anthracene, chrysene, benzo[b]flouranthene, benzo[k]flouranthene, benzo[a]pyrene and benzo[ghi]perylene (IARC, 1991).

In order to understand more clearly the behaviour of these pollutants on a global scale and to prepare the future environmental policies, a baseline study is aimed to determine the occurrence of POPs in sediment and fish samples collected from the western coast of Alexandria.

Materials and Methods

Sampling

The sampling stations were located along EL Max Bay and at Kilo 21 during 2005 (Figure 14.1). Surface sediment (0-3 cm) samples were collected using a stainless steel grab from 14 sites during March 2005 and stored in pre-cleaned aluminum containers and frozen in a deep freezer at –20°C till analysis. The samples were analyzed for aliphatic and aromatic hydrocarbons following well-established techniques (UNEP/IOC/IAEA, 1989, 1991). In addition, 5 fish species; *Euthynnus alleferatus, Scomberomorus commerson, Sphyraena sphyraena, Diplodus vulgaris, Alepes djedaba* were collected during the same period to study the accumulative organic pollutants.

Procedures

The sediments were freeze-dried, dry/wet ratios determined and then sieved through a stainless steel mesh (250 mm). Each sediment sample (30 g) was Soxhlet extracted with 250 ml of hexane for 8 hours and then re-extracted for 8 hours into 250 ml of dichloromethane (Villeneuve *et al.*, 1999). Then the extracts were combined and concentrated down using rotary evaporation at 30°C followed by concentration with nitrogen gas stream down to a volume 2 ml. Sulphur was removed by shaking the extracts with mercury. Clean-up and fractionation was performed by passing the extract through a silica/alumina column. The 1st mL of the extracted volume was passed through the silica column prepared by slurry packing 20 mL (10 g) of silica, followed by 10 mL (10 g) of alumina and finally 1g of anhydrous sodium sulphate. Elution was performed using 40 mL of hexane/dichloromethane (90:10) followed by 20 mL of hexane/dichloromethane (50:50) (which combined contain PAHs). Finally, eluted samples were concentrated under a gentle stream of purified nitrogen to about 0.2 mL, prior to be injected into GC/FID for PAHs analysis.

The 2nd mL of the extracted volume was passed through florisil column prepared by slurry packing 20 g of florisil, followed by 10 g of alumina and finally 1g of anhydrous sodium sulphate. Elution was performed using 50 mL mixture containing 70 per cent hexane and 30 per cent dichloromethane for pesticide fractions. Finally, eluted samples were concentrated under a gentle stream of purified nitrogen to about 0.3 mL, prior to be injected into GC/ECD for pesticides analysis. All samples were analysed by a Hewlett Packard 5890 series II GC gas chromatograph equipped with a flame ionization detector (FID) and electron capture detector. For hydrocarbons analysis the instrument was operated in splitless mode (3µL splitless injection) with the injection port maintained at 290 °C and the detector maintained at 300°C. Samples were analysed on a fused silica capillary column HP-1; 100 per cent dimethyl polysiloxane (30 m length, 0.32 mm i.d., 0.17 µm film thickness). The oven temperature was programmed from 60-290°C, changing at a rate of 3°C/min and maintained at 290°C for 25 min. The carrier gas was nitrogen flowing at 1.2 mL/min. However, HP-5 capillary column (50m x 0.32mm x 0.52µm film thickness) with ^{63}Ni-electron capture detector (ECD) was used for pesticides analysis. The oven temperature was programmed from 90°C-140°C at rate of 5°C/min, maintained at 140°C for 1 min, then from 140°C–250°C at rate of 3 °C/min, maintained at 250°C for 1 min, then from 250°C -300°C at rate of 20°C/min and maintained at 300°C for 1 min, the carrier gas was nitrogen flowing at 1.5 mL/min.

A stock solution containing the following PAHs with their alkyllaed derivatives was used for quantification of hydrocarbons: naphthalene, acenaphthylene, acenaphthene, fluorene, phenathrene, anthracene, fluoranthene, benzo(a)anthracene, chrysene, benzo(b)fluoranthene, benzo(k)fluoranthene, pyrene, benzo(a)pyrene, dibenzo(a,h)anthracene, benzo(ghi)perylene and indeno(1,2,3-cd) pyrene by dilution to create a series of calibration standards of PAHs at 0.1, 0.25, 0.5, 0.75, 1.0, 2.0, 5.0, and 10 µg/mL. The detection limit was approximately 0.01 µg/mL for each PAH. For analytical reliability and recovery efficiency of the results, 6 analyses were conducted on PAH reference materials, HS-5 and 2974 (provided by EIMP-IAEA). The laboratory results showed recovery efficiency ranged from 92-111 per cent with coefficient of variation (CV) 8-14 per cent for all studied pollutants (16 PAHs fractions with their derivatives).

Organochlorine pesticides were quantified from individually resolved peak areas with the corresponding peak areas of the external standards (POC mixture provided by IAEA). They includes: α, β and γ-Hexachlorocyclohexanes, Heptachlor, Aldrin, Heptachloroepoxide, Dieldrin, DDT and its isomer DDE. To control the analytical reliability and assure recovery efficiency and accuracy of the results, four analyses were conducted on organochlorine compounds reference material SRM-1974 provided by EIMP-IAEA. The laboratory results showed recovery efficiency ranged from 95-105 per cent for pesticides reference material NO. 2974 with coefficient of variation of 15-20 per cent for all pesticide congeners. All solvents were pesticide grade purchased from Merck and appropriate blanks (1,000 fold concentrates) were analyzed.

Results and Discussion

Aliphatic and Aromatic Hydrocarbons in Sediment and Fish

Concentrations of aliphatic hydrocarbons in sediment samples are presented in Table 14.1. The results represent average concentrations of nineteen detectable n-alkanes (n-C10 to n-C34) ranged from 542.87 ng/g at El Dekhela harbour (St. 6) to 3265.1 ng/g at El Max Bay (St. 3) with an average of 1304.51 ng/g. The presence of n-alkane C26 (n-hexacosane) in all studied sediment samples suggests possible bacterial

Table 14.1: Concentrations of Aliphatic Hydrocarbons (ng/g) in Sediment Samples Collected from the Western Part of Alexandria during March, 2005

Compound	*Station*										
	3	*4*	*5*	*6*	*7*	*8*	*9*	*10*	*Mohandseen*	*Marakia*	*Marabila*
C10	ND	6.84	3.51	3.57	6.37	3.28	ND	ND	4.08	5.58	8.36
C12	8.68	38.00	13.01	12.29	37.40	11.04	21.70	ND	14.32	18.98	16.67
C14	42.12	32.52	36.00	17.99	39.85	31.63	25.78	5.63	23.38	27.40	22.67
C16	39.88	39.80	31.24	14.53	29.48	28.25	21.12	9.78	20.72	20.25	16.40
C17	74.96	5.15	11.49	9.02	7.49	3.88	4.29	4.61	10.50	18.67	5.83
Pristane	91.04	2.66	11.24	5.48	8.44	1.41	17.85	0.88	6.98	12.08	8.49
Octadecene	61.48	5.84	9.88	4.99	11.76	3.76	4.96	3.23	7.48	23.72	4.65
C18	98.34	2.64	38.48	16.48	36.57	7.15	ND	8.47	13.12	30.07	12.51
Phytane	58.70	2.09	11.84	4.71	3.44	ND	9.75	1.01	4.38	13.65	1.35
C20	144	12.14	28.76	13.97	29.56	7.63	8.26	4.46	13.10	39.37	7.47
C21	102	2.25	16.76	7.87	3.87	1.52	3.05	0.80	4.08	41.60	2.04
C22	132	8.03	18.84	13.93	14.16	4.71	ND	3.81	9.74	32.65	6.99
C24	162	5.06	20.49	8.68	4.36	27.28	3.49	1.75	4.92	53.00	3.09
C26	281	1.59	51.43	15.89	32.81	33.07	22.05	7.81	32.68	102.78	45.33
Squalene	256	7.47	65.12	17.45	15.17	11.16	7.01	22	14.10	74.17	8.35
C28	268	9.17	43.80	93.64	8.77	2.65	211.1	5.46	34.54	91.00	383.73
C30	646	444	227	157	162	486	1751	625	384	578.97	1138.5
C32	391	58.88	212	108	58.91	33.04	250	5.76	47.06	250.15	129.87
C34	409	16.07	106	21.33	103	46.08	ND	25	ND	509.48	53.87
Total	3265	693	954	543	607	740	2362	735	645	1937.98	1867.8
Pri/Phyt	1.6	1.3	0.95	1.16	2.45	NR	1.8	0.87	1.59	0.9	6.3

ND: < detection limit, NR: Not recorded.

contamination (Neff 1979). Pristane (C19) and Phytane (C20) are common isoprenoids detected in coastal marine sediments.

They are present in most petroleum products, usually as the major constituents within a much wider range of isoprenoid alkanes and usually considered as good indicators of petroleum contamination (Readman *et al.*, 2002). The present study show that the Sum of C16–C34 of Aliphatic fractions was < 4000 ng/g (Figure 14.2) indicating the presence of fresh petroleum source as stated by Villeneuve *et al.* (1999). Also, the ratio of Pristane/Phytane (Pr/Ph) was >1 indicating fresh petroleum source.

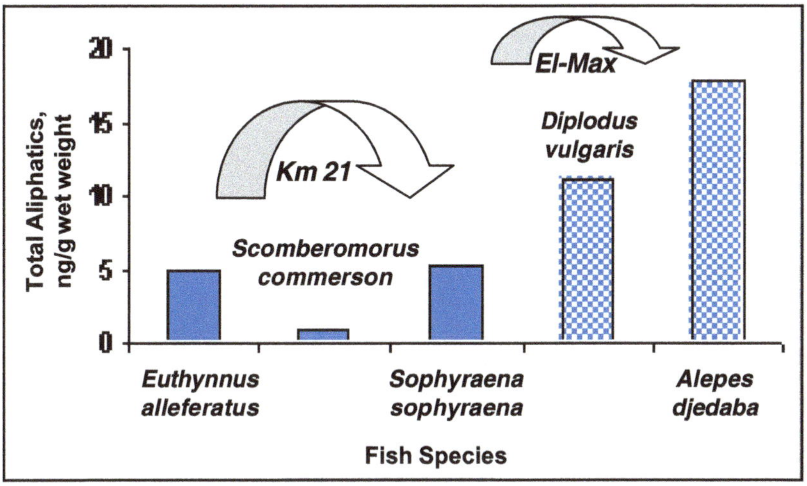

Figure 14.2: Total Aliphatic Concentrations (ng/g; wet wt) in Fish Species Collected from Regions; Kilo 21 and El Max of Alexandria Coast

Twenty-six PAH members were identified in the marine sediments where important qualitative differences among the different sampling sites were observed (Table 14.2 and Figure 14.3). The range of PAHs in sediment ranged between 38.5 ng/g at El Mohandseen village and 31405.0 ng/g at El Max Bay (St. 3) with an average of 7921.25 ng/g. Lower concentrations were detected at the stations of El Mohandseen, Marakia and Marabilla with average values of 125.87 ng/g. These values are similar to those already published. Baumard *et al.* (1998) have measured total PAHs from 0.3 to 100 ng/g in the Western Mediterranean. Concentrations from 48.1 to 86.5 ng/g and the total of 14 PAHs have been reported for coastal areas in the northwest Mediterranean Sea (Benlahcen *et al.*, 1997). The polycyclic aromatic hydrocarbon concentration (PAH) patterns differ according to the emission sources (Rogge *et al.*, 1993a,b,c,d; Sicre *et al.*, 1997; Wang *et al.*, 1999) and several PAH concentration diagnostic ratios have been extensively used in order to identify and quantify the contribution of each source of contamination to the specific compartments (Gogou *et al.*, 2000; Kavouras *et al.*, 2001). The reported ratios should be treated with caution as biogeochemical and physical process may alter the PAH profiles during transport. Differences in reactivity, volatility, water solubility and in sorption rates among the PAHs could lead to the incorrect designation of the sources. The general

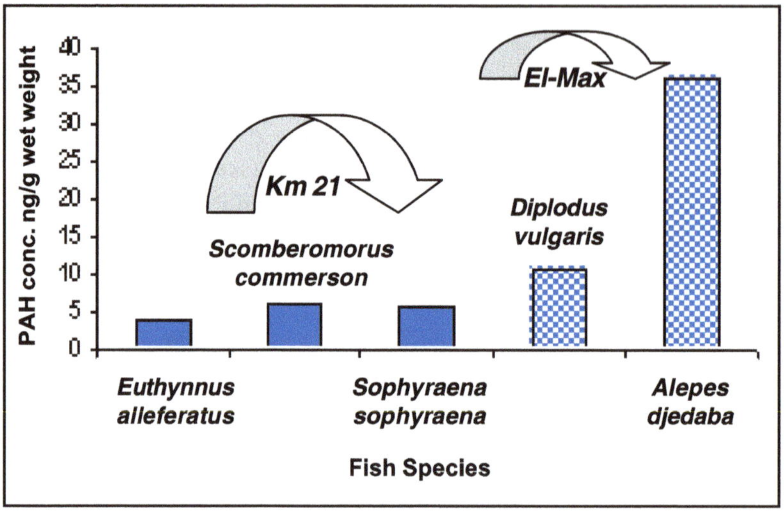

Figure 14.3: Total PAHs Concentrations (ng/g; wet wt) in Fish Species Collected from Regions; Kilo 21 and El Max of Alexandria Coast

distribution of the PAHs in all sediments reflects the high contribution of pyrolytic sources because of the predominance of parent PAHs over their alkylated derivatives (Prahl and Carpenter, 1983; Garrigues *et al.*, 1995). The ratios of methyl phenanthrene to phenanthrene (MePh/Ph) was < 1 for all stations (Table 14.2). Yunker *et al.* (2002) stated that anthracene to phenanthrene ratio was < 0.1 is considered as an indication of fossil fuel inputs while a ratio > 0.10 indicates a dominance of combustion source. For all sediments in our study, ANT/PHE ratio was > 0.1 (Table 14.2). The sum of five carcinogenic PAHs (IARC 1991) was highest at stations 9 and 10 of El Max Bay showing concentrations of 26757.0 and 12554.0 ng/g, respectively. The major inputs of hydrocarbon pollutants in the Western coast of Alexandria are from combustion of organic matter, anthropogenic industrial activity and/or natural fires. These sources give rise to complex mixtures of PAHs characterized by higher abundance of parent PAHs than their alkylated derivatives. The range of total hydrocarbons (aliphatic + PAHs) in sediment was 683.8–34670.1 ng/g with an average of 9286.88 ng/g. The total aliphatic concentrations were higher than the corresponding aromatic fraction for all sediment samples. In addition, the recorded total n-alkanes were lower than that recorded in Black Sea ranged from 1200 to 240000 ng/g for sediment reported by Readman *et al.* (2002).

The concentration of total n-alkanes in various fish species was varied in the range of 253 to 11132 ng/g of wet weight and the concentration of total PAHs was ranged from 3862 to 35746 ng/g with average of 4472 ng/g (Table 14.3). The average concentration of total PAHs was lower than that measured in mussels of the Egyptian Red Sea coast (average 5452 ng/g; wet weight) recorded by El Nemr *et al.*, 2003, and also lower than that recorded in bivalves from the Egyptian Mediterranean Sea (average 8180 ng/g wet weight) recorded by El Sikaily *et al.* (2002). The concentration range of total PAHs was lower than the acceptable level of total PAHs (100,000 ng/g) on the

Table 14.2: Concentrations of Aromatic Hydrocarbons (ng/g) in Sediment Samples Collected from the Western Part of Alexandria During March, 2005

Compound	*Station*									
	3	*5*	*6*	*7*	*8*	*9*	*10*	*Mohandseen*	*Marakia*	*Marabila*
Naphthalene	121	102	53	208	78	40	2.53	0.93	0.70	152
Triethyl benzene	0.67	0.36	ND	ND	0.17	ND	0.18	ND	0.08	0.27
1-Methyl naphthalene	0.67	ND	ND	0.97	0.36	0.28	0.37	0.43	0.32	0.40
1-Ethyl naphthalene	0.08	ND	ND	ND	ND	ND	ND	ND	ND	ND
2,6-Dimethyl naphthalene	0.08	ND	ND	ND	0.11	ND	ND	ND	ND	ND
Acenaphthene	1.04	ND	ND	ND	18.4	ND	0.12	ND	ND	ND
2,3,6-Trimethyl naphthalene	0.54	ND	ND	ND	ND	ND	ND	ND	ND	ND
Fluorene	131	43.5	ND	43.7	21.9	50.5	20.6	ND	27.6	40.4
Phenanthrene	288	162	ND	ND	21.3	ND	36.4	ND	ND	ND
2-Methyl phenanthrene	1.17	0.27	ND	0.33	0.09	0.45	0.13	ND	ND	0.32
1-Methyl phenanthrene	2.80	ND	ND	ND	ND	ND	ND	ND	ND	ND
3,6-Dimethyl phenanthrene	0.93	ND	ND	ND	0.25	0.33	ND	ND	ND	ND
Anthracene	217	51.8	ND	60.8	72.5	83.1	38.6	ND	35.6	59.7
Fluoranthene	53.6	116	ND	ND	12.1	0.13	ND	ND	ND	ND
Pyrene	234	222	ND	ND	19.7	ND	ND	ND	ND	ND
1-Methyl pyrene	72.5	ND	ND	ND	2.32	ND	ND	ND	ND	ND
Benzyl butyl phthalate	0.57	1.25	ND	ND	0.11	ND	ND	ND	ND	ND
Benzo(a)anthracene	201	ND	ND	ND	ND	ND	ND	ND	ND	ND
Chrysene	141	ND	ND	ND	ND	ND	ND	ND	ND	ND
Did n-Octyl phthalate	6.46	ND	ND	ND	ND	ND	ND	ND	ND	ND

Contd...

Table 14.2–Contd...

Compound	*Station*									
	3	*5*	*6*	*7*	*8*	*9*	*10*	*Mohandseen*	*Marakia*	*Marabila*
Benzo(b)fluoranthene	290	ND	73	ND	ND	ND	ND	ND	ND	ND
Benzo(k)fluoranthene	238	51	ND	ND	30	ND	ND	ND	ND	ND
Benzo(a)pyrene	669	57	54	1441	29	4129	12554	36	22	ND
Dibenzo(a,h)anthracene	624	0.99	328	42	23	ND	ND	0.63	0.20	ND
Benz(ghi)perylene	25376	ND	2487	ND	ND	ND	ND	ND	ND	ND
Indeno(1,2,3-cd)pyrene	2733	ND	1854	ND	62	22628	ND	ND	ND	ND
Total PAHs	31405	809	4849	1796	392	26932	12653	38.5	86.1	253
ANT/PHE	0.7518	0.32	NR	NR	3.41	NR	1.059	NR	NR	NR
PAHs (Carc)	4517.7	58.3	2309	1483	114	26757	12554	37.1	21.78	NR

basis of wet weight in edible fish tissue suggested by EPA (1979). The fish species collected from El Max Bay were generally more polluted than that collected from Kilo 21 (Figures 14.2 and 14.3). The average concentration of PAHs in the present study (12112 ng/g; wet weight) was higher than the average of PAHs (295 ng/g; wet weight) in mussels and clams reported by Law and Andruleniez (1983) for the southern Baltic Sea.

Table 14.3: Concentrations of PAHs and Aliphatic Hydrocarbons (ng/g; dry wet) in Fish Samples Collected from the Western Coast of Alexandria During 2005

Compound	*Species*				
	Euthynnus alleferatus	*Scombero-morus commerson*	*Sophyraena sophyraena*	*Diplodus vulgaris*	*Alepes djedaba*
Naphthalene	123.48	570.44	726.47	47.94	149.15
Triethyl benzene	ND	0.26	0.71	0.32	ND
1-Methyl naphthalene	ND	1.92	1.55	2.56	ND
1-Ethyl naphthalene	ND	ND	0.21	ND	ND
2,6-Dimethyl naphthalene	ND	0.62	0.85	ND	ND
Acenaphthylene	ND	ND	ND	156.70	ND
Acenaphthene	ND	105.04	143.64	ND	ND
2,3,6-Trimethyl naphthalene	ND	ND	0.17	ND	ND
Fluorene	ND	60.44	50.11	175.12	59.92
Phenanthrene	ND	0.61	338.76	ND	ND
2-Methyl phenanthrene	ND	ND	1.31	1.51	0.54
1-Methyl phenanthrene	ND	ND	1.44	0.93	ND
3,6-Dimethyl phenanthrene	ND	0.73	0.98	0.99	ND
Anthracene	ND	201.59	243.48	281.40	101.34
Fluoranthene	ND	0.62	179.00	0.41	ND
Pyrene	ND	ND	427.57	ND	ND
1-Methyl pyrene	ND	ND	ND	ND	ND
Benzyl butyl phthalate	ND	ND	2.40	ND	ND
Benzo(a)anthracene	106.89	39.95	ND	ND	ND
Chrysene	461.56	ND	ND	ND	ND
Di n- Octyl phthalate	ND	ND	12.39	ND	46.71
Benzo(b)fluoranthene	ND	ND	ND	ND	155.18
Benzo(k)fluoranthene	449.24	266.33	331.73	691.50	1514.41
Benzo(a)pyrene	5693.38	2511.81	1902.72	4309.32	32905.49
Dibenz(a,h)anthracene	98.56	34.85	168.76	74.60	272.63
Benz(ghi)perylene	2.19	27.99	3465.59	ND	ND
Indeno(1,2,3-cd)pyrene	ND	39.31	276.71	ND	540.88
Total PAHs	6935.30	3862.50	8276.53	5743.30	35746.25

Contd...

Table 14.3–Contd...

Compound	*Species*				
	Euthynnus alleferatus	*Scombero-morus commerson*	*Sophyraena sophyraena*	*Diplodus vulgaris*	*Alepes djedaba*
n-C10	20	38	28	43	42
n-C12	46	47	49	58	64
n-C14	63	49	70	68	63
n-C16	54	42	56	50	42
n-C17	49	ND	60	25	ND
Pristane	20	ND	32	10	ND
n-Octadecene	ND	ND	4	ND	ND
n-C18	27	19	39	26	21
Phytane	ND	ND	3	ND	ND
n-C20	25	ND	19	14	ND
n-C21	ND	ND	3	ND	ND
n-C22	6	ND	10	8	ND
n-C24	7	ND	4	2	14
n-C26	24	ND	20	20	202
Squalene	75	ND	6	ND	162
n-C28	58	ND	25	119	ND
n-C30	4263	58	449	4859	10522
n-C32	71	ND	14	ND	ND
Total Aliphatics	4808	253.3	891	5302	11132

PCBs and pesticides in sediment and fish.

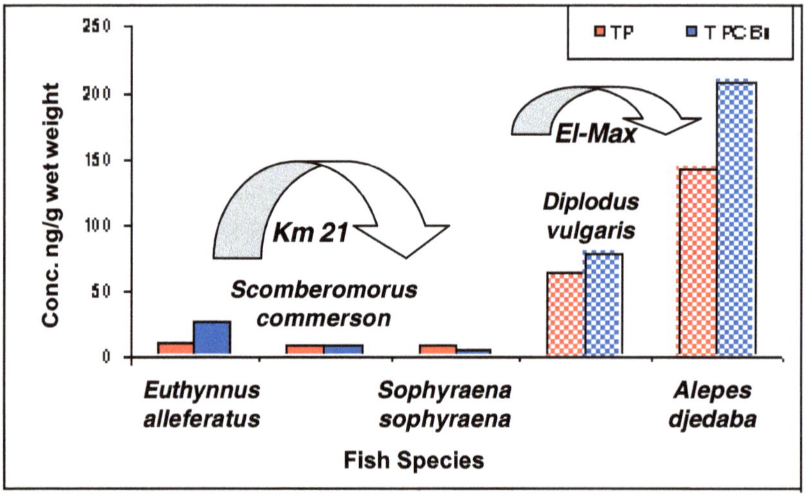

Figure 14.4: Distribution of Total Pesticides and Total PCBs in ng/g Wet Weight in Fish Samples Collected from Alexandria Coast During 2005

Table 14.4: Concentration of Chlorinated Pesticides and PCBs (ng/g) in Sediment Samples Collected from the Western Coast of Alexandria During March 2005

Compound	*Station*										
	3	*4*	*5*	*6*	*7*	*8*	*9*	*10*	*11*	*12*	*13*
Alpha-HCH	0.16	0.04	ND	0.06	0.03	0.07	0.05	0.05	0.03	0.10	0.04
Beta and Gama-HCH	0.15	0.92	ND	2.19	0.19	0.80	0.50	0.37	0.38	0.12	0.17
PCB 28	6.98	0.22	0.15	7.58	0.28	0.34	0.24	0.16	0.09	0.35	0.20
PCB 52	57.31	9.54	0.64	34.61	0.83	3.94	1.39	1.72	1.62	1.57	1.00
Aldrin	8.00	0.07	0.06	2.87	0.67	0.16	0.06	0.17	0.04	0.05	0.12
PCB 101	0.25	0.18	ND	0.21	0.13	0.11	0.14	0.11	0.03	0.25	0.05
o,p-DDE	ND	0.07	ND	0.09	0.05	0.07	0.05	0.03	ND	0.09	0.01
Dieldrin	0.04	0.09	ND	0.07	0.03	0.14	0.04	0.07	0.03	0.13	0.01
p,p-DDE	2.46	0.71	0.02	0.81	0.71	0.77	0.65	0.42	0.11	0.91	0.30
o,p-DDD	0.27	0.96	ND	0.05	ND	ND	ND	ND	ND	ND	ND
Endrin	ND	0.06	ND	0.01	0.04	ND	0.05	0.02	ND	0.04	0.01
PCB 118	0.06	0.07	ND	0.02	0.04	0.06	0.05	0.04	0.01	0.11	0.02
p,p-DDD & o,p-DDT	0.40	0.04	ND	0.06	0.14	0.14	0.04	0.01	0.01	0.04	0.02
PCB 153	0.11	0.04	ND	0.01	0.04	0.06	0.06	0.04	0.01	0.06	0.02
p,p-DDT	0.05	ND	ND	ND	0.03	0.04	0.04	0.01	ND	ND	ND
PCB 138	0.13	ND	ND	0.01	0.03	ND	0.03	ND	ND	ND	ND
PCB 180	0.10	ND	ND	ND	0.03	0.01	0.04	0.01	0.01	0.01	ND
Total PCBs (TPCBs)	64.94	10.05	0.79	42.44	1.38	4.51	1.95	2.09	1.76	2.34	1.29
Total pesticides (TP)	11.53	0.11	0.08	6.21	1.89	2.19	1.49	1.13	0.61	1.49	0.68
Total (TPCBs+TP)	76.47	10.16	0.87	48.65	3.27	6.70	3.44	3.22	2.36	3.83	1.98

ND: Below the detection limit.

Polychlorinated biphenyls (PCBs) concentrations were ranged from 0.79–64.9 ng/g with an average 12.14 ng/g of wet weight (Table 14.4). The PCB concentrations were higher at St. No. 3; 64.94 ng/g, St. No. 6, 42.44 ng/g. The concentrations of organochlorine in fish species (*Euthynnus alleferatus, Scomberomorus commerson, Sphyraena Sphyraena, Diplodus vulgaris and Alepes djedaba*) were decreased in the order: PCBs>DDTs>HCHs>total cyclodienes (aldrin, dieldrin and endrin) (Table 14.5). The concentrations of total HCHs (alpha-, beta- and gama-HCH) were ranged from 0.3 to 65.7 ng/g with an average of 16.35 ng/g which is higher than that recorded in coral reef samples of the Red Sea (average 4 ng/g) recorded by El Nemr *et al.* (2004). The concentrations of total cyclodienes were ranged from 0.99 to 45.74 ng/g wet weight. Although the use of HCHs in agriculture has been greater than cyclodienes and DDTs, the relatively low concentrations of HCHs in fish tissues reflect lower potential for bioaccumulation. Furthermore, higher vapour pressures for HCHs than DDTs facilitate relatively rapid atmospheric dissipation in the tropics, leaving fewer residues in sediment (Kannan *et al.*, 1995). This was clearly found for lower HCHs concentrations than DDTs concentrations are recorded in Tables 14.4 and 14.5).

Table 14.5: Concentration of Chlorinated Pesticides and PCBs (ng/g) in Fish Samples Collected from Western Coast of Alexandria During February 2005

Compound	Species				
	Euthynnus alleferatus	*Scombero-morus commerson*	*Sphyraena sphyraena*	*Diplodus vulgaris*	*Alepes djedaba*
Alpha-HCH	0.104	0.18	0.15	2.67	11.33
Beta and Gama-HCH	1.944	1.32	0.15	9.52	54.37
PCB28	4.309	1.21	0.75	13.57	15.71
PCB 52	20.803	5.85	3.92	48.11	168.27
Aldrin	1.838	1.65	0.65	12.67	29.90
PCB 101	0.285	0.60	0.56	4.50	5.33
o,p-DDE	0.153	0.23	0.27	1.32	2.47
Dieldrin	0.102	0.15	0.16	1.64	15.11
p,p-DDE	5.088	3.85	5.79	25.60	18.01
o,p-DDD	ND	ND	ND	ND	3.86
Endrin	0.092	0.11	0.18	1.71	0.73
PCB 118	0.165	0.20	0.30	2.58	10.79
p,p-DDD& o,p-DDT	0.328	0.60	0.69	5.12	3.63
PCB 153	0.496	0.22	0.51	2.89	4.20
p,p-DDT	0.296	0.21	0.75	4.33	3.28
PCB 138	0.184	0.07	0.10	0.55	1.11
PCB 180	0.181	0.08	0.25	6.48	2.63
TP	9.945	8.30	8.79	64.59	142.68
T PCBs	26.423	8.22	6.40	78.67	208.04
Total	36.368	16.52	15.19	143.26	350.72

ND: Below the detection limit.

Concentrations of DDTs in fish tissues ranged from 4.89 to 36.37 ng/g with an average of 16.4 ng/g of wet weight. These values were much lower than that recorded in fish samples collected from the Suez Canal (total DDTs ranged from 161-3100 ng/g with an average of 1087 ng/g; wet weight) by Said and Hamed (2005), and also lower than that recorded in mussels of the Egyptian Red Sea coast (ranged between 125 and 722 ng/g wet weight of Total DDTs) by Khaled *et al.* (2004). The metabolic transformation of DDT under oxidative conditions which lead to p,p-DDE was detected in clearly for sediment samples (Table 14.4). The same was true for p,p-DDE detected in fish tissues (Table 14.5).

The persistent half-life ($T_{1/2}$ = 5 years) of DDT in marine systems (Carvalho 1994) and recent work on the dechlorination of DDE in anaerobic sediments $T_{1/2}$ = 6 years are similar (Quensen *et al.*, 1998). Assuming that after 1974 there have been no further releases of DDT, these half-life values would allow for an estimated reduction of DDT in the coastal environment. Nevertheless, despite the ban of DDT there are still continuous inputs into the coastal environment, mainly by atmospheric deposition of DDT (Villeneuve and Cattini 1986) and DDT leaching from agricultural soils followed by discharges into estuarine areas (Claisse 1989). These inputs would help maintain DDT presence in the coastal environment, as the concentrations of total DDTs were nearly the major component of total pesticides (Table 14.4 and 14.5). Other chlorinated pesticides results indicate either a more rapid disappearance from the coastal environment than DDT (Ilmer 1975; Villeneuve *et al.*, 1999) or a lower use there in comparison with DDT. The acceptable daily intake for PCB is 10-30 µg/person every day (WHO 1989), DDT is 1.4 pg/person every day (FAO/WHO 1985) and HCH is 1.8 mg/person every day (EPA 1988).

The PCB congeners and pesticides can cause toxic symptoms similar to those caused by dioxin exposure, including developmental abnormalities and growth suppression, disruption of the endocrine system, impairment of immune function and cancer promotion. The maximum permissible levels of toxic DDTs, PCBs and cyclodienes recommended by the National Academy of Sciences and National Academy of Engineering (NAS-NAE 1972), for the protection of aquatic biota are 1000 ng/g to 5000 ng/g for PCBs and 100 ng/g for cyclodienes (all as weight concentrations in whole-body tissue). In Sweden, the recommendations are 5000 ng/g for DDTs, 2000 ng/g for PCBs and 20 ng/g for HCH (Swedish Food Regulation, 1983).

Conclusion

Data from the current study indicates that the average of hydrocarbon concentrations in the collected fish species of Alexandria coast was lower than that recorded in mussels of the Egyptian Red Sea coast and lower than that recorded in bivalves from the Egyptian Mediterranean Sea. However, the average concentration of PAHs in the present study was higher than that measured in mussels and clams of the southern Baltic Sea. Furthermore, the concentration range of total PAHs was lower than the acceptable level recorded in edible fish tissue, so there is no adverse effect on biota in the investigated area. Both of pesticides and PCBs recorded in different fish species of the study area were lower than the acceptable levels.

This study provides a preliminary assessment of the level of POPs, in a restricted area of the Egyptian Mediterranean coast; Alexandria coast, which serve as a baseline study for future comparisons. Repeating this study on a large scale in the near future may be useful as a quantitative integration of pollution by the most persistent organic pollutants and in identifying the sources of pollution.

Acknowledgements

The author would like to express his sincere gratitude to Prof. Dr. Soliman Hamed the president of NIOF and Prof. Dr. Mohamed Shreadah the Vice president of NIOF for their financial support. Also deep appreciation for Prof. Dr. Zinab El Sherif for her kind assistance and effective orientations during this study.

References

Abu El-Dahab, O.M.T., 1980. Oil pollution of the marine environment in the area of Alexandria. M.Sc. Thesis. Alexandria University- Egypt, pp. 1-123.

Baumard, P., Budzinski, H., Garrigues, P., 1998. Determination of polycyclic aromatic hydrocarbons (PAHs) in sediments and mussels of the Western Mediterranean Sea. Environ Toxicol and Chem, 17, pp. 765-776.

Benlahcen, K. T, Chaoui, A, Budzinski, H., Belloca, J., Garrigues, P. 1997. Distribution and sources of polycyclic aromatic hydrocarbons in some Mediterranean Coastal sediment. Mar Pollu Bull, 34, pp. 298-305.

Carvalho, F.P., Flower, S.W., Readman, J.W., Mee, L.D., 1994. Pesticides residues in tropical coastal lagoons-use of ^{14}C labeled compounds to study the cycling and fate of organ chemicals. Int. Symp. Applications of Isotopes and Radiation in conservation of the environment, 9-13 March 1992, Karlsruche, Germany. Vienna, Austria: International Atomic Energy Agency (IAEA), 94, pp. 637-653.

Claisse, D., 1989. Chemical contamination of French Coast. The results of ten year mussel watch. Mar Pollu Bull, 20, pp. 523-528.

El Nemr, A., El Sikaily, A., Khaled, A., Said, T.O., Abd-Alla, A.M.A., 2004. Determination of hydrocarbons in mussels from the Egyptian Red Sea Coast. Environ Monitor Assess, 96, pp. 251-261.

El Nemr, A., El Sikaily, A., Khaled, A., Said, T.O., Abd-Alla, A.M.A., 2004. Chlorinated pesticides and polychlorinated biphenyls in the Coral reef skeleton of the Egyptian Red Sea Coast. Bull. Environ. Contam., 72, pp. 1195-1202.

El Sikaily, A., Khaled, A. El Nemr, A., Said, T.O., Abd-Alla, A.M.A., 2002. Determination of hydrocarbons in bivalves from the Egyptian Mediterranean Coast. Medit Mar Sci, 3(20), pp. 123-131.

Emara, H. I., Shriadah, M.A., 1995. Aliphatic and aromatic petroleum hydrocarbons in coastal seawater of Alexandria. Report of the third FAO/UNEP. Regional Workshop on the monitoring of chemical contaminants in marine biota for trends, 6–9 November, Alexandria, Egypt.

EPA, 1988. Integrated risk information system (RIS), Environmental Criteria and assessment office. Cincinnati, Ohio.

FAO/WHO, 1985. Pesticides residues in food 1984 evaluations. FAO Plant Production and Protection Paper 67, ROME.

Garrigues, P., Budzinski, H., Manitz, M.P., Wise, S.A., 1995. Pyrolytic and petrogenic inputs in recent sediments: a definitive signature through phenanthrene and chrysene compound distribution. Polycyclic Aromatic Compounds, 7, pp. 75–284.

GESAMP, 1993. 'Impact of Oil and Related Chemicals and Wastes on the Marine Environment'. Report Studies, 50, pp. 1-180.

Gogou, A., Bouloubasi, I., Stephanou, *E.G.*, 2000. Marine organic geochemistry of the Eastern Mediterranean: 1. Aliphatic and polyaromatic hydrocarbons in the Cretan Sea surficial sediments. Mar Chem, 68, pp. 265–282.

IARC, 1991. Monographs on Evaluation of Polynuclear Aromatic Compounds, Part 1, Chemical, Environmental, and Experimental Data (Internal Agency for Research on Cancer), Lyon.

ILMR, 1975. Biennial activity report of international laboratory of marine radioactivity. Monaco: IAEA-ILMR.

Kannan K., Tanabe S., Tatsukawa, R., 1995. Geographical distribution and accumulation features of organochlorine residues in fish in tropical Asia and Oceania. Environ. Sci. Technol., 29, pp. 2673-2683.

Kavouras, I.G., Koutrakis, P., Tsapakis, M., Lagoudaki, E., Stephanou, *E.G.*, Baer, D., Oyola, P., 2001. Source apportionment of urban aliphatic and polyaromatic aromatic hydrocarbons (PAHs) using multivariate methods. Environ Sci Technol, 35, pp. 2288–2294.

Khaled, A., El Nemr, A., Said, T.O., El Sikaily, A., Abd-Alla, A.M.A., 2004. Polychlorinated biphenyls and chlorinated pesticides in mussels from the Egyptian Red Sea coast. Chemosphere 54, pp. 1407-1412.

Laflamme, R.E. and Hites, R.A., 1988. The global distribution of polycyclic aromatic hydrocarbons in recent sediments. Geochem Cosmochim Acta, 42, pp. 289–303.

Law R. and Andruleuicz E. 1983. Hydrocarbons in water, sediment and mussels from the southern Baltic Sea. Mar. Poll. Bull., 14, pp. 289-293.

Mohmoud, Th. H., 1985. Phosphorus and nitrogen dynamics in the polluted coastal waters off Alexandria. Ph.D. Thesis, Faculty of Science, Alexandria University.

National Academy of Sciences and National Academy of Engineering (NAS-NAE), 1972. Ecol Research Ser Environmental, Washington DC, pp. 1-62.

Neff, J.M., 1979. Polycyclic aromatic in aquatic environment sources. Fates and biological effects. Applied Science, London, UK.

Nelson, T.R., Eisenreich, S.J., 2001. Dynamic air–water exchange of polychlorinated biphenyls in the New York–New Jersey Harbor Estuary. Environ Sci Technol, 20, pp. 810–816.

NRC, 1989. Oil in the Sea: Inputs, Fates and Effects. National Academy Press, Washington, DC.

Prahl, F.G., Carpenter, R., 1983. Polycyclic aromatic hydrocarbon (PAH)-phase associations in Washington coastal sediments. Geochim Gosmochim Acta 47, pp. 1013–1023.

Quensen, J.F., Mueller, S.A., Jain M.K., Tiedje, J.M., 1998. Reductive dechlorination of DDE in marine sediment microcosms. Science, 280, pp. 722-724.

Readman, J.W, Fillmann, G, Tolosa, I., Bartoccim J., Villeneuve, J.P., Catinni, C., Mee, L.D., 2002. Petroleum and PAH contamination of the Black Sea. Mar Pollu Bull, 44, pp. 48-62.

Rogge, W.F., Hildemann, L., Mazurek, M.A., Cass, G.R., Simoneit, B.R.T., 1993a. Sources of fine organic aerosol: 2. Noncatalyst and catalyst-equipped automobiles and heavy duty diesel trucks. Environ Sci Technol, 27, pp. 636–651.

Rogge, W.F., Hildemann, L., Mazurek, M.A., Cass, G.R., Simoneit, B.R.T., 1993b. Sources of fine organic aerosol: 3. Road dust, tire debris and organ metallic brake lining dust: roads as sources and sinks. Environ Sci Technol, 27, pp. 1892–1904.

Rogge, W.F., Hildemann, L., Mazurek, M.A., Cass, G.R., Simoneit, B.R.T., 1993c. Sources of fine organic aerosol: 4. Particulate abrasion products from leaf surfaces of urban plants. Environ Sci Technol, 27, pp. 2700–2711.

Rogge, W.F., Hildemann, L., Mazurek, M.A., Cass, G.R., Simoneit, B.R.T., 1993d. Sources of fine organic aerosol: 5. Natural-gas home appliances. Environ Sci Technol, 27, pp. 2736–2744.

Said, T. O., Hamed, M.A., 2005. Distribution of chlorinated pesticides in surface water and fish of El Temsah and Bitter lakes, Suez Canal. Egypt J Aqaut Res, 31(1), pp. 200-213.

Shreadah, M.A. Tayel, T.R., 1992. Environmental conditional the western harbor of Alexandria. The Bulletin of the High Institute of Public Health, XXLL (1), pp. 213-228.

Sicre, S., M.A., Bayona, J. M., Grimalt, J.O., Saliot, A., Albaiges, J., 1997. Mass balance and dynamics of polycyclic aroma tic hydrocarbons in the Mediterranean Sea. Deep-Sea Research 44, pp. 881–905.

Svobodova, Z., Loyd, R., Machova, J. and Vykusova, B., 1993. Water quality and fish health', EIFAC technical paper, No. 54 ROME, FAO, pp. 59-64.

Swedish Food Regulation (SFR), 1983. Foreign substances in food. The National Administration SLVFS: 1.

UNEP/ECE/UNIDO/FAO/UNESCO/WHO/IAEA, 1984. Pollutants from land based sources in the Mediterranean Sea. UNEP Regional Seas Reports and Studies No. 32, UNEP, Geneva.

UNEP/IOC/IAEA, 1989. Determination of DDTs and PCBs in selected marine organisms by capillary column gas chromatography, reference methods for marine pollution studies No. 40. Nairobi: United Nations Environment Program, 18.

UNEP/IOC/IAEA, 1991. Sampling of selected marine Organisms and sample preparation for the analysis of chlorinated hydrocarbons. Reference methods for marine pollution studies no. 12, revision 2. Nairobi: United Nations Environment Program 17.

Villeneuve, J.P., Cattini, C., 1986. Input of chlorinated hydrocarbons through dry and wet deposition to the Western Mediterranean Coast. Chemosphere, 28, pp. 897-903.

Villeneuve, J.P., Carvalho, F.P., Fowler, S.W., Cattini, C., 1999. Levels and trends of PCBs, Chlorinated pesticides and petroleum hydrocarbons in mussels from the NW Mediterranean Coast: Comparison of Concentrations in 1973/1974 and 1988/1989. The science of the total Environment, 2371(238), pp. 57-65.

Wahby, S.D., El Deeb, K.Z., 1980. A study of the state of pollution by petroleum hydrocarbons along the Alexandria Coast. Ves Journees Etud. Pollutions, Cagliari, C.*I.E.*S.M., pp. 257-262

Wang, Z., Fingas M, Shu, Y.Y., Sigouin L., Landriault, M., Lambert, P., Turpin, R., Campagna, P., Mullin, J., 1999. Quantitative characterization of PAHs in burn residue and soot samples and differentiation of pyrogenic PAHs from petrogenic PAHs—the 1994 mobile burn study. Environ Sci Technol, 33, pp. 3100–3109.

WHO, 1989. Polychlorinated dibenzo-para-dioxins and dibenzofurans. Environ. Health Criteria No. 88, International Programme on Chemical Safety, WHO, Geneva.

Yunker, M.B., Macdonald, R.W., Brewer, R., Mitchell, R.H., Goyette, D., Sylvester, S., 2002. PAHs in the Fraser River basin: a critical appraisal of PAH ratios as indicators of PAH source and composition. Organic Geochemistry, 33, pp. 489–515.

Chapter 15

The 2001 Fish Kill in Kuwait and the Outcomes Following the Phenomena

Abdullah Al-Zaidan
National Biodiversity Committee, Environment Public Authority, State of Kuwait
E-mail: alzaidan.abdullah@gmail.com

Fish Kill

What is Fish Kill?

The death of aquatic organisms in massive numbers within an aquatic ecosystem (Rivers, streams, lakes and seas ect.) by natural causes or human activities.

Fish kill occur naturally by Temperature fluctuation which lead to low oxygen levels, and can occur by the influence of human activities such as, coastal activities which include, economic activities, effluent or sewage water, Industrial waste, Dredging and land filling, Oil spills, aquaculture and desalination plants etc. Any human activity that leads to the alteration of the physical and chemical composition of the aquatic ecosystem.

Fish kill occur worldwide both in fresh and seawater ecosystems. Examples of neighboring countries are in Oman August 2000, which was said to be from pollution and harmful algal bloom but finally they concluded that the main cause was from Natural environment changes that reduced low oxygen surface waters. http://earthobservatory.nasa.gov/Study/oman/CV

Introduction

In August and September 2001 Kuwait Bay experienced a massive fish kill resulting in the death of nearly 3000 tons of Mullet. The 2001 fish kill had an ecological, economical and political effect. The ecological impacts include massive amounts of dead fish on the beaches with the disposal and unpleasant odors of dead and rotten fish. The loss of fish from the Fishkill is estimated to be roughly four years of Kuwait's annual consumption.

As the conditions worsened a ban was placed on fishing as well as the import, display and sale of all kinds of fresh fish, and Kuwait fish market was subsequently closed (Arab Times 5 Sept 2001). Fisherman around 4000 continually fish the waters of Kuwait and their loss of jobs caused tremendous hardship. The shrimp industry was also impacted and was shut down during the crisis. Another effect of economic value was that many gulf states banned the import of fish and shrimp from Kuwait making the recovery from the crisis even more difficult.

As the city and country of Kuwait that is largely dependant on the sea, the closures of the beaches for swimming and fishing resulted in intense fear to the public.

Kuwait Bay

Kuwait Bay is a moderately large (850 km^2) sub-tropical, semi-enclosed embayment located in the northwest region of the Arabian gulf.

Kuwait Bay is shallow with a maximum depth of 20 m and mean depth of <3 m (Anderlini *et al.*, 1982), this means that the bay and its bottom sediments receive sunlight year round (Sheppard 1993; Carpenter *et al.*, 1997). Mixing and circulation are largely wind driven (Lehr 1984; Aabo-Seida and Al-Sarawi 1990; Reynolds 1993), the net surface flow is counterclockwise at the head of the Arabian Gulf.

Kuwait Bay supports a thriving fishing industry, with a range of species typically harvested, and also is an important nursery ground for many families of such. Kuwait Bay is also the site of an aquaculture facility for the production of sea bream.

The fish farm owned and operated by Bubiyan Fisheries Co. is near Doha Port in waters of approximately 11 m depth. There are 80 cages and they produce around 500-600 tones per year. Cultivated Species include Sobiaty seabream (*Acanthopagus cuvieri*), Gilthead sea bream (*Sparus auratus*) and yellow-finned black sea bream (*A.latus*).

The shoreline of Kuwait Bay undergone extensive construction and development over the past twenty years. Construction projects include the Doha Marina, Green Island, and artificial beach enhancement near Kuwait Towers and other new projects.

Physical and Ambient Conditions

Sea temperatures in Kuwait Bay typically range from about 16°C to 31°C (Wright 1988,1989) but EPA Kuwait data suggest that record temperatures were reached in 2001. Calm conditions and low winds also persisted, leading to little circulation during this time. Additionally, salinities were higher than normal limits (F. Al-Yamani).

Some measures of water quality were also outside normal limits during the periods before and during the fish kill. One of these was PH, which reached as high as 8.5 in late July and early August, before the first fish kill were noted (F. Al-Yamani).

Such high levels are not uncommon during conditions of high productivity and high temperatures, and are produced *bt* draw-down of CO_2 by photosynthetic organisms. Levels of chlorophyll in the water column were also lower than normal during the period of the fish kill. There are several possible reasons for this. First, the elevation in the water temperature may have been sufficient to exceed the temperature limits of phytoplankton which typically proliferate. Second, the spread of the bacteria and its high levels may have conditioned the water such that phytoplankton growth was not favored over that of bacteria.

The Aquaculture Fish Kill

Water temperatures during early August were exceedingly warm even for Kuwait Bay exceeding 35°C. Alone this is stressful for fish for several reasons. Their metabolism may be functioning at or near their temperature limit. Furthermore, such temperatures reduce the saturation of oxygen so that even without a high bacterial level, oxygen saturation levels would be lower than normal.

The first fish kills were observed in the aquaculture cages. These began with a relatively slow rate of death, from just a few hundred dead fish per cage per day, and gradually increased in severity as the first week of August progressed. This fish kill was coincident with an outbreak of the dinoflagellate *Ceratium furca*, as observed by personnel from the EPA Kuwait. This species typically blooms in high abundance under conditions of high nutrients, including sewage. It is not thought to be toxic, and therefore is not likely kill the fish in the cages.

However, because of its high accumulation, it most likely resulted in the further reduction of oxygen, compounding the effects of oxygen saturation due to heat. *Ceratium furca* has been responsible for fish kills around the world as a result of low oxygen that develops following its accumulation. Due to the confined conditions within the cages, fish would not be able to escape the effects of low oxygen.

The numbers of sea bream that died in the fish kill in the cages in early August was not large in relation to the total harvest of these systems. There are no samples of fish from this period, so the presence of *Streptococcus* bacteria cannot be established. However, if these fish were carriers of the bacteria, their infection and subsequent death may have promoted, through cohabitation with mullet and the subsequent infection in mullet. Mullet are found in and near the pens, feeding on the added feed or the accumulation of organic matter that is found around such cages. As these are highly productive systems, fish have ample sources of food if they aggregate around the cages. The first mullet deaths were reported several days after the sea bream mortality was observed.

The Wild Mullet Fish Kill

Dead mullet were observed in increasing numbers from mid- through late August. The first reports of lesioned fish were from around 18 August 2001. By this stage, the *Ceratium furca* bloom in and around the net pens had apparently collapsed, and further fish kills within the pens were not reported. Overall, chlorophyll levels were considered lower than normal for this time of year by personnel from the Kuwait

Institute for Scientific Research (KISR), but dissolved oxygen levels continued to be lower than normal due to the continued warm temperatures of the water and the apparently high levels of bacteria, which would consume oxygen through their respiration. Low oxygen levels would be most apparent at night; midday measurements may not fully reflect the oxygen conditions to which the fish would be exposed.

By mid-August, mullet were observed to display a number of signs of disease that are characteristic of infection. Abnormal behavior included erratic swimming, swimming on the surface and "C"-shaped body curvature while at the surface. Reddening and hemorrhage were common findings in the integumental and muscoskeletal system. Lesions were noted at Bubiyan Fisheries, and at the Scientific Centre of Kuwait. Collections were made on 2-3 September, 2001. Frozen fish from 18 August, 2001, were also examined.

In order to fully diagnose what was causing the fish to die, 44 fish samples were analyzed. Brain, eye, head kidney and/or intestines were sampled from 39 mullet (mean length=151 millimeters), and 5 sea bream (mean length=266 millimeters). Also, 1 white shrimp was collected for microbiological analyses. Of these, seven mullet were collected without disease signs as controls. Lastly, an isolate of *Strepotococcus*, prepared by scientists from KISR was also analyzed and compared with *Strepotococcus* isolated from both fresh and frozen fish.

Identification of the Bacterium

Based on the analyses above, the isolates from both the mullet and the sea bream fish collected conformed to Group B *Streptococcus agalactiae.*

Streptococcal isolation from mullet and sea brean conformed to (Group B) *Streptococcus agalactiae* microbiologically using conventional phenotypic characteristics, API Rapid ID 32 strep, and Slidex strept grouping latex agglutination tests (Evans *et al.*, submitted manuscript). The phenotypic characteristics of the fish isolates displayed similar characteristics to the American Type Culture Collection (ATCC) reference strains of *S. agalactiae* and differed from *S. iniae, S. pyogenes, S. uberis,* and *S. zooepidemicus*. The fish isolates differed from ATCC type isolates in some tests depending on the methodology. Rapid API tests gave a positive Voges Proskauer reaction for fish and ATCC isolates, while conventional tests gave a negative reaction. Fish isolates were negative for tagatose fermentation by Rapid API tests, although ATCC isolates were positive for this compound. Fish isolates were positive for pululan, although ATCC isolates were negative. Considerable variability was noted in the fish isolates' fermentation reactions of many carbohydrates tested in the API system.

Sources of the Bacterial Infection

It may never be possible to determine with certainty the source of *S. agalactiae* to Kuwait Bay. However, it is worth exploring the various sources and their likelihood as sources of the disease.

There are four possible scenarios for the occurrence of this bacteria in the Bay. First, *Streptococcus* may have been present for a long period of time within the ecosystem of Kuwait Bay, but not in virulent form. It is known that this bacterium can be harbored within organisms, including humans, for years without any signs of disease. In support of this theory, the mullet sampled from the Scientific Centre of Kuwait and which had originally been collected in Kuwait Bay, showed no external signs of infection, nor did it yield any sign of infection upon testing with dry swabs, but did produce a positive test in the brain and in the eye upon further analysis. This might suggest that even fish that were not symptomatic were carriers. Once the environmental conditions were such that the bacteria could grow well, it was able to spread and cause infection. Reports of *Streptococcus* in fish in other Arabian Gulf states (Bahrain, Saudi Arabia) also lend credence to this possibility. However, this is the first report of the presence of *S. agalactiae* in the region; the previous reports suggest the presence of *S. iniae.*

Second, *Streptococcus* may have also been introduced from the net pen aquaculture system. There are many pathways by which this may have occurred. The initial fish stock itself may have been infected. Knowledge of the exact source of the stock, and further testing of that operation would be required to substantiate this possibility. In Greece, one of the source countries of the fingerlings for Kuwait, streptococcal infections have been common. In a related manner, the feed may also have been a source of infection. The possibility should also be considered that the bacterium was introduced by this mechanism, but at some point in the past. This may have led to the harboring of the bacterium in the sediments, and its release only upon the establishment of the right environmental conditions. With this possibility, even samples collected from the sources of the stock and feed might not indicate the presence of this bacteria.

The third possibility of the source of the infection is the sewage outfalls. Introduction of the bacteria by sewage is a possibility, and the high prevalence of infection of fish near the outfalls may support this. Alternatively, the high nutrients from the sewage plume may also provide an excellent broth for growth of the bacteria, so its association with outfalls may be related to the nutrient enrichment.

The fourth possibility is introduction via ballast water discharge. As with all the other potential sources, if this was the source, it might have been a single introduction, or it may have been a more continuous source.

Thus, identification of the source of the bacteria is extremely difficult. The source may have been via a single introduction, rather than a continuous source, either from sewage or from the aquaculture system, or from ballast exchange, but current testing of that source may not reveal the presence of the bacteria at this time. Alternatively, the fish may have been carriers of the bacteria for a long period of time. Furthermore, from a management perspective, identification of the source will not reduce the likelihood of future outbreaks. In the future, the entire Bay will be a source, as this microorganism has spread and will, with all probability, remain within the system. The key will be to reduce the conditions that promote its growth and virulence.

Previous and Current Program Following the Fishkill Event

Following the Fish kill event monitoring programs, legislations and research projects on behalf of the EPA with different organizations were carried out to help in the preservation and rehabilitation of Kuwait Bay.

The Environment Public Authority has a monthly sampling program that handles the collection and analysis of Water sample for measuring nutrients, Meiofuna (organisms that live interstitially and range between 0.5 mm to 0.063 mm) samples this sampling program is one of the most oldest (20 years) and deals with the collection of near the shore lines to measure the environmental stress and pollutants (if any) on the organisms. And finally a sampling program in 2000 was established following the 1999 fish kill for collection of phytoplankton samples to determine the presence of Harmful algal blooms (HAB) and there toxicity according to species. Also a Monitoring program with the Ministry of defense is organized for air survey during summer and out breaks to monitor the whole Bay.

In 1998 a legislation concerning the banned and movement of the Stable fishing traps (locally know as the Hathra) to the southern parts of Kuwait, this was implemented in 2001 after the fish kill,to reduce there impact from Kuwait bay knowing that in 2001 the first signs of stress were shown on the fish located with in the fish traps. Since the aggregation of large numbers of fish leads to a reduction in oxygen levels. Another legislation is under consideration that states the movement of the fish cages (Aquaculture) out form Kuwait Bay and be placed to the southern parts of Kuwait.

National committee in relation to the fish kill event was formed after the event 2001. It consists of representatives from the Environment public authority, Public Authority for Agriculture Affairs and Fisheries Resources, Kuwait Institute for Scientific research, Ministry of Defense, Ministry of Health etc. The committee specializes in the follow-up and determining the effect and sources of the fish kill phenomena.

Followed by the installment of the oceanographic Buoy system in January 2005 which are located at five stations within Kuwait Bay and help in measuring the following parameters. Wind speed and direction, Atmospheric temperature, Relative humidity, Atmospheric pressure, oil spill, dissolved oxygen, current velocity and direction, chlorophyll, solar radiation, water temperature, salinity, pH, heave, turbidity, trace metals and nutrients. The data afterwards is transmitted to an on-land computer where the data is analyzed.

A joint project from 2003 till 2004 was set up between the Environment Public authority (Kuwait) and The Centre of Environment, Fisheries and Aquaculture Science (United Kingdom) for the assessment of environmental quality of Kuwait Bay. This project helped form a clear understanding concerning the over all inland hazards and their effect biologically on the Biota in the bay especially FISH.

Finally a new project is established in 2006 with Japan External Trade organization which deals and is titled the rehabilitation of Kuwait Bay this was briefly discussed during the presentation.

Chapter 16

The Distribution of Heavy Metal Pb, Cr and Cd in Water, Sediment and Mud Crab, *Scylla serrata* of Donan River, Segara Anakan Central Java, Indonesia

Annisa Rahmawati Jumhan, Asrul Sahri Siregar, Edy Yuwono and Purnama Sukardi

Program Perikanan dan Kelautan, Komplek GOR Unsoed, Karangwangkal, Purwokerto 53123, Indonesia

ABSTRACT

Heavy metal is non-biodegradable pollutant material, which can be accumulated, toxic, and can persistent in aquatic environment. In aquatic environment, heavy metals are distributed in water, sediment and in organism. The aim of the study was to know heavy metals Pb, Cr and Cd distribution in water and sediment; to know their concentration in mud crabs (*Scylla serrata* Forskal).

A survey was applied using Group Random Sampling technique with three replications at one-week interval. This study was done in August until November 2005. Data was analyzed qualitatively using F-test.

The result showed that heavy metals distribution in water, for Pb undetected–0,0338 mg/L, for Cr 0,0032–0,0308 mg/L and for Cd 0,0015–0,0658 mg/L; heavy metals distribution in sediment, for Pb 2,7221–4,3580 mg/kg, for Cr 1,3495–4,7811 mg/kg, and for Cd 1,3845–6,8580 mg/L; heavy metals concentration in mud crabs, for Pb 0,0448–2,2365 mg/kg, for Cr 0,0262–2,2968 mg/kg and for Cd 0,0398–2,6611 mg/kg. It can be concluded that heavy metals Pb, Cr and Cd distribution in water and sediment on Donan River increased from upstream to downstream; the heavy metals concentration in mud crabs varying between its stations.

Introduction

Donan river is an estuarine within Segara Anakan Lagoon situated at 07°38′45″ LS and 108°01′45″ BT to 07°43′56″ LS and 109°59′18″ BT, ± 6.900 meter in length and ± 150 meter in width (Muhayat, 2002). In the end, Donan river, Kembang Kuning river and Sapuregel river meet in the junction of Nusakambangan strait (Ambarwati, 2002). The river used for many human activities including navigation for oil refinery and cement industry, fisheries and recreation. Settlement, aquaculture, industries and harbor are catchments area from which organic and inorganic, solid and liquid waste originally.

Extend and degree of water pollution, are increasing a rapid rate. Industries are dumping poison and complex toxic mixtures into natural bodies of water. Added to this is the ever-increasing use of other poisons by agriculture. Long term action of numerous water pollutants occurs in low concentration but which potentially dangerous on a chronic basis.

Heavy materials because non bio-degradable, can be accumulated, toxic, and persistent in aquatic environment. In aquatic environment, heavy metals are distributed in water, sediment and in organism.

The aimed of the study was to know heavy metals Pb, Cr and Cd distribution in water, sediment and mud crabs (*Scylla serrata* Forskal).

Method

A survey was applied using *Group Random Sampling* technique with three replications and one week interval. This study was done in August until November 2005. Data was analyzed descriptively. ANOVA was applied to compare means.

Sample was taken from four stations from which each site represent mangrove area, cement industry, oil refinery and harbor area.

Water sample was taken using water sampler (± 250 ml) and to embedded metal water pH was dropped to 1 or 2 added dilute HNO_3 (0.75 mL), then stored in the ice-box for analysis *Eckman grabb* was used to collect sediment sample (± 250 gram) for examination. Sample was stored in plastic bag, then stored in ice-box. A cage ("wadong") was used to trap mud crab from which meat was taken (± 250 gram) and then stored in ice-box for analysis. (Dolaria, 2004).

Table 16.1: Sampling Sites

Station	*Location*	*Coordinate*
I	Mangrove area	(07°39'355" LS–109°01'840" BT)
II	Cement industry area	(07°41'476" LS–108°59'578" BT)
III	Oil refinery area	(07°41'550" LS–108°59'653" BT)
IV	Harbour area	(07°43'500" LS–108°59'610" BT)

Atomic Absorbtion Spectro-meter with accuracy of 10^{-4} ppm was applied to determine Pb. Wave length was set 217 nm using current 3.5 mA. *Neutron Activation*

Analysis with accuracy of 10^{-4} ppb was used to measure cr and cd in water, sediment and mud crab (Susetyo, 1995).

Results

In the water, the concentration of lead in industrial and harbor area was significantly higher (P<0.05) compared to mangrove area. In the harbor and oil refinery area, however, the concentration was significantly higher (P<0.05) compared to cement industry area.

Table 16.2: The Average of Pb, Cr and Cd in Water in Each Station

Station	*Pb (mg/L)*		*Cr (mg/L)*		*Cd (mg/kg)*	
	Range	*Average*	*Range*	*Average*	*Range*	*Average*
I	*Ud–0,0039	0,0021[a]	0,0015–0,0026	0,0020[a]	0,0032–0,0057	0,0044[a]
II	0,0083–0,0095	0,0089[b]	0,0103–0,0203	0,0148[b]	0,0514–0,0652	0,0608[b]
III	0,0291–0,0338	0,0309[c]	0,0225–0,0301	0,0259[c]	0,0304–0,0402	0,0341[c]
IV	0,0248–0,0319	0,0284[c]	0,0281–0,0308	0,0297[c]	0,0125–0,0263	0,0184[ac]

*Ud: Undetected .

The concentration of chromates was also significantly higher (P<0.05) in harbor and industries area compared to mangrove area. However, concentration in oil refinery and harbor areas was significantly higher than that of cement factory area.

Water discharge around cement industry area was significantly highest cadmium concentration (P<0.05) compared to mangrove, oil refinery and harbor area. Harbor area was not significantly different compare to oil refinery and mangrove area, but oil refinery area was significantly higher than that of mangrove area.

Table 16.3: Pb, Cr, and Cd in Sedimen Taken Compare to US-EPA (1976)

Station	*Pb (mg/L)*		*Cr (mg/L)*		*Cd (mg/kg)*	
	Average	*Allowable Concentration (AC) US-EPA*	*Average*	*Allowable Concentration US-EPA*	*Average*	*Allowable Concentration US-EPA*
I	*2,9701*	8,0	*1,4722*	22,0	*1,8346*	6,0
II	*3,1105*		*2,5842*		**6,6152**	
III	*4,0246*		*3,2875*		*4,1102*	
IV	*3,6861*		*4,5183*		*5,3358*	

Keterangan: *Italic*: Under AC; **Bold**: Upper of AC.

The concentration of Pb, Cr and Cd in water in station I (mangrove area) was relatively lower compare to station II (cement industry area), III (oil refinery area) and IV (harbor area). It is far from industrial and harbor activities. Compare to Indonesian Maximum Allowable Level Act (MALA) PP. 82, 2001 of Water Quality for Aquaculture and Fishery, the concentration of Pb in station I and II was in the range of allowed concentration, however in the station III and IV where industrial sites operated, it was higher than the range permitted. Chromium was in the range allowable concentration in every station. Only in the station I cadmium concentration was in the range of allowable, whilst other stations were not allowable.

All of Pb and Cr were below the allowable concentration (AC) of US-EPA mean that concentration of these metals was safe around Donan river. In the station I, III and IV the concentration of cd was below threshold of US-EPA, however in the station II from which cement industry activity influenced the concentration of lead was high in the sediment. It indicated that a cement factory had a significant lead waste.

Table 16.4: Pb, Cr and Cd in Mud Crab, *Scylla serrata* Forskal Catched in Each Station

Stasiun	*Logam Pb (mg/kg)*		*Cr (mg/kg)*		*Cd (mg/kg)*	
	Range	*Average*	*Range*	*Average*	*Range*	*Average*
I	0,0448–0,0927	0,0612[a]	0,0262–0,0358	0,0294[a]	0,0398–0,0936	0,0701[a]
II	0,6413–0,9380	0,7901[b]	1,0326–1,0581	1,0431[b]	1,8597–2,6611	2,3651[b]
III	1,9332–2,2365	2,0858[c]	1,2991–2,0889	1,5732b[c]	1,0679–1,7432	1,3289[c]
IV	0,9381–1,5330	1,1861[b]	1,5582–2,2968	2,0416[c]	0,9852–1,5829	1,3712[c]

*: Values represent pooled means (± SD). Means in each column with different letters are significantly different ($p<0.05$).

The lead concentration in *Scylla* in the discharge of oil refinery area was significant higher ($P<0.05$) compared to mangrove, cement industry and harbor area, whilst cement and harbor area was significant higher ($P<0.05$) than that of mangrove area. Mangrove area was significantly lowest ($P<0.05$) compared to industrial and harbor area.

Scylla serrata caught in harbor and oil refinery area had the concentration of chromates significantly higher ($P<0.05$) compared to cement industry and mangrove areas. The lowest concentration of Cr was crabs caught in the mangrove area.

The highest concentration of cadmium in crabs caught was in the area of cement industry. It was significantly highest ($P<0.05$) compare to that in harbor, oil refinery and mangrove area and the lowest concentration of Cd was in crabs caught in mangrove area.

Discussion

Cement industry activity will produce waste include Cr, Cd, Pb and Hg (Alloway and Ayres, 1998). Water in mangrove area had lowest concentration of lead compare to industrial and harbor area. It seems to have capability to absorb metal materials.

Salami and Oktaviatun (2005), reported that metals were absorbed by sediment and biota, include plankton, alga, invertebrates and fish. *Sonneratia alba* has capability to absorb lead from zero to 2.8487 mg/kg Mulyadi (2006) and Norvianda (2006). The distributions of lead also influenced by distance where source of waste come from, the longest distance will the smallest concentration (Fardiaz, 1992). Lead poisoning in animals was known to be associated with cellular alteration of erythrocytes, including swollen mitochondria, simple and compound vacuoles cluster of ferritin in cells. Alterations of mitochondria have been ascribes to a blockage of heme synthesis (Wilber, 1971). The concentration Cr in the water around Donan River increased compared to previous research in March 2001 (Sudibyaningsih and Siregar, 2001). In the station II (cement industry area) concentration increased about 0.0001 mg/L, in station III (oil refinery area) about 0.0008 mg/L and in station IV about 0.0005 mg/L compared to present study.

Cadmium concentration of water in cement industry area was significantly higher ($P<0.05$) compared to oil refinery, harbor and mangrove area. In the harbor and oil refinery area was also significantly higher ($P<0.05$) compare to mangrove area. The lowest concentration was in mangrove area. The ion of Cd^{2+} is in the form of bioavailability means that it is ready absorbed by organism (Soegianto *et al.*, 2002) and after absorbed and accumulated in body of organisms then Cadmium will process biotransformation and bioaccumulation (Darmono, 1995 and Darmono, 2001). Cadmium and zinc act additively to produce their more toxic effect.

References

Alloway, B. J. 1990. *Heavy Metal in Soil*. John Wiley and Sons. Inc, New York. 339 hal

Ambarwati, D.T. 2002. *Kandungan Logam Berat Pb dan Cu Cacing Diopatra* sp. *sebagai Biomonitor Pencemaran di Sungai Donan Segara Anakan Cilacap*. Skripsi. Fakultas Biologi. Universitas Jenderal Soedirman, Purwokerto. 51 hal.

Darmono. 1995. Logam dalam Sistem Biologi Makhluk Hidup. UI Press, Jakarta. 140 hal.

Dolaria, N. 2004. Analisis Logam Berat Merkuri (Hg) dalam Air Laut, Sedimen, dan Biota di Laboratorium. *Buletin Teknik Litkayasa Akuakultur*, Vol. 3, No. 1: 39–42.

Fardiaz, S. 1992. Polusi Air dan Udara. Kanisius, Yogyakarta. 190 hal.

Muhayat, 2002) Muhayat, A. 2002. *Sebaran Larva Crustacea Planktonik di Sungai Donan Cilacap*. Skripsi. Fakultas Perikanan dan Ilmu Kelautan. Universitas Muhammadiyah Purwokerto, Purwokerto. 69 hal.

Mulyadi, H. A. 2006. *Kandungan Logam Berat Pb di Kawasan Perairan Sungai Donan Cilacap dan Pengaruhnya Terhadap Morfometrik Daun Sonneratia alba Jsm*. Skripsi. Program Sarjana Perikanan dan Kelautan. Universitas Jenderal, Purwokerto. 74 hal.

Norvianda. 2006. *Hubungan Kinerja Reproduktif Ikan Belanak (Mugil Dussumieri) dengan Kandungan Logam Pb di Perairan Sungai Donan, Cilacap*. Skripsi. Program Sarjana Perikanan dan Kelautan. Universitas Jenderal Soedirman, Purwokerto. 114 hal.

Salami, I. R. S dan Oktaviatun. 2005. Uptake dan Depurasi Logam Pb pada Ikan Nila. *Jurnal Kimia Lingkungan*, Vol. 6, No. 2: 75–81.

Soegianto, A; Nurtiati; S. I. Wardhani. 2002. Toksisitas Kadmium Terhadap Udang *Macrobrachium sitangese* de Man. (Crustacea: Decapoda) yang Dipelihara pada Medium dengan Salinitas Berbeda. *Jurnal Matematika dan Ilmu Pengetahuan Alam Universitas Airlangga*, Vol. VII, No. 1: 15–18.

Sudibyaningsih, Th dan A. S. Siregar. 2001. Pemanfaatan Kerang *Polymesoda* sp. dan Cacing *Diopatra* sp. Sebagai Bioindikator dalam Sistem Peringatan Dini di Perairan Segara Anakan Cilacap. Laporan Hasi Penelitian. DUE-Project Batch II. Universitas Jenderal Soedirman, Purwokerto. 28 hal.

Susetyo, W. 1995. Spektrometri Gamma dan Penerapannya dalam Analisis Pengaktifan Neutron. Gajah Mada University Press, Yogyakarta.172 hal.

Wilber, C.G. 1971. The Biological Aspects of Water Pollution. Pp. 296. Charles C. Thomas Publisher. USA.

Part III

General Coastal Management Issues Including Use of GIS and Remote Sensing Technologies

Chapter 17

Coastal Ecosystems Status in Bottom-Up vs. Top-Down Pressures

Luis O. Duarte
Laboratorio de Investigaciones Pesqueras Tropicales,
Universidad del Magdalena, Santa Marta, Colombia
E-mail: luis.o.duarte@gmail.com

ABSTRACT

Colombia is the only South American country having coast on both the Caribbean Sea and the Pacific Ocean. A total shoreline of nearly 3000 km includes the great majority of the tropical special habitats with high biodiversity, several goods and services exploited by humans, and vulnerabilities and conservation needs to be considered. The, coral reefs occupy nearly 1000 km^2 (77 per cent are situated in Caribbean oceanic islands); mangrove wetlands represent a total of 3790 km^2 (2927 in the Pacific coast); seagrass meadows coverage is estimated to be 433 km^2 (only developed in the Caribbean); lagoon and estuary systems occupy an extension of nearly 1650 km^2; beach systems comprise 1300 km of the Pacific shoreline and 1650 km of the Caribbean one; and no more than 15 per cent of the shoreline consist in rocky shores. From a functional point of view, impacts on coastal areas can be focused as bottom-up (*i.e.* natural) and top-down (*i.e.* anthropogenic) forces. Coastal ecosystems are deeply disturbed by combinations of bottom-up and top-down perturbations with consequences in the ecological, economic and social functions. This report analyses some cases of such kind of interacting forces in Colombia, examines the current capacity of the country to deal with natural and anthropogenic stressors and discuss the necessity of an integrated management approach to counter natural hazards and to avoid deleterious human actions in the coastal ecosystems. Overall, convergence of bottom-up and to-down factors is recognized in coastal ecosystems of Colombia; however the establishment of an adequate hazards prevention and rehabilitation policy is an ongoing task. Although there are some efforts for biological monitoring and international agreements for El Niño and tsunamis warning systems at operational level but still there is need for further studies for integrated programs for detailed data acquisition and modelling of ecosystem processes and feedbacks, as well as studies focusing on the ecological role of coastal ecosystems for suitable management measures.

Keywords: *Coastal ecosystems, Hazards, Natural, Anthropogenic, Ecosystem management, Caribbean Sea, Eastern Pacific Ocean, Colombia*

Introduction

The coastal seas are one of the most valuable and vulnerable of the Earths habitats. Humans use these areas as renewable and non-renewable resources (*e.g.* almost all of the world's fish catch coming from coastal waters including upwellings), and for recreation, waste disposal, power generation and transportation. In fact, coastal ecosystems were assigned a higher value than the whole terrestrial or open ocean system, despite their much smaller area (Constanza *et al.*, 1997). Colombia is the only South American country having coast on both the Caribbean Sea and the Pacific Ocean (Figure 17.1). A total shoreline of nearly 3000 km includes the majority of the tropical special habitats with high biodiversity (Díaz and Acero 2003), and exploited by the humans for several goods and services (Botero and Alvarez-León 2000; Marrugo-González *et al.*, 2000). We thus need to seriously consider vulnerabilities and conservation of the shoreline (Steer *et al.*, 1997).

Coral reefs occupy nearly 1000 km^2, 77 per cent are situated in the Caribbean oceanic Archipielago of San Andrés and Providencia, including two barrier reefs, five atolls and several small coral banks. Along the Caribbean coast small fringing reefs and patch reefs are found in the Tayrona Natural Park near Santa Marta and the Rosario and San Bernardo Islands Natural Park near Cartagena. In the Pacific coast, reefs are small and patchy mainly located in the Gorgona Island Natural Park and the Malpelo Island Natural Park (Díaz *et al.*, 2000).

Mangrove wetlands represent a total of 3790 km^2. In the Pacific coast, extensive mangrove forests occupy about 2927 km^2 and up to 30 k inland in some areas between Buenaventura and Tumaco. Cliffs in the northern region of the Pacific coast cause smaller forests fringing the mouths of the major rivers. Mangroves in the Caribbean coast are found in the deltas of major rivers and smaller forest are found in bays of the La Guajira Peninsula, the Tayrona Natural Park, the Rosario and San Bernardo Islands Natural Park and the Archipielago of San Andrés and Providencia (Marrugo-González *et al.*, 2000).

Seagrass meadows coverage is estimated to be 433 km^2, occurring only in the Caribbean coast since wide tide fluctuations avoid their development in the Pacific coast. Almost 80 per cent of the seagrasses are located in the shallow waters off the La Guajira Peninsula. Some beds occur in nearshore waters of the continental shelf and around the Archipielago of San Andrés and Providencia (Díaz *et al.*, 2003).

Lagoon and estuary systems occupy an extension of nearly 1650 km^2. Discharges of the Magdalena, Sinú and Atrato rivers shape the four major estuaries of the

Figure 17.1: Coastline and Marine Boundaries of Colombia

Caribbean coast. The Cienaga Grande de Santa Marta, located within the Magdalena River flood plain, is the largest of the 59 coastal lagoons distributed in the Colombian Caribbean. Several rivers flow into the Pacific Ocean, thus numerous estuaries are found in this region (Marín, 2004).

Beach systems comprise around 1300 km of the Pacific shoreline and 1650 km of the Caribbean one (Ceballos, 2002), and no more than 15 per cent of the total shorelines consist in rocky shores. Cliffs arise in the northern of the La Guajira Peninsula; in the Santa Marta area where the Sierra Nevada, the world's highest coastal peak reaching an altitude of 5,775 m above sea level just 42 km from the Caribbean coast, descend abruptly to the sea; in the southernmost area of the Caribbean coast; in the northern zone of the Pacific coast, and in the Gorgona and Malpelo Islands (López-Victoria, 2004).

Wind driven upwelling processes characterize coastal waters off the La Guajira Peninsula (Andrade and Barton, 2005) and off northern Pacific coast of Colombia (Rodriguez-Rubio and Stuardo, 2002). The enhanced biological productivity observed in these areas shows seasonal variability according with the climatic forcing.

This wide variety of coastal ecosystems in Colombia is impacted by a range of both natural and human perturbations. Some attempts to assess biological effects of such perturbations have been carried out, but usually focusing natural and anthropogenic factors separately and with short term and small scale perspectives. From a functional point of view, impacts on coastal areas can be focused as bottom-up (*i.e.* natural) and top-down (*i.e.* anthropogenic) forces, since coastal ecosystems are deeply disturbed by combinations of bottom-up and top-down perturbations with consequences in the ecological, economic and social functions. This work analyses some cases of such kind of interacting forces in Colombia, examines the current capacity of the country to deal with natural and anthropogenic stressors and discuss the necessity of an integrated management approach to counter natural hazards and to avoid deleterious human actions in the coastal ecosystems.

Natural and Anthropogenic Hazards

Several kinds of hazards have been recognized for the coastal ecosystems of Colombia. Table 17.1 shows a list of natural and anthropogenic pressures identified in the major coastal ecosystems in Colombia. Hazards for coral reefs are derived from Garzón-Ferreira *et al.* (2000); for mangrove wetlands the main bibliographic source is Ocampo-Aguirre (1997); Barrios and Rodriguez (2002) provide information of seagrass meadows; perturbations on lagoon and estuarine systems are reported by Marín (2004); hazards for beach systems are evaluated by Ceballos (2002). The Pacific region is occasionally affected by natural disturbances like El Niño events, tsunamis, and earthquakes. The Caribbean coast is less impacted (or indirectly impacted) by natural phenomena with exception of hurricanes threatening oceanic islands. Long term disturbances like global warming (sea level rise) or low frequency climate variability is expected to impact both the Caribbean and Pacific coasts. Human actions impact coastal zones, ranging high are ecosystems destruction, damage fishing practices, pollution and alterations in the hydrology. These concurring bottom-up and top-down stressors are exemplified by a selected set of documented cases in the Colombian coasts.

Case 1

The hydrology of the largest estuarine lagoonal complex in Colombia, known as Ciénaga Grande de Santa Marta, was altered by road works since the 1950's, by channelization and by increased sedimentation product of deforestation in the drainage basin of the Magdalena River (Botero and Mancera, 1996); reduced freshwater supply increased extremely the salinity of the lagoon, that is top-down or anthropogenic pressures. Climate variability (El Niño Southern Oscillation–ENSO cycle) was responsible for changes in rainfall patterns, streamflows and salinity increments of the lagoon (Blanco *et al.*, 2006), *i.e.* a bottom-up or natural perturbations. During El Niño events severe droughts are recorded in the northern Colombia which increases salinity levels of the Ciénaga Grande de Santa Marta; conversely during la Niña conditions high rainfall and high Magdalena River streamflow prevailed. The consequence is a profound environmental stress, reflected in mass fish kills and mangrove deaths. In addition, the efficiency of the hydraulic management program (reopening of channels connecting the lagoon with the Magdalena River) proposed by national and international entities is modulated by the flooding prompted by global climate variability.

Table 17.1: List of Natural and Anthropogenic Hazards Identified in the Major Coastal Ecosystems in the Caribbean and Pacific Coasts of Colombia

		Caribbean	*Pacific*
	Coral Reefs		
Natural	El Niño events		X
	Hurricanes	X	
	Global warming	X	X
	Bleaching	X	X
	Sea urchin mortality	X	
	Earthquakes and Tsunamis		X
	Harmful algal blooms		X
	Macroalgae proliferations	X	
Anthropogenic	Overfishing	X	X
	Coral extraction	X	X
	Fishing with dynamite	X	X
	Deforestation	X	X
	Channelization	X	X
	Pollution	X	
	Tourism and diving	X	X
	Increased sedimentation	X	X
	Mangrove Wetlands		
Natural	El Niño events	X	X
	Changes in rainfall patterns	X	X
	Hurricanes	X	
	Earthquakes and Tsunamis		X
	Erosion	X	X
Anthropogenic	Destruction and exploitation	X	X
	Alterations in the hydrology	X	
	Urbanization	X	X
	Wastewater disposal	X	X
	Agriculture	X	X

Contd...

Table 17.1–Contd...

		Caribbean	Pacific
	Seagrass Meadows		
Natural	River discharges	X	
	Epiphytes	X	
	Hurricanes	X	
Anthropogenic	Wastewater disposal	X	
	Tourism	X	
	Channelization	X	
	Waste disposal	X	
	Bottom trawl fishing	X	
	Fishing with dynamite	X	
	Lagoon and Estuary Systems		
Natural	Sedimentation	X	X
	Changes in rainfall patterns	X	
	Global warming	X	X
	El Niño events	X	X
	Earthquakes and Tsunamis		X
Anthropogenic	Alterations in the hydrology	X	
	Overfishing	X	X
	Wastewater disposal	X	X
	Waste disposal	X	X
	Channelization	X	
	Urbanization	X	X
	Agriculture	X	X
	Pollution	X	X
	Deforestation	X	X
	Beach Systems		
Natural	Sea level rise	X	X
	Earthquakes and Tsunamis		X
	Erosion	X	X
	Flooding	X	X
	Subsidence	X	
	El Niño events	X	X
Anthropogenic	Sand extraction	X	X
	Channelization	X	X
	Deficient engineering defences	X	
	Pollution	X	X
	Compaction	X	X
	Deforestation	X	X

Case 2

Historical reductions of coral reefs and seagrass meadows in the coastal areas influenced by the Magdalena River are greatly explained by the temporal variability of sediment loads and freshwater discharges (Restrepo *et al.*, 2006). An ENSO atmospheric teleconnection affect the hydrological processes of the Magdalena River promoting interannual variability of its runoff: increased streamflow during La Niña and reduced streamflow during El Niño events (Poveda and Mesa, 1997), while human actions like destruction of mangroves, river diverting and deforestation of drainage basin increase sediment flux (Restrepo *et al.*, 2006). Mixed and prolonged actions of these stressors contribute to the partial disappearance of corals and seagrasses.

Case 3

Low-frequency atmospheric signal (North Atlantic Oscillation–NAO) reaches the wind fields of the northern Colombian Caribbean Sea modulating the upwelling intensity off La Guajira Peninsula and related conditions for biological productivity. A study of the Caribbean sediments provides evidence of the known decadal mode of Atlantic variability in the tropical upwelling spanning the past eight centuries (Black *et al.*, 1999). Bottom trawl of the shrimp fishery established in such upwelling area since the late of the 1960's removes a huge range of biota and reduces habitat complexity (Duarte *et al.*, 2003; Duarte *et al.*, 2006). Combinations of these factors threat the structure and functioning of the ecosystem as is reflected in the sustained decline of marine resources biomass during the last 40 years (Garcia *et al.*, 2003). Mitigation strategies have been explored to reduce fishing impacts (Criales *et al.*, 2006), but climate variability should be included if management programs are to be effective.

Case 4

Local manifestations of shore erosion in the Caribbean coast of Colombia can be attributed to human impacts, associated to beach sand extraction, destruction of mangroves, deficient engineering defences, channelization and improper use of wadis; however the generalized phenomenon also suggest a probably relation with natural factors like sea level rise and neotectonic activity. In critical areas natural erosive trends are accelerated by anthropogenic actions, thus coordinating efforts are required to cope with littoral land losses (Correa and Vernette, 2004).

Case 5

Coral reefs of the Pacific coast of Colombia are affected by recurrent natural perturbations like El Niño events (Zapata and Vargas-Angel, 2003) and by catastrophic events like tsunamis (Rodriguez, 2003), but also are impacted by damage fishing practices and deforestation. As a consequence coral cover losses have been observed.

Discussion

Overall, convergence of bottom-up and top-down factors is recognized in coastal ecosystems of Colombia. Signals of ecosystem deterioration by natural and anthropogenic; local and global; temporal and permanent perturbations are detected; however, the establishment of an adequate hazards prevention and rehabilitation policy is an ongoing task.

There is a widespread consensus that marine ecosystems respond to climatic events (Durant *et al.*, 1998; Parsons and Lear, 2001; Kennedy *et al.*, 2002). However, biological consequences of climate variability on oceans are greatly unknown because limitations in spatial and temporal coverage of ecosystem research. Usually atmospheric and hydrographic processes are much better sampled than biological ones (McGowan *et al.*, 1998), thus, large, low-frequency changes are simply not well detected by short, small-scale studies (Duarte *et al.*, 1992). Similarly, ecosystems responses to anthropogenic actions must be assessed in the long term. Thus, it is necessary to document historical performance of ecosystems to define appropriated reference points and for identifying targets for rehabilitation measures avoiding the so-called shifting baseline syndrome, which, in a general form, can be defined as the tendency of each generation of scientist to accept as a baseline the ecosystem status occurred at the beginning of their careers and uses it to evaluate changes (Pauly, 1995). These considerations are particularly necessary in the Colombian coastal ecosystems. Marine science in general is a very young discipline in Colombia, only developed in the last 35 years (Diaz and Acero, 2003). Although there are some efforts for biological monitoring (*e.g.* Garzón-Ferreira *et al.*, 2002; Rodríguez-Ramírez and Garzón-Ferreira, 2003) and international agreements for the "Regional Study of the El Niño Phenomenon" (ERFEN are its initials in Spanish) and for tsunamis warning systems are operating in Colombia, integrated programs for detailed data acquisition and modelling of ecosystem processes and feedbacks, as well as studies focusing on the ecological role of coastal ecosystems are required for suitable management measures.

A system of natural parks is operating in Colombia as a conservation tool (PNNC, 2005) with relative success due to conflicts with local communities and few controls for the use of natural resources. A network of marine reserves to protect species, ecological structure, and functioning of costal ecosystems of human disturbances is under development.

The current capacity of the country to deal with natural and anthropogenic hazards is limited by the lack of comprehensive systems for climate and ecosystem monitoring in the coastal zones, including mitigation and rehabilitation schemes. The Colombian "National System for the Prevention and Attention to Disasters" is focused in the reduction of the catastrophic effects of natural and anthropogenic disasters in the human populations. Coastal ecosystems require a homologue system.

The institutional policy framework for coastal zone management recently developed in Colombia (Steer 1997) can be effective if ecosystem-based (integrated) management approaches and long-term assessments of coastal ecosystems are implemented. Urgent implementation of adequate management schemes is required to avoid further catastrophic damage. Mitigation and rehabilitation tools for hazards linked with climate variability and human activities are to be developed in order to reduce ecosystem alterations and protect human populations.

Acknowledgement

I am grateful to the Centre for Science and Technology of the Non-Aligned and Other Developing Countries (NAM S&T Centre), the Centre for Tropical Marine Ecology (ZMT–Zentrum für Marine Tropenökologie), and Jenderal Soedirman University (UNSOED) for the invitation to attend to the Workshop-cum-training course on coastal ecosystems: hazards management and rehabilitation. I thank the Columbian Institute for the Development of Science and Technology–COLCIENCIAS for the nomination to participate in the Workshop and Ivanna Haag at Internationalisation of ST&I Division of COLCIENCIAS for her assistance in the nomination process. This report was supported by COLCIENCIAS grant 1117-09-13723 to the Project "Valoración biológico-pesquera y ecológica de la pesca industrial de arrastre camaronero e impacto de la introducción de dispositivos reductores de fauna acompañante, en el Mar Caribe colombiano" and by Universidad del Magdalena, Universidad Nacional de Colombia, FAO and INCODER. I also acknowledge additional support by European Community grant 003739 to the project "Integrating multiple demands on coastal zones with emphasis on aquatic ecosystems and fisheries, INCOFISH" and by Deutscher Akademisher Austauschdienst–DAAD under a PhD scholarship.

References

Andrade, C.A., Barton, E.D., 2005. The Guajira Upwelling. Continental Shelf Research, 25(9), pp. 1003-1022.

Barrios, L.M., Rodríguez, A., 2002. Estado de las praderas de pastos marinos en Colombia. In: Informe del estado de los ambientes marinos y costeros en Colombia: año 2002. Invemar, Santa Marta, pp. 115-136.

Black, D.E., Peterson, L.C., Overpeck, J.T., Kaplan, A., Evans, M.N., Kashgarian, M., 1999. Eight centuries of North Atlantic Ocean Atmosphere Variability. Science 286, pp. 1709-1713.

Blanco, J.A., Vitoria, E.A., Narváez, J.C., 2006. ENSO and salinity changes in teh Ciénaga Grande de Santa Marta coastal lagoon system, Colombian Caribbean. Estuarine, Coastal and Shelf Science 66, pp. 157-167.

Botero, L., Alvarez-León, R., 2000. The Caribbean coast of Colombia. In: Sheppar C (Ed.) Seas at the Millenium, an Environmental Evaluation. Pergamon, Ámsterdam.

Botero, L., Mancera, J.E., 1996. Síntesis de los cambios de origen antrópico ocurridos en los últimos 40 años en la Ciénaga Grande de Santa Marta (Colombia). Revista de la Academia Colombiana de Ciencias Exactas, Físicas y Naturales 20(78), pp. 465-474.

Ceballos, C., 2002. Estado de las playas en Colombia. In: Informe del estado de los ambientes marinos y costeros en Colombia: año 2002. Invemar, Santa Marta, pp. 149-156.

Costanza, R., d'Arge, R., de Groot, R., Farber, S., Grasso, M., Hannon, B., Limburg, K., Naeem, S., O'Neill, R.V., Paruelo, J., Raskin, R.G., Suttony, P., van den Belt, M.,

1997. The value of the world's ecosystem services and natural capital. Nature, 387, pp. 253-260.

Correa, I., Vernette, G., 2004. Introducción al problema de la erosión litoral en Urabá (Sector Arboletes–Turbo) costa Caribe colombiana. Boletín de Investigaciones Marinas y Costeras 33, pp. 7-28.

Criales-Hernandez, M.I., Duarte, L.O., Garcia, C.B., Manjarres, L., 2006. Ecosystem impacts of the introduction of bycatch reduction devices in a tropical shrimp trawl fishery: insights through simulation. Fisheries Research 77(3): 333-342.

Díaz, J.M., Acero, A., 2003. Marine biodiversity in Colombia: Achievements, status of knoledge, and challenges. Gayana 67(2), pp. 261-274.

Díaz, J.M., Barrios, L.M., Cendales, M.H., Garzón-Ferreira, J., Geister, J., López-Victoria, M., Ospina, G.H., Parra-Velandia, F., Pinzon, J., Vargas-Angel, F., Zapata, F., Zea, S., 2000. Areas coralinas de Colombia. Publicación Especial 5, INVEMAR, Santa Marta.

Díaz, J.M., Barrios, L.M., Gómez-López D.I. (Eds.) 2003. Praderas de pastos marinos en Colombia: Estructura y distribución de un ecosistema estratégico. INVEMAR, Serie Publicaciones Especiales No. 10. Santa Marta.

Duarte, C.M., Cebrian, J., Marbá, N., 1992. Global change in the marine environment: the life-history of long-term monitoring programmes in Europe. Nature 356, pp. 190.

Duarte, L.O., García, C.B., Altamar, J., Manjarrés, L., 2003. Bottom trawl by-catch of the shrimp fishery in the upwelling area off Colombia, Caribbean Sea. Current status and historical insights. In: Manjarrés, *et al.* (Eds.) Dinámica espacio-temporal del ecosistema de afloramiento del área Bocas de Ceniza–Punta Espada (Caribe colombiano) y sus implicaiones para un regimen de pesca responsable. Technical Report. COLCIENCIAS, Universidad del Magdalena, INCODER, INPA, Universidad Nacional de Colombia. Santa Marta.

Duarte, L.O., Gómez-Canchong, P., Manjarrés, L., García, C.B., Escobar, F.D., Altamar, J., Viaña, J.E., Tejada, K., Sanchez, J., Cuello, F., 2006. Variabilidad circadiana de la tasa de captura y la estructura de tallas en camarones e ictiofauna acompañante en la pesquería de arrastre del Mar Caribe de Colombia. Investigaciones Marinas 34(1), pp. 23-42.

Durand, M-H., Cury, P., Mendelssohn, R., Roy, C., Bakun, A., Pauly, D., 1998. Global versus local changes in upwelling systems, Orstom, Paris.

García, C., Duarte, L.O., Altamar, J., 2003. Demersal fish density in teh upwelling ecosystem off Colômbia, Caribbean Sea: historic outlook. In: Manjarrés, *et al.* (Eds.) Dinámica espacio-temporal del ecosistema de afloramiento del área Bocas de Ceniza–Punta Espada (Caribe colombiano) y sus implicaiones para un regimen de pesca responsable. Technical Report. COLCIENCIAS, Universidad del Magdalena, INCODER, INPA, Universidad Nacional de Colombia. Santa Marta.

Garzón-Ferreira, J., Cortés, J., Croquer, A., Guzmán, H., Leao, Z., Rodríguez-Ramíres, A., 2000, Status of coral reefs in Southern Tropial America: Brazil, Colombia, Costa Rica, Panama and Venezuela. In: Wilkinson, C. (Ed.) Status of Coral Reefs of the World: 2000. Australian Institute of Marine Science, Cape Ferguson, pp. 331-348.

Garzón-Ferreira, J., Reyes-Nivia, M.C., Rodríeguez-Ramíres, A., 2002. Manual de métodos del SIMAC Sistema Nacional de Monitoreo de Arrecifes Coralinos en Colombia. INVEMAR, Santa Marta.

Kennedy, V.S., Twilley, R.R., Kleypas, J.A., Cowan, J.H., Hare, S.R., 2002. Coastal and marine ecosystems and Global Climate Change. Potential effects on U.S. Resources. Pew Centre on Global Climate Change, Arlington.

López-Victoria, M., 2004. Estado de los litorales rocosos en Colombia. In: Invemar. Informe del estado de los ambientes marinos y costeros en Colombia: año 2004. Invemar, Santa Marta, pp. 147-152.

Marín, B., 2004. Lagunas Costeras y Estuarios. En: Invemar. Informe del Estado de los Ambientes Marinos y Costeros en Colombia: año 2003. Invemar, Santa Marta, pp. 199-222.

Marrugo-González, A.J., Fernández-Maestre, R., Alm, A.A., 2000. The Pacific coast of Colombia. In: Sheppar C (Ed.) Seas at the Millenium, an Environmental Evaluation. Pergamon, Amsterdam.

McGowan, J.A., Cayan, D.R., Dorman, L.M., 1998. Climate-Ocean variability and Ecosystem Response in the Northeast Pacific. Science 281, pp. 210-217.

Ocampo-Aguirre, P., 1997. Evaluation of the mangrove resources in Colombia–their status, management and future. MSc Thesis, Centre for Tropical Coastal Management. University of Newcastle upon Tyne.

Parsons, L.S., Lear, W.H., 2001. Climate variability and marine ecosystem impacts: a North Atlantic perspective. Progress in Oceanography 49, pp. 167-188.

Pauly, D., 1995. Anecdotes and the shifting baseline syndrome of fisheries. Trends in Ecology and Evolution 10(10), pp. 430.

PNNC (Parques Nacionales Naturales de Colombia), 2005. Lineamientos para la coordinación del Sistema Nacional de Areas Protegidas. PNNC, Bogotá.

Poveda, G., Mesa, O.J., 1997. Feedbacks between hydrological processes in tropical South America and large-scale ocean-atmospheric phenomena. Journal of Climate 10, pp. 2690-2700.

Restrepo, J.D., Zapata, P., Díaz, J.M., Garzón-Ferreira, J., García, C.B., 2006. Fluvial fluxes into the Caribbean Sea and their impact on coastal ecosystems: The Magdalena River, Colombia. Global and Planetary Change 50, pp. 33-49.

Rodríguez, D.Y. (Ed.) 2003. Informe del estado de los ambientes marinos y costeros en Colombia: Año 2002. Serie de publicaciones periódicas No 8, INVEMAR, Medellín.

Rodríguez-Ramírez, A., Garzón-Ferreira, J., 2003. Monitoreo de arrecifes coralinos, pastos marinos y manglares en la bahía de Chengue (Caribe colombiano) 1993-1999. INVEMAR, Serie de Publicaciones Especiales No. 8, Santa Marta, pp 1-170.

Rodriguez-Rubio, E., Stuardo J., 2002. Variability of photosynthetic pigments in the Colombian Pacic Ocean and its relationship with the wind field using ADEOS-I data. Proceedings of the Indian Academy of Science (Earth and Planetary Science), 111(3), pp. 227-236.

Steer, R., Arias-Isaza, F., Ramos, A., Sierra-Correa, P., Alonso, D., Ocampo, P., 1997. Documento base para la elaboración de la Política Nacional de Ordenamiento Integrado de las Zonas Costeras Colombianas. Technical Report. Ministerio del Medio Ambiente. Serie publicaciones especiales No.6.

Zapata, F.A., Vargas-Angel, B., 2003. Corals and coral reefs of the Pacific coast of Colombia: In: Cortes J (Ed.) Coral reefs in Latin America. Elsevier, Amsterdam.

Chapter 18

Kenya Coastal Ecosystems: Sustainable Resource and Hazards Management

J.N. Kamau and J.M. Kazungu
Kenya Marine and Fisheries Research Institute, Mombasa, Kenya
E-mail: jkamau@kmfri.co.ke

ABSTRACT

The Kenyan coastline is about 600 kilometers in length and forms part of the western border of the Indian Ocean. Kenya is a signatory to the Arusha resolution on Integrated Coastal Area Management (ICAM) in Eastern Africa including the island states of 1993. The government established the Coastal development Authority in 1990 to be responsible for planning, initiating, coordinating and integrating development activities in the administrative coastal areas of Kenya including the Exclusive Economic Zone (EEZ).

Coral reefs exist along most of the Kenyan coast. They occur as coral flats, lagoons, reef platforms and as fringing reefs. The total area of coral reef in Kenyan waters is estimated at 50,000ha and the coral types include the families Portidae (*Porites* and *Goniopora* ssp.) and Faviidae (*Meandrina* and *Favia* species). There are four Marine National Parks in total–those at Malindi, Watamu, Kisite and Mombasa. Their total area is $54km^2$ and all are managed by the Kenya Wildlife Services. Fauna and flora in the Marine National Parks are fully protected and the introduction of species is prohibited.

The role of mangroves as sediment traps is observed when considering the ratios of water discharge to that of sediments, in the two main Kenyan Rivers. The Sabaki river discharges 2,000 million m^3 of freshwater and 2 million tonnes of sediment annually, while the Tana River discharges an average of 4,000 million m^3 of freshwater and 3 million tonnes of sediment annually. The ratios are 1:1 for Sabaki i.e for every 1 m^3 of water there is 1 kg of sediment and 4:3 for Tana, thus Tana river discharges less sediments, for every 1 m^3 of water there is 0.75 kg of sediment. The Tana system is 25 per cent more efficient in trapping sediments; this can be attributed to the presence of Mangrove forests at the Tana delta.

Kenya Marine and Fisheries Research Institute through community participatory forestry rehabilitation has overseen the reforestation of deforested mangrove areas at Gazi bay and Shimoni.

Kenya Marine and Fisheries Research Institute (KMFRI) has developed the Kenya Sensitivity Atlas and its GIS database, to provide the much needed support in effective decision process for the management of the Marine and Coastal Area of Kenya in the event of oil spill.

Introduction

The Kenyan coastline is about 600 kilometers in length and forms part of the western border of the Indian Ocean. Its most distinctive feature is the almost continuous fringing coral reef usually running parallel to the coast. Other features include the Lamu Archipelago with its extensive mangrove forests; the Tana River Kenya's longest, which discharges through a complex wetland system into Ungwana bay; the Sabaki river discharges just north of Malindi. The Tana river delta is Kenya's only major ocean delta, it is a low-lying area composed largely of sediments brought down by the river. It is subject to frequent flooding and changes in the network of channels and canals. The input of water is almost exclusively from the river itself because of the net outward flow of water. The delta maintains high levels of productivity in a dynamic balance that revolves around the frequency, extent and duration of flooding. Water circulation transports nutrients, influences a wide variety of habitat types, flushes away wastes, controls salinity and disperses and nurtures larval stages of a number of coastal organisms. Close to one million people inhabit the Kenyan coastal areas at an overall density of 100-200 persons/km^2 of theses about 400,000 live in Mombassa. Which is Kenya's major seaport and second largest urban area. The coast provides these people with food, shelter and employment. On land, the coastal corridor is important for its agricultural crops such as cassava, maize, cowpeas, rice, mango, banana, pineapple, cashew nut, coconut as well as sisal and cotton. Coastal forests, though not necessarily all truly coastal, (some extend beyond the strip along river valley), they are of great interest in view of their protective role against environmental degradation especially soil erosion. Coastal forests, therefore, play a big role in the conservation of other coastal ecosystems against siltation and erosion. However, centuries of human residence and interaction with the resource have reduced this role. The coastal forest communities of Kenya exist as isolated blocks comprising about 83,800 ha. These forests are however under heavy utilization pressure by surrounding communities which extract a variety of resources especially charcoal, timber and poles for income. Some communities obtain medicines, honey and mushrooms. Uses such as charcoal burning and logging maintain very high pressure, which threatens sustainability of the forest resources as well as the secondary values such as soil stabilization, which helps reduce soil erosion and sedimentation of coastal water resources. In the inter-tidal and sub-tidal areas, the coast provides finfish shellfish, marine algae and mangrove poles.

Status of Major Kenyan Rivers

The Tana River is the longest in Kenya being approximately 850 km in length and it has a catchment area of 95,000 km^2. It discharges an average of 4,000 million m^3

Figure 18.1: Map of Kenya Showing the Location of Marine Parks and Reserves Along the Coastline

of freshwater annually with peak flows occurring between April and June and a shorter high flow period during November/December. The river also discharges some 3 million tonnes of sediment per year. The Sabaki River is the second longest with length of 650 Km and a catchment area of 70,000 km^3 and discharges 2,000 million m^3 of freshwater and 2 million tonnes of sediment annually into southern Ungwana bay through the Sabaki estuary north of Malindi (UNEP, 1998).

The high sediment loads carried by the Tana and Sabaki Rivers are partly attributed to poor land use practices in the upper catchments that are important agricultural lands. Such a high rate of sediment discharge is threatening the sustainability of marine and coastal ecological biotopes such as mangroves sea grass meadows and coral reefs.

Groundwater Resources

The coastal region of Kenya has enormous potential in terms of ground-water resources. This is a result of its geological structure that promotes rapid infiltration and percolation of surface runoff to recharge. Hot water springs with temperatures ranging between 65^{0}C and 75^{0}C are found near Mkongani and Mwananyamala in Kwale district. However ground water quality varies depending on the depth of the borehole or well, nearness to the ocean and proximity to human settlements. Boreholes located near the sea have problems with seawater intrusion and this problem is exacerbated by over extraction. Boreholes and wells located in urban areas such as Mombasa have the added threat of pollution originating from pit latrines and septic tanks.

The Kenyan coastal zone has large deposits of quality sand for building. Building sand continues to be harvested with no control measures in place, mainly at Mazeras, the south coast and in areas around Ngomeni sites where the Sabaki changed course. Sand deposits are an important resource in enabling groundwater recharge and should be exploited sustainably. The Northern coastline is fringed with sand dunes some of which are recent in formation and migrate under prevailing wind conditions. A resent occurrence is the near burial of Mumbrui, a coastal tourist resort located in the North coast close to Malindi. The situation however reversed just in time avoiding the relocation of the tourist resort town.

Mangrove Resources

Mangrove ecosystems are important habitats for a variety of faunal species; they act as nursery grounds, providing shelter from predators and food to the juvenile species. They are rich in organic matter and can thus support diverse benthos communities. The mangrove benthic system is a closed system as reported by Kazungu J.M (1996). It thus appears that nitrogen introduced into the mangrove sedimentary compartment is mainly left for (I) recycling through an active bacterial production system (II) uptake by the mangrove root system and (III) accumulation as refractory nitrogen. These different observations indicate that mangrove ecosystems are probably not important exporters of organic nitrogen and dissolved inorganic nutrients to the coastal ecosystem. Instead they function as rather efficient traps and appear to efficiently recycle nitrogen in order to primarily satisfy nitrogen requirements of the mangrove trees themselves. As also reported by Silva *et al.* (1990) Mangrove forests are efficient barriers of heavy metal transport in tropical coastal areas. The above observations qualify mangrove forests as efficient buffer zones protecting other nearshore regimes from pollutants and sediment runoff, this is a very important function that tends to be assumed. The role of mangroves as sediment traps is observed when considering the ratios of water discharge to that of sediments, in the two main Kenyan Rivers. The Sabaki river discharges 2,000 million m^3 of freshwater and 2 million tonnes of sediment annually, while the Tana River discharges an average of 4,000 million m^3 of freshwater and 3 million tonnes of sediment annually. The ratios are 1:1 for Sabaki i.e for every 1 m^3 of water there is 1 kg of sediment and 4:3 for Tana, thus Tana river discharges less sediments, for every 1 m^3 of water there is 0.75 kg of sediment. The Tana system is 25 per cent more efficient in trapping sediments; this can be attributed to the presence of Mangrove forests at the Tana delta (3045 ha, Abuodha and Kairo, 2001) and the lack of mangroves at Sabaki estuary.

Effects of Mangrove Overexploitation

Clear-cutting of Mangroves causes hydrodynamic changes in inshore circulation that tend to increase erosion of shorelines. Losses of mangroves are assumed to affect fisheries and other faunal components of the food web but little documentation exists. The decline of fisheries resources in Lamu district, Kenya, has been associated to loss of mangroves (Tiensongrusmee, 1991). Degradation of mangrove forests not only depletes the resources available within it's boundary, but also affects the productivity of the adjacent waters.

Along the northern Kenyan coast, conversion of mangrove areas for pond culture and salt works is localized in Ngomeni. During the period 1921-1976, some 9922 ha of tidal swamps between Ngomeni and Kurawa were converted for salt works (Yap and Landoy, 1986). Six of the salt works are operational, producing an estimated 71 400 metric tonnes of raw salt. Such a vast amount of salt production has a harmful effect on the mangroves (Kigomo, 1991). The destruction of such a higher biological system, without intensive utilization of the converted area illustrates the lack of resource management within an ecological framework. In Lamu, close to 100 ha of mangrove forest was destroyed by dumping of dredged up sediment deposited during the construction of Mokowe sea jet (Abuodha and Kairo, 2001).

Mangroves develop ecosystems with adaptation to exist in hostile coastal river environment mainly by trapping terrestrial sediments, litter and nutrients (*e.g.* in deltas) to build platforms using their root systems, which then absorb generated nutrients for growth. The above mechanism cleans the water environment making it suitable for the establishment of other near-shore ecosystems such as sea-grass beds and coral reefs. Mangrove forests hence form nutrient rich environments, which promote a variety of food chains, and therefore function as feeding and nursery ground to many species of finfish, shellfish, prawns and crabs. The natural embankment formed by Mangrove forests act as self-regenerating barriers against coastal erosion, believed to be, in some ways, superior to manmade structures that have to be rebuilt if broken down. Construction of illegal structures along the coastline has resulted in coastal erosion and shoreline change.

Resource Management

Anthropogenic and Natural Stress on Coastal Ecosystems: Action Taken

The exploitation of mangrove ecosystems for poles and charcoal caused political concern. In 1975, the Government of Kenya imposed a ban on the use of mangrove poles for charcoal production. This was followed by a ban on export of mangrove poles in 1982, a step that seriously affected the economy of Lamu district. Despite the ban, the actual average harvest per year from Lamu for domestic use remained more or less equal to the 31,734 scores of mangrove poles extracted up until 1983 (Forest Department of Kenya, 1983).

Coastal ecosystems are exposed to both natural and human hazards; natural hazards with significant effect on coastal ecosystems have been experienced once in Kenya, this was during the El-Nino rains in 1997-98. These rains affected estuarine mangroves and more so those upstream; the major cause of death was sedimentation. Tons of sediments were deposited on the mangrove beds leading to suffocation on the arial roots. These large amounts of sediment influx affected also the nearshore ecosystems, some effect was observed on the seagrass beds but significantly affected were the coral beds causing widespread coral bleaching.

The important and heavily fished reef zone close to shore is particularly vulnerable to pollution and silting. Oil is a major pollution threat to coastal ecosystems; owing to the heavy use of the tanker route along the East African coast. On any given day there are hundreds of tankers in the Region, many of them Very Large Crude Carriers (VLCCs). Slicks are brought in from spills in the open ocean by coastal currents, while operational discharges from ships and refineries add to the load. Oil spill is one of the major causes of loss in coastal biodiversity, due to negligence or accidental.

Two major oil spills have been experienced in Kenya the most resent being that of 2005 where a single haul ship was damage while docking, spilling about 20,000 tons of light crude oil. The total area affected was approximately 234 ha, spreading for about 7.5 km long. Kenya lies within the international shipping route and is therefore at risk of offshore oil spill. It is in this connection that the government of Kenya through Kenya Marine and fisheries Research Institute embarked on the formation of an Environmental Sensitivity Atlas for the Coastal area of Kenya.

In recent decades, the growth of industry has brought an increasing volume of effluents to coastal waters. The use of agricultural chemicals has continued to grow, and sewage treatment continues to be inadequate in many parts of the region. Eastern Africa is also undergoing an extraordinary rate of urbanization. As the cities have become overcrowded, water supplies have proven insufficient, and systems for drainage, sewerage and refuse disposal inadequate. Domestic sewage is discharged directly into rivers and in some cases the sea. Although industrialization remains slow relative to other parts of the world, it takes place without proper environmental impact assessments legislative controls, leading to further pressure on the environment. Rivers, creeks and the sea have become dumping sites for industrial wastes. Industries of major environmental concern in the region include textiles, tanneries, paper and pulp mills, breweries, chemical factories, cement factories, sugar factories, fertilizer factories, and oil refineries.

A pilot project is underway to investigate the role played by mangroves in the phyto-remediation of domestic sewage. This is a regional project acronym PUMPSEA funded by the EU and comprising Kenya (KMFRI), Tanzania, Mozambique and other development partners; Sweden, Portugal, Italy, Denmark, Germany and Belgium. The ultimate product is a monitored wetland system that can be used for sewage treatment.

Some species of marine animals are already endangered as a result of human activities, particularly the dugong or manatee, which is often caught in fishing nets and drowned. Marine turtles continue to decrease in numbers as their eggs are poached and the adults are killed for their meat and decorative shells. Kenya led Africa with the establishment of the continent's first marine protected areas in 1968. These areas are primarily designed to conserve Kenya's coral reefs which run along most of the coastline and which form a biodiversity hotspot second only to the tropical rainforests. The larger protected areas also enclose important breeding sites for migratory marine birds, marine mammals and three species of turtle.

There are four Marine National Parks in total–those at Malindi, Watamu, Kisite and Mombasa. Their total area is 54 km^2 and all are managed by the Kenya Wildlife Services. Fauna and flora in the Marine National Parks are fully protected and the introduction of species is prohibited.

In addition there are five Marine National Reserve–Malindi, Watamu, Mpunguti, Mombasa and Kiunga. Their total area is 706 km^2 and they are administered by the Kenya Wildlife Services with traditional fishing being allowed within their boundaries.

Hazard Management Tools

Environmental Sensitivity Atlas for Coastal Area of Kenya

The problem of oil spill response can be complex if a decision has to be made in short timelines of less than one day as is often the case. The knowledge of the spread and distribution of natural resources along the Kenyan coast would be quite challenging to a command centre grappling with operational and political pressure from various sectarian interests all seeking satisfactory response during a spill incidence. In order to arrive at objective compromise decision, information must be available in a speedy manner.

The development of the atlas by Kenya Marine and Fisheries Research Institute (KMFRI) was an integral part of the National Marine Oil Spill Contingency Plan (NMOSCP). It provides an environmental data dictionary to be utilized as a tool in risk assessment, clean up prioritisation as well as in selection of appropriate methods and tools of response.

The GIS datasets include information on environmental parameters and socio-economics data along the coast of Kenya thereby providing vast opportunities for secondary use. Information has been used to generate hard copy and electronic maps showing the degree of sensitivity to oil spill.

This product is part of the Kenya Government's policy on poverty alleviation whereby the Government ensures it has contingency plans and resources in place to combat disasters such as forest fires, floods, drought and oil spills threatening to ruin the natural resource base which provides income to its citizens.

In addition the atlas represents the government of Kenya's efforts in fulfilling its obligations with respect to various conventions on prevention of pollution including the International Convention on the Prevention of Pollution by Ships (MARPOL 73/78), the safety of Life at Seas (SOLAS), the International Convention on Oil Pollution Preparedness and Response Cooperation (OPRC90), the Civil Liability Convention (CLC92) and the International Oil Pollution Compensation Fund (IOPC92).

Institutional Structures

Oil Spill Mutual Aid Group (OSMAG)

OSMAG is a voluntary association drawing membership from oil companies, the petroleum refinery, Kenya Ports Authority and other government agencies dealing with wildlife, maritime activities and environmental conservation. The Ministry of Transport in conjunction with OSMAG is involved in the preparation of a contingency plan for responding to oil spills and to oversee oil spill surveillance duties. These include Kenya Wildlife Services, Kenya Marine and Fisheries Research Institute and the Fisheries Department. The Kenya Ports Authority has the repository for oil spill response equipment.

Legislation and Regulations

The management of the coastal environment and resources in Kenya is governed by various legislation covering different sectors and issues. Similarly, no single policy document has been prepared specifically on conservation, development or management of the coastal zone. The National Environment Action Plan identifies 77 statutes relating to the management and conservation of the environment (Annex I), most of which apply to the marine coastal environment. In the last few years, several task forces have been formed to review some Kenyan laws, including some related to the management of the coastal zone. Several policy documents relating to the marine areas have recently been finalised or are under review. The new Environment Policy and Forestry Policy documents were finalised in 1996. The policy on Wildlife Conservation and Management is also being reviewed to take into account the conservation of biological diversity and representative indigenous ecosystems; the promotion of environmentally sustainable tourism and promotion of compatible land use in priority biodiversity areas, and channelling the benefits thereof to the local communities. The Wildlife Conservation and Management Act (Cap 376 of the Laws of Kenya) will be amended in line with the revised policy. According to the Policy Framework Paper 1996-1998, comprehensive environmental legislation will be presented to parliament for approval.

The Maritime Zones Act of 1989 consolidates the law relating to the territorial waters; provides for the establishment and delimitation of the Exclusive Economic Zone (EEZ) of Kenya; and provides for the exploration, exploitation, conservation and management of resources in the maritime zones. The Continental Shelf Act of

1975 vests the rights in the government in respect of the continental shelf, and the resources thereon, therein and thereunder. Whereas the holder of a title to land is free to use resources on the land, the "..water or every body of water.... is vested in the government" under the Water Act. Similarly, all unextracted minerals, other than common minerals (*e.g.* clay, murram, limestone) are vested in the government by the Mining Act (Cap 306 of the Laws of Kenya). Sand extraction is governed by local government by-laws. Land ownership and land use is a particularly touchy subject in the coastal region because of the relatively large number of squatters (people staying in land which they do not legally own). In the Policy Framework Paper, the government recognizes this and indicates that a Land Use Commission will be created to address land tenure and land use policy issues. Other issues to be addressed in proposed land legislation is the ownerships/allocation of land created by recession of the sea.

The Fisheries Act (Cap 378 of the Laws of Kenya) provides for the development, management, exploitation, utilisation and conservation of fisheries and for connected purposes. Under the Act, the Fisheries Department, in collaboration with other appropriate agencies and other government departments shall promote the development of traditional and industrial fisheries, fish culture and related industries; and may impose measures necessary for the proper management of any fishery.

The Act also declares Kenya fishery waters a pollution prevention zone for the purposes of protecting the aquatic environment and ecology. Regulations on pollution and its control are spread in several Acts, with different enforcing agencies. These include the Factories Act, the Public Health Act, Merchant Shipping Act and the Local Government Act. The Environmental legislation currently being prepared will consolidate statutes on pollution of the environment, especially the coastal environment.

Kenya is party to several international conventions including the Law of the Sea Convention, Framework Convention on Climate Change; Biodiversity Convention; Convention for the Protection; Management and Development of the Coastal Environment of the Eastern African Region (Nairobi Convention); and the Ramsar Convention. Kenya is the depository for the Nairobi Convention.

The laws relating to the coastal areas have not been adequately enforced by the relevant authorised institutions, owing to a number of reasons such as poor or weak administrative structures; absence of provisions to specify standards of performance; inadequate deference and inadequate incentives; generally low levels of active and participative awareness among the majority of the population; preference for short term gains at the expense of more sustainable alternatives in policy making and planning; gaps and overlaps in institutional responsibilities making enforcement difficult; and poverty which promotes unsuitable use of resources. Some of the enforcing agencies do not have the necessary resources to adequately enforce regulations and follow-up on complaints and violations.

Early Natural Hazard Warning Tools

Tide Gauge

Kenya Marine and Fisheries Research Institute (KMFRI) in collaboration with the University of Hawaii Sea Level Centre have installed two tide gauges, one located at Mombasa Port area and the other at Lamu about 200 km north of Mombasa. The tide gauges were commissioned in 1998 and have been operational to date. Data obtained is recorded electronically and monitored for indications of sea level rise. The data obtained from the two stations is feed to a model for the processing of tide tables used for navigation and research purposes. Historical data can be used to monitor sea level rise as a management tool to regulate coastal infrastructure rules.

The reference port for tidal observations in Kenya is Kilindini (Port Mombasa) where the maximum tidal range does not usually exceed 3.8 m but may occasionally go beyond this. Tidal range for Malindi is 2.0 m for neap tide and 2.9 m for spring tide. There is a lag in the tidal state which increases with distance north along the Kenyan coast. Malindi is normally 5 minutes after Kilindini while Lamu is about 40 minutes behind.

Western Indian Ocean Satellite Application Project (WIOSAP)

A one year pilot project was initiated by Kenya Marine and Fisheries Research Institute (KMFRI) in August 2004 to develop useful dynamic Meteorological Second Generation (MSG) data products and to broadcast (disseminate) them to the industrial fishing community. One of the specific objectives within the project was to develop a front following algorithm for the detection and tracking of thermal fronts in collaboration with the National Oceanographic Centre, Southampton. Such information is important as an early warning tool.

References

Abuodha, P. A. W, Kairo, J. G. 2001. Human-induced stresses on mangrove swamps along. the Kenyan coast. Hydrobiologia, 458: 255-265

Carlos Augosto, R. S., Lacerda, L. D and Rezende, C. E. 1990. Metals Reservoir in a Red Mangrove forest. Biotropica 22: 339-345.

Forest Department, Kenya. 1983. Operational Cruise on Lamu Mangroves. Nairobi. Cited in: FAO. 1991. Management and Conservation of Mangrove Forest in Kenya.

Kazungu, J. M. 1996. Nitrogen-transformation processes in a tropical mangrove ecosystem (Gazi bay, Kenya). Ph.D. thesis. Vrije Universiteit Brussel, Brussels,

Kigomo, B.N. 1991. Forest ecosystems, rare germplasm, cultural sites and conservation efforts under development. Report No.4 ECE Group, KFMP, Forest Department, Nairobi.

Silva, C. A. R, Lacerda LD and Rezende CE 1990. Metals reservoir in a red mangrove forest. Biotropica 22(4): 339-345.

Tiensongrusmee, B. 1991. *Mangrove conservation and management*. Washington, DC. World Bank.

UNEP, 1998. East African Atlas of Coastal Resources: Kenya.

Yap, WG and Landoy, RJ: 1986, Report on survey of the Coastal Area of Kenya for Shrimp Farm. Development, Rome, FAO.

Annex I
Statutes Relating to Coastal Zone and Enforcement Agencies

Issue/Concern	*Legislation*	*Enforcement/Implementation*
Security	Maritime Zones Act Continental Shelf Act	Kenya Navy
Land Tenure	Government Land Act Land Titles Act Registration of Titles Act Land (Group Representatives Act) Trust Land Act Mazrui Land Trust Act Registered Land Act Land Planning Act	Commissioner of Lands Local Government Authorities Physical Planning Department
Water Use and Conservation	Water Act National Water Conservation Pipeline Corporation Act	Ministry of Land Reclamation, Regional and Water Development/NWCPC
Environment and Conservation, including pollution	Chiefs Act Local Government Act Kenya Ports Authority Act Public Health Act Factories Act Fisheries Act Wildlife Management and Conservation Act National Museums Act Petroleum Act	Provincial Administration Municipal, Town, County Kenya Ports Authority Ministry of Health Ministry of Labour Fisheries Deapartment KWS, Ministry of Tourism and Wildlife NMK
Tourism	Tourism Industry Act Tourism Development Corporation Act	Ministry of Tourism and Wildlife/KTDC
Industrial Development	Coast Development Authority Act Factories Act Export Processing Zones Act Land Planning Act Town Planning Act	CDA Ministry of Labour EPZA Physical Planning Department Municipal, Town, County
Shipping	Kenya Pots Authority Act Merchant Shipping Act Carriage of Goods at Sea Act	KPA/Ministry of Transport and Communications
Agriculture	Agriculture Act Crop Production and Livestock Development Act Plant Protection Act Seeds and Plants Varieties Act Coconut Preservation Act Irrigation Act Pests Control Act	Ministry of Agriculture and Livestock Development
Forestry	Forestry Act	Ministry of Environment and Natural Resources
Research	Science and Technology Act	Ministry of Research Technical Training and Technology/ Research Institutes
Fisheries	Fisheries Act	Fisheries Department

Chapter 19

The Role of Remote Sensing and Geographic Information Systems (GIS) in Hazard Prediction and Management of the Coastal Mangroves of Kenya

Ogalleh Sarah Ayeri
Department of Environmental Science, Egerton University, Kenya
E-mail: sarahayeri@yahoo.com

ABSTRACT

The mangroves ecosystems are subjected to intense land-sea interaction as a result of human activities. Human activities that affect the coastal zones of Kenya include fishing, aquaculture, agriculture, logging, tourism and recreation. These activities threaten the mangrove forests' habitats and they impair their ecological functions. Efforts are needed to create better understanding of the structure and functioning of coastal systems. This will include appreciation of their complexity and interactions, their response to natural and anthropogenic pressures and devising ways to restore their functional capacity in order to provide services and products. Remote Sensing (RS) techniques and Geographical Information Systems (GIS) are important tools, which can be used in both hazard prediction and management of mangrove ecosystems.

Although large amount of information on the subject exists, it is widely dispersed across the countries among various organizations, especially research institutions. Moreover, some of the information is not accessible or available in a readily usable electronic form, yet individuals and organizations involved in mangrove forests management are in dire need of such information for making fundamental decisions. Kenya, therefore, urgently needs comprehensive quality information on the distribution, status, utilization and threats of mangrove forests for efficient hazard prediction, planning and management. Remote Sensing (RS) and Geographic Information Systems (GIS) technology will provide a platform, which will make it possible for electronic management of data on mangroves for utilization by forest managers, policy makers and would be users.

This paper assesses the potential role of RS and GIS in integrating, analyzing data on mangroves, predicting and documenting all forms of mangroves forests information for sustainable management of mangrove forests in Kenya.

Keywords: *GIS, Remote Sensing (RS), Mangrove forests, Hazard, Management, Conservation.*

Introduction

Mangroves are high-value aquatic resources expected to deliver multiple uses to the public, including the simultaneous provision of recreational opportunities and of "ecosystem" services or functions, as well as maintenance of regional biodiversity and the possibility of income-making opportunities for local communities.

At the same time, the mangroves are commonly subjected to multiple social, economic and ecological threats, commonly being exposed to rising salinity, altered water regime and the introduction of exotic species. Human activities that affect the coastal zones of Kenya include fishing, aquaculture, agriculture, logging, tourism and recreation. In the past, managers have had to manage such aquatic resources without general protocols to guide them. In particular, the significance of interactions across multiple threats is poorly understood, as is the appropriate spatial and temporal scale on which to base management decisions. Data visualization uses maps, graphs and statistics to process and facilitate understanding of the enormous amount of data that can be derived from analysis of mangrove forests utilization. Therefore GIS software becomes of paramount importance. Mironga, (2004) reports that GIS software is being underutilized in Kenya, hence the need to enhance its utilization.

This review paper seeks to address the multiple socio-economic and ecological threats, how they compound each other and produce impacts that can be analyzed, interpreted and predicted based on implementation of the remote sensing and geographical information systems. This paper has utilized two case studies that show both the application and the benefits of the RS and GIS technology, which can be emulated and applied in the management of the mangrove ecosystems in Kenya. Figure 19.1 gives the conceptual framework of the application of RS and GIS.

The recommendations given can be used by the policy makers, decision makers, development agents and the local communities to manage their natural resources.

The Mangroves

One of the most biologically diverse forests is the mangrove forest. They are known as the "rainforests by the sea". Mangroves are woody, specialized types of trees of the tropics that can live on the edge, where rainforests meet oceans. Mangrove trees look as if they grow on stilts. The stilts are their specialized aerial roots, which hold the trunk and leaves above the water line. The rising and falling of the ocean's tides affect mangrove forests. Mangroves have very specialized adaptations that

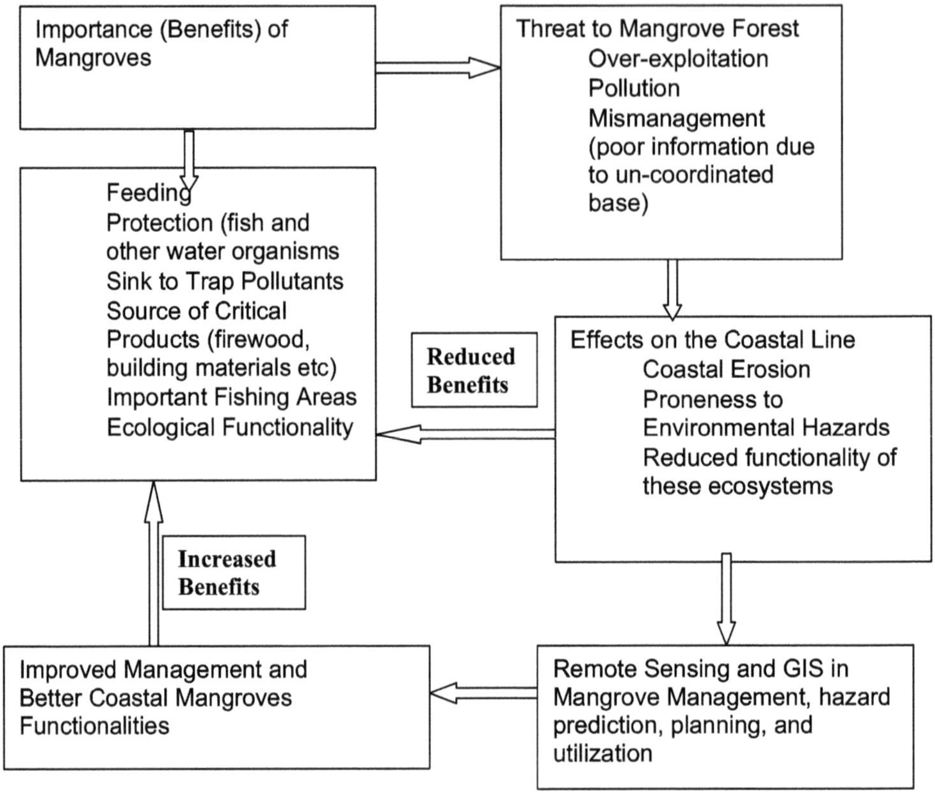

Figure 19.1: Conceptual Framework

enable them to live in salty waters. Breathing roots allow them to survive in anaerobic sediments. Buttresses and aboveground roots enable them to grow in unstable mud flats. Their foliage removes excess salt from the sap, and they conserve water to cope with periods of high salinity. Their seeds are buoyant to allow them to disperse and establish themselves in new areas. Figure 19.2 depicts mangroves of the Kenya coast.

Importance of Mangroves

The mangrove forests are the breeding grounds for fish, shrimp, prawns, crabs, shellfish and snails.

They protect the coastline and prevent erosion by collecting sediment from the rivers and streams and slowing downs the flow of water. Mangroves act as sinks, which concentrate pollutants like sewage, toxic minerals and pesticides and herbicides.

Threat to Mangrove Forests

Today mangrove forests are one of the most threatened habitats in the world. Mangrove roots are very susceptible to pollutants like crude oil clogging their lenticels, and continual flooding from artificial dikes and sea walls. Mangroves act as sinks,

Figure 19.2: Mangroves at the Kenyan Coast

which concentrate pollutants like sewage, toxic minerals and pesticides and herbicides. Over time the stress of the pollutants and reduced light kill large areas of mangroves forests. Mangrove wood also makes a superior kind of charcoal and many trees are being cut down to sustain local charcoal and timber industries. Mangrove forests are also being filled in for developments and as a form of mosquito control.

All these activities impair the ecological functioning of the mangroves and these impairment leads to proneness to disasters and hazards of the coastal areas that further deteriorate the mangroves ecosystems and its environs.

Role of Remote Sensing in Mangrove Management

Remote Sensing is the science and art of obtaining information about an object, area, or phenomenon through analysis of data acquired by a device that is not in contact with the object, area, or phenomenon under investigation. The most frequent data used in this discipline are aerial photographs and satellite images. While Remote sensing technology provides tools for the analysis of these images and the understanding of underlying spatial patterns and processes, GIS can easily integrate raw or processed images its environment (CDE, 2006). Therefore, the RS will be a crucial source of data to be utilized by the GIS software.

Role of GIS and Mangrove Data

An aspect of nature conservation that deserves special attention in the context of GIS analysis, measurement, planning and management is related to forests. GIS can play a key role in the management of the mangrove forests in terms of stock taking.

GIS stores both the geographic and numerical structure of the forest stands and links that spatial database to the planning models. It allows the manager to effectively add both the important temporal and spatial dimensions to the management planning process. Within the limits of the inventory and model, the manager (decision makers, conservationist, users) can then map what the forest will look like five, 10, 25, or 100 years in the future (Arc News, 2003-2004). Therefore, GIS will play an important role as a tool for environmental management particularly with current concern for sustainable use of resources particularly mangrove forests, conservation and monitoring of these ecosystems.

Case Studies Using GIS

Example 1: Case Study of the East Mau Forest in Kenya

Mau Forest complex is the largest forest system in Kenya, representing over 25 per cent of Kenya's forest reserves (Akotsi and Gachanja, 2004). It covers over 290,000 hectares and is one of the largest remaining blocks of mountain forests in East Africa. Over the years, degradation of Mau forest has occurred due to a number of factors including:

1. Human settlements in the Eastern Mau Escarpment leading to an increased population
2. Indiscriminate excision of the Mau forest (Kenya Forests Working Group, 2001); and
3. Change in land use

The displaying potential of the GIS (Figure 19.3) enabled the mapping out on the Mau forest complex that allows for easy interpretation of the map. For example, from the map, the Elburgon and Njoro areas have undergone high deforestation compared to other areas. Policy makers and decision makers can therefore take advantage to read and make clear-cut decisions on how to conserve the forest based on the map. Also, future prediction of the forest cover over the years especially when no action is taken to curb deforestation, can easily be done based on the map.

The GIS technology can be employed therefore, to map out and display mangrove forest areas based on the land use (logging, fishing, pollution etc), the rate of destruction, and impacts of the land use on the mangrove forests as well as the threatened areas of the mangrove forests in Kenya. Such information can be utilized to predict associated impacts of a particular hazard and appropriate measures taken to arrest or prevent the disaster from occurring.

Example 2: Case Study of Tropical Forest in East Kalimantan, Indonesia

The Tropical Forest in East Kalimantan, Indonesia is characterized by high botanical diversity of the tree species. Large parts of natural forest in East Kalimantan have been logged. An organization in Indonesia, known as Lembaga Ekolabel Indonesia (LEI) developed some ecological criteria and indicators through a study. The objective of the study was to demonstrate how remotely sensed data could support sustainable forest management by verifying some ecological criteria and indicators for tropical timber certification. Within the framework of the study, one ecological criterion, two indicators and five verifiers were selected from a list already developed by LEI. Specifically, the study aimed at using optical satellite remotely sensed images (*i.e.* Landsat-TM data) to detect deforestation in protected area in comparison to the

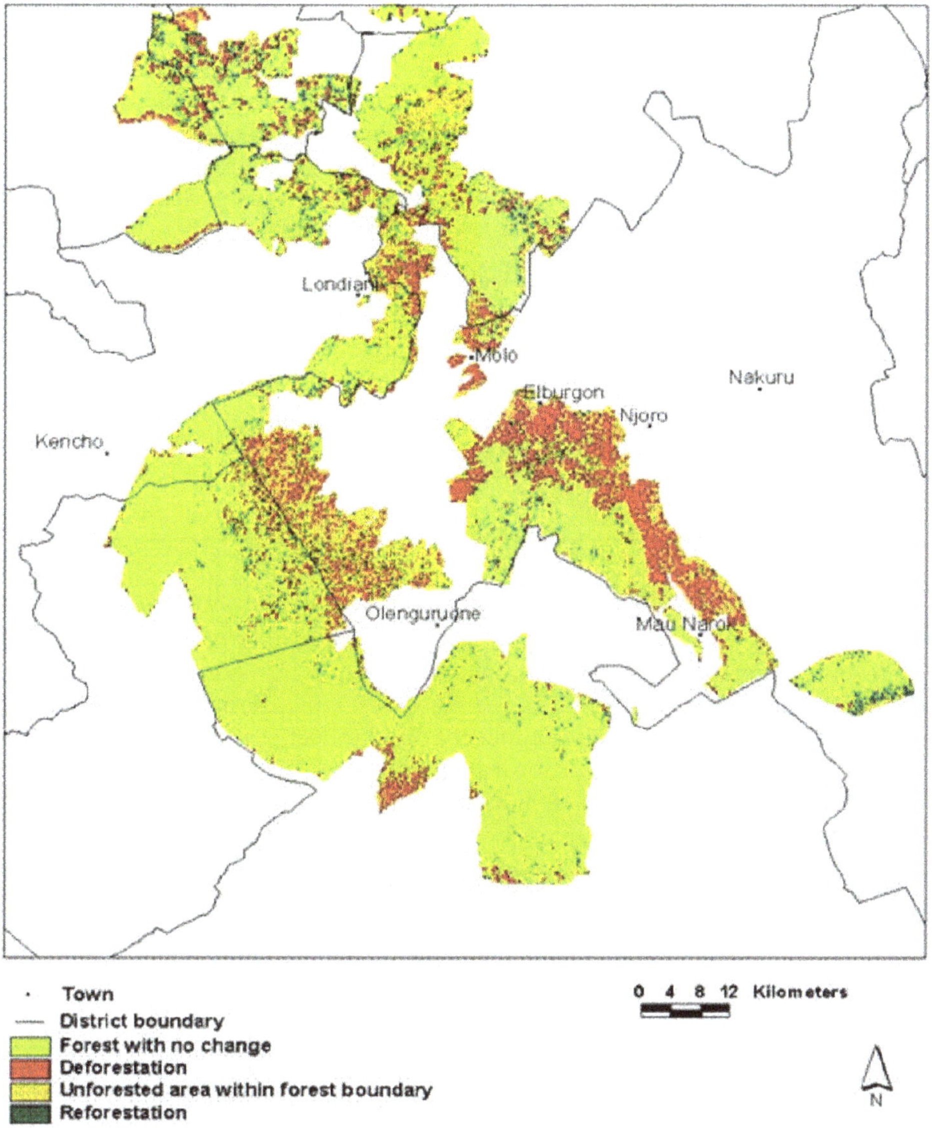

Figure 19.3: Mau Forest Cover Changes between 1995 to 2003
***Source*: Inter-Ministerial Task Force, Government of Kenya (2005)**

overall forestland in the study site (Aguma and Hussin, 2004). Figure 19.4 shows the protected area around the river in 2000.

From (Table 19.1), forest encroachment increased drastically from 1996 to the year 2000 in the Lebanon concession. Using the GIS technology, the information was used to map encroachment along the river. Increase in colour (red) on the map in (Figure 19.4) confirms thorough encroachment that could culminate into further forest destruction. This therefore serves as a warning to all forest stakeholders. Such information can be used in mapping out disturbed and undisturbed areas of mangrove forests. This can then be used in the prediction of any consequences in case no action is taken to reverse the negative trend. Also the information can be used to predict the hazards likely to take place and the appropriate measures taken to prevent such disasters along the coasts line. Therefore, the GIS can be of great help to the policy makers, the decision makers and most important, the forest users in that appropriate measures can be taken by simply studying the GIS generated map.

Table 19.1: Showing the Size of Encroached Area in 1996 and 2000 Within the River Protected Area

Year	*Encroached Area (Ha)*	*Total River Protected Area*	*Area Encroached Expressed as a Percentage of Total Protected Area*
1996	22	1144	1.92 per cent
2000	86	1144	7.52 per cent

GIS Implementation for Mangrove Forests and Hazard Management

Hazards are a serious disruption of the functioning of a society, causing widespread human, material or environmental losses, which exceed the ability of the affected society to cope using only its own resources. While the occurrence of disasters triggered by natural hazards, the number of people affected and economic losses are rising in Kenya. The disasters on the increase in Kenya include Epidemics, drought and floods worst disasters in Kenya. Hence the need to develop strategies to arrest these hazards.

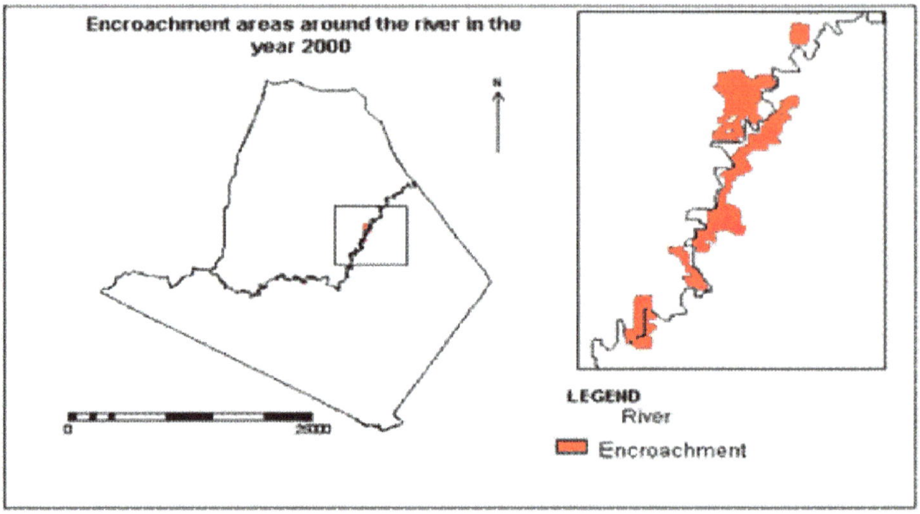

Figure 19.4: Map Showing Encroachment Areas Around the River-protected Areas in Lebanon Concession in the Year 2000

The implementation of GIS for hazard management of forests requires coordination of software, hardware, data, personnel and an overall information management strategy. Software and hardware are generally available for many research organizations in Kenya and thus not an issue, (Mironga, 2004) although cost implications cannot be underrated. The use of GIS for mangrove forests management no longer requires expensive high-end computer hardware; the modern desktop computer can manipulate and analyze these data within an acceptable time period. The modern desktop computer now has a hard drive space over 80 Gbytes and can easily store a wide range of data sets for GIS analysis. However, the price of GIS software may still be a financial barrier for many researchers, but not for the organizations. The forest data and skilled personnel are limiting progress with forest management.

Once the GIS technology is implemented, it will allow for risk assessment and analysis for the mangrove forests hence the GIS technology will be useful is assessing the risks associated with the destruction of these forests, for example over-harvesting of the mangrove forests will lead to the impairment of the ecological function of mangroves against protection of the coastline against coastal erosion and flooding.

Once the risk assessment has been developed, an awareness of the risks associated with the destruction will ensue. Such awareness will lead to the development of an early warning system that is likely to alert the people of any forthcoming expected disasters/hazards against the coastal lines and the accompanying consequences to the humans and their environment. This will lead to hazard management strategies that are supposed to protect and guard the coastal systems from any natural hazards, and hence protect the coastline against any hazards.

Participatory GIS can be incorporated particularly to enable mapping of the mangrove forest areas susceptible to risk of any hazard such as flooding and subsequent, vulnerable populations etc.

Barriers to Application of GIS in Kenya

Use of GIS in Kenya for forest planning, management and conservation is still exploratory and adhoc in nature, although there are signs of growing GIS application in specific areas such as ecological monitoring, wildlife movement, tourism, surveys and mapping (Mironga, 2004).

There are considerable barriers to be dealt with if GIS is to be fully applied in the management of indigenous forests. Major issues concerning awareness and understanding, data quality, data sharing and management need to be addressed by all those involved in one way or the other in forest management. Lack of suitably qualified personnel to manage GIS is another obstacle to GIS implementation. Whereas GIS is now taught in secondary schools, colleges and universities in developed countries, Kenya has only limited teaching of GIS in tertiary institutions (Mironga, 2004).

More so, the dilemma that these systems are often established without any inter-institutional coordination and thus results in highly incompatible (with each other) hardware, software and data coding procedures. Also such practices duplicate data and waste considerable effort and funds (Qadri, 1995).

Conclusion

This paper has discussed some application of GIS for forest management and conservation, which will hopefully encourage forest conservationists in Kenya to use GIS as their research tool. Given that GIS requires coordination and calibration at a strategic management level, it is important that high level conservation managers also have a general understanding of GIS functionality and application. There is need for a coordination of environmental information collection and dissemination throughout the country as a matter of urgency. Topological identification of arcs that connect at a node, within a linear network, the from- and to-node numbers for each arc define connectivity. So, arcs that share a common node are connected. Researchers also require that data to be shared and results and findings to be made known. Most research organizations need to know what their counterpart institutions are doing in terms of research. This will certainly reduce the duplication of common data and save on our meagre resources.

This paper suggests that a single well-linked Geographic Information System is needed in Kenya. It is proposed that a centralized, well-managed platform be established with various nodes at different locations. This is the only and the best way that a GIS can be a valuable and long-lasting asset in countries like Kenya where developing and updating an information bank may become an impractical task.

The use of GIS and remote sensing will thus be able to achieve the following:

1. Reducing vulnerability through integrated management that involves all the stakeholders in a participatory approach
2. Monitoring key indicators of hazards using GIS/RS
3. Improving the model inputs from RS
4. Mapping hazard/potential
5. Modeling hazard potential to guide land use decisions.

Recommendations

Forest management planning involves making forecasts about what the future forest will look like relative to alternative management activities. This ability is crucial to nearly all aspects of management forecasting, particularly long-term forest management and conservation. GIS can play a key role in this analysis. GIS stores both the geographic and numerical structure of the forest stands and links that spatial database to the planning models. It allows the manager to effectively add both the important temporal and spatial dimensions to the management planning process. Within the limits of the inventory and model, the manager can then map what the forest will look like five, 10, 25, or 100 years in the future.

GIS expertise: resource managers, waders and decision makers should be trained in basic GIS knowledge that includes functions and management applications of GIS. Once GIS operators understand the functions of GIS, they should explore ways of using it to provide forest ecosystems with information and recommendations needed for their sustainable management. The managers should encourage GIS specialists to participate in decision-making and preparation of reports.

Public awareness and education about the mangrove forests may help protect them. It is necessary to involve local communities in managing and protecting their mangroves. Designated conservation areas may also save some of the forests. However, national governments have not been able to enforce laws and regulate development and industry to save the wetlands. Adequate provisions for conservation and restoration have not been included in development of the mangrove wetlands.

References

Aguma Robert and Hussin Yousif Ali. 2004. The use of Remote Sensing and GIS to Support Sustainable Management of Tropical Forest in East Kalimantan, Indonesia. In: Geospatial Application Papers>Environment>Conservation and Monitoring

Centre for Development and Environment (CDE) 2006. Limitations of Use: Declaration of Version and Sources. Eastern and Southern Africa Partnership Programme

http.//www.gisdevelopment.net/application/index.htm

http://www.blueplanetbiomes.org/mangrove_forests.htm

Inter-Ministerial Task Force, Government of Kenya 2005. *Assessment of the Mau Forest Complex*. Nairobi: Government of Kenya Publications.

"Everything You Ever Wanted to Know About Mangroves", http://mbgnet.mobot.org/salt/mudflats/mangrove.htm

Kenya Forests Working Group 2001. *Excision and Settlement in The Mau Forest: Report of A Fact-Finding Mission on The Status of the Eastern Mau Forest*. Nairobi: KFWG.

Mironga, M.J. 2004. Biodiversity Conservation in Kenya: The Dilemma of Data Handling and Information Flow. OSSREA Bulletin Vol I, No. 3.

Mironga, M.J. 2004. Geographic Information Systems (GIS) and Remote Sensing in the Management of Shallow Tropical Lakes. Journal of Applied Ecology and Environmental Research 2(1): 83-163.

Ogalleh S.A. 2005. The Role of Gis in Promoting Information Management for Indigenous Forest Conservation in Kenya. In conference programme. Kenya ESRI-Supported User Group, Second Annual Conference. Egerton University, Kenya. 5-6th October 2005.

Ogalleh, S, A., and Nato, G.N. 2005. Human Settlement around Mau Forest and its Implications on Forest Succession. Unpublished paper, Egerton University

Chapter 20

Environmental Impacts of Tourism Development in Fragile Coastal Ecosystems: Operational Planning for Environmental Rehabilitation in Cuba

Lourdes Ruiz Gutiérrez
Higher Institute of Technologies and Applied Sciences, Havana City, Cuba
E-mail: lruiz@apache.isctn.edu.cu

Introduction

Cuba participated in the United Nations Conference on the Environment held in Rio de Janeiro, signed Agenda 21 and has signed and ratified both Global Conventions (CBD, UNFCCC). Cuba also participated in the World Conference on Sustainable Development of Small Island States, and is a signatory and active member of an important number of international and regional agreements, conventions, and protocols relating to the environment, (Convention on International Trade in Endangered Species of Wild Fauna and Flora-CITES, Basel Convention on the Control of Transboundary Movements of Hazardous Wastes and their Disposal, MARPOL, CEPOL, SPAW, Protocol of Kyoto, etc.). In 1994, Cuba developed a National Program for Environment and Development as a national Agenda 21, which became the National Environmental Strategy in 1996.

In 1994, there was a restructuring of the state apparatus in search of more functional efficiency, and in the environmental arena, the functions of COMARNA and the Academy of Sciences were absorbed by the new Ministry of Science, Technology and Environment (CITMA). COMARNA disappeared as such, but not its functions: they were strengthened and perfected in the new institutional structure.

The name of the Academy of Sciences of Cuba has been assumed, as originally, by a new organization of members selected among outstanding scientists to assist in resolving policy problems and to provide scientific guidance to the country, and to stimulate and reward excellent scientific work, among other functions.(1)

In 1994, as a result of increasing awareness of the crucial importance of the environmental dimensions of economic development and sustainability, Cuba created the Ministry of Science, Technology and Environment (CITMA) with its Environment Agency, the Environment Units in CITMA's Provincial Delegations, the Environment Control and Management Centre, the National Centre for Protected Areas and the Environmental Information and Education Centre. Just before, the National Centre of Biodiversity (belonging to the Institute of Ecology and Systematics) was created.(2).

Law, Policy and Institutional Arrangements in Cuba

To incorporate the environmental dimension into the country's development, considerable supportive or enabling legislation has been formulated and approved: the Law on the Environment (July 1997); the Resolution on Environmental Impact Assessment, and Resolution 168/95 "Rules for the Realization and Approval of Environmental Impact Assessments" (October 1995); the Law on the Tax System (October 1994) introducing taxes to be levied on use or exploitation of natural resources, and for use in environmental protection; the Resolution on Inspection by Government of the Environment and Toxic Chemical Products (1995) and the creation of CITMA's Environmental State Inspection Body; the Law on Foreign Investment highlighting the principle of sustainable development and addressing environment protection in the process of foreign investment; the Decree-Law 164 "Rules for Fisheries"(May 1996) and the creation of the Inspection Body of the Ministry of Fisheries; and Resolution 111/96: Regulations on Biological Diversity (1996). Currently, the Law on Soil Use, Territorial Ordering and Urbanism, the Decree-Law on Protected Areas, and the Decree-Law on Management of the Coastal Zone are under discussion and pending approval. Finally, Cuba is currently preparing a National Biodiversity Strategy and Action Plan (UNEP-GEF).

The National Centre for Control and Management of CITMA, together with the Provincial Delegations of CITMA and the organizations of People's Power of the Provinces and Municipalities will be charged with the gradual implementation of the vast EIA Programs in Cuba. The Environmental Units of CITMA will play an important and direct role in each one of the provinces.

The Institute of Physical Planning (Ministry of Economics and Planning) and its Provincial Directorates of Physical Planning will collaborate with other relevant institutions of the project to establish a clear delineation between areas to be protected and areas prioritized for development. The Institute will also continue conceptual strategic environmental planning for other important keys where biodiversity could be seriously damaged by tourism development if adequate planning is overlooked.

Environment monitoring will be carried out through a network of small Environment Monitoring Stations to be set up in the provinces. CITMA will be responsible for the implementation of this network through its Provincial Delegations and with the support of the Popular Power bodies (local government) in the relevant provinces and municipalities. CITMA's Environment Units will play an important and direct role in this effort.

Institutional and financial sustainability of activities is guaranteed by the Government's recognition of the vital importance of protecting biodiversity as part of the country's natural capital on which its national sustainable development depends. The Government will guarantee the functioning of the Protected Area System and the Council for Integrated Coastal Management upon completion of the project, including the long-term monitoring programme. Despite the huge economic difficulties the country is facing, the Government continues to invest considerable resources in scientific research, institution building and environmental legislation and policymaking, among other elements.

Cuba is currently creating the National Fund for the Environment, supported by the recently approved Law on the Environment. This fund will permit continued financing of activities aimed at protecting biodiversity in fundamental ecosystems. The Fund will be capitalized from a variety of sources, including entrance and users fees in protected areas, as well as charges for ecosystem services. By current law, forty percent of income generated from entrance and users fees is expected to remain within protected area administrations and will be managed and redistributed according to the possibilities of each protected area to cover its needs with profits it has generated. Design and legal establishment of the fund is expected to become final later this year. The Environmental Impact Assessment (EIA) will play an advisory role to the Fund and will provide important inputs towards capitalization of the Fund.

Sustainability will be enhanced over time with an increasingly comprehensive legal framework; the Resolution on Environmental Impact Assessment (EIA) and the Law on the Environment, as well as the Decree-Law on Coastal Zone Management, currently being approved, and other new laws, will contribute to the EIA process.

CITMA acts as rector of scientific, technical and environmental policy. Within the environmental branch, its operative function is executed through the new Environmental Agency, which, in turn, is integrated with the Centre of Environmental Control and Management (CICA), the National Protected Area Centre (CNAP), the Centre of Environmental Information, and Education (CIGEA), and several scientific institutes (that belonged to the former Academy of Sciences of Cuba), such as the Institute of Ecology and Systematic, Institute of Oceanology, Institute of Tropical Geography, Institute of Meteorology, and the Coastal Ecosystems Research Centre. The National Zoological Park, the National Aquarium, and the National Museum of Natural History also belong to the Environmental Agency. This Ministry maintains its representation at the level of provinces in the Provincial Delegations of CITMA (DPCITMA) whose activity related to environment is carried out by the Environmental Units (UMA) in each province.

Cuba is maintaining a sustained trend towards refining the decision making process regarding land use and physical planning as well as environmental issues in development at policy level. This can be seen in the expeditious strengthening of the institutional framework and legal system, and the increased awareness, experience and know-how of Government. The Institute of Physical Planning (of the Ministry of Economics and Planning) is the governing agency for physical planning and regulation of land use and its work is closely coordinated with CITMA concerning environmental issues.

Operational Planning for Environmental Rehabilitation

CITMA's Environment Units together with the National Centre for Control and Management, Protected Areas, together with the participation of other relevant institutions, (Fisheries Ministry, National Institute of Hydraulic Resources, Ministry of Sugar, Ministry of Construction, Ministry of Tourism, etc.) will lead the operational planning for environmental rehabilitation and recovery actions (affected species, habitats and landscapes identified by the pilot project). This will involve measures to control organic loading and pollution, regulate dams and reservoirs, construct culverts and bridges in marine roads (bermroads), reintroduce species, monitor and protect threatened species populations, as well as carry out environmental education, research, and monitoring.

Human Resources and Capacity

Inspite of being a developing country, Cuba possesses the human resources required to carry out research and biodiversity conservation through integrated management of ecosystems. This is the result of more than a century of scientific tradition of the country, and the emphasis and investment of resources of the Government in education and scientific development of the country. The community of scientists that work on aspects of biodiversity and environmental management at present includes internationally renowned specialists in their respective disciplines. Also, the country possesses herbaria, botanical gardens, natural history museums, and scientific libraries that were founded more than 150 years ago.

Brief Summary of Tourism Cases

Sustainable Tourism Development in Sabana Camaguey Archipielago (SCA)

Tourist development plans were presenting a development stereotype that would be in conflict with important biological resources, and not in agreement with the evolution of the tourism market in the Caribbean. Tourism is recognized as an important source of foreign currencies, being the second industry in Cuba. Given the current status of economy, and the relatively good health of natural environment in the Sabana-Camaguey, tourism based on the environment is the most compatible option of development. This kind of tourism is not only sensitive to the marine and coastal habitats and species, but also encourages the quality of these. Linking tourism with nature conservation through the establishment of a protected areas system in this ecosystem, not only offers an economic incentive to execute environmental standards for tourist facilities, but also ensures that large natural areas are indeed protected.

Traditional construction methods in fragile ecosystems have produced high levels of negative environmental impacts during the stages of project and construction, these impacts increase in the operation and abandonment phases. The conceptions of urban design and architectural typology not linked to natural landscape and wild life and based on the design of anthropic landscapes cause loss of important natural and scenic values, as well as environmental stress. It is necessary to eradicate the tendency to subordinate nature to architectural design, instead of acting the other way around.

A strategy for construction includes very important actions where new criteria for architectural design of facilities and landscape, construction and work making, and appropriate and alternative engineering systems are defined. Sustainable development of constructions in cays implies preserving and interpreting tangible and intangible natural and cultural resources of sites; and it designs solutions that are compatible with environmental planning techniques, permissible carrying capacities, and acceptable limits to change the project area. It also uses clean engineering systems, appropriate constructive technologies and materials that get adapted to the extreme climatic conditions existing, thus providing a visual and aesthetic comfort in the dialogue with and the interpretation of nature.

More of littler island (keys) are presently the focus of an expanding, large-scale tourism development program. The development of "sun and sand" tourism requires basic environmental measures such as preservation of water quality, adequate waste treatment, appropriate design measures regarding infrastructure and location, etc. Biodiversity protection has not yet been a significant factor in tourism development planning. Human exploitation of the keys to date has been very limited and is restricted mainly to more conventional tourism development on Coco and Guillermo keys. Currently planned tourism must be developed in an ecologically responsible manner, compatible with biodiversity conservation. Otherwise, planned infrastructure and activities will pose a serious threat to the flora, fauna, habitats and landscapes of the keys.

Key government staff lack adequate training and capacities in environmental monitoring.

Territorial and sectoral planning by provincial staff does not always incorporate environmental considerations in a systematic and comprehensive manner.

There is limited awareness and understanding of sustainable development concepts amongst decision-makers.

The level of environmental awareness among communities and civil society is limited, thus hindering broad-based participation, ownership and pride from civil society in sustainable development.

There is limited access to and exchange of information on sustainable development practice and methodologies.

The risk of failing to achieve the targeted basic or incremental investment is minimal, and it is unlikely that the Government should give conscious priority to a non-sustainable development plan with short-term goals. It is highly improbable that there be a shortage of suitable staff for the actions proposed given the wide availability of highly trained people from different educational levels. Participation of the provinces and sectors involved has been secured since the implementation of the GEF pilot project and because of their participation in project design.

Effectiveness, Opportunities and Constraints

The Centre for Environmental Control and Management, the National Protected Area Centre, the Centre for Environmental Information and Education, and the Environmental Units (provincial), all from the Ministry of Science, Technology and Environment constitute a strong foundation of environmental institutions.

The level of training and hands-on experience of personnel from relevant institutions and agencies is generally high, are well qualified and familiar with the latest theoretical and applied methodologies in their respective disciplines. Supported with appropriate technical assistance, they can undertake research and monitoring programs, integrated coastal management, protected area management and environmental planning, undoubtedly at the level of international standards. However, national experts will need targeted support in certain specialized areas, as well as in other areas where there is a need to keep abreast with recent developments, inspite of the existence of a good knowledge base, due to our great economic restrictions. Related to this targeted expertise is the need for bolstering infrastructural facilities that can assist in the execution of relevant professional activities. Such activities include monitoring (environmental monitoring stations that are in the process of remodeling) and the administration of the protected areas system (interpretive centres, administrative and control facilities, basic services, etc.).

References

CITMA, Ministry of Science, Technology and Environment Law 81 of the Environment. Havana, 1997 (44 pp.)

Ruiz Lourdes., Diaz Gisela, Ferro Sergio. Sustainable Architecture Desing in the Kays of Cuba. Project Document. Constructions for the Tourism in Cuba. Havana, 1996. (42 pp.)

Ruiz Lourdes. Environmental Impact Assessment for Tourism in Cuban Coastal Zones. P.H.D. Thesis. Havana, 1999 (99 pp.)

U.N.D.P./G.E.F. Project Document. Protecting Biodiversity and Sustainable Development of the Sabana-Camaguey Ecosystem. Havana, 1997. (254 pp)

Chapter 21

Human Security vs. Mangrove Destruction: Case Studies from East Coast of India

Rita Chauhan[1], Gurmeet Singh and A.L. Ramanathan[2]
School of Environmental Sciences, Jawaharlal Nehru University, New Delhi, India
[1]E-mail: rita.env@gmail.com; [2]E-mail: enviroguru@yahoo.co.in

Introduction

Ecosystems of coastal areas, small islands and mangroves in particular have an important economic value for local populations living from both terrestrial and marine resources. More than 60 per cent of the world's population lives within 60 km of the coast and more than two-thirds of the cities with over 2.5 million inhabitants are situated in these areas. Due to population growth and persistent migration of people towards large urban centres, the number of people living in coastal zones is constantly growing. Moreover, coastal zones and islands are under permanent pressure in the name of development. This situation leads to constant degradation of natural resources in coastal zones and islands, linked with social conflicts and environmental degradation. Furthermore, the loss of traditional land use and fishery practices leads to overfishing, erosion and loss of coastal land. Here we have attempted to synthesis the effect of over-exploitation of these already fragile ecosystems leading to the conflict between the man and these two mangroves (Sundarban Mangrove ecosystem and Bhitarkanika Ecosystem) in the context of the human security and management.

Mangrove Forest as "Ecosystem"

Mangrove forest ecosystems support important wetland communities of plants and animals. They are characterized by unique species of trees and shrubs that fringe the intertidal zone along sheltered coastal, estuarine and riverine areas in tropical and subtropical latitudes.

The mangrove ecosystem has important direct and indirect economic, ecological and social values to man. Mangrove ecosystems have consistently been undervalued, usually because only their direct goods and services have been included in economic calculations (*e.g.* forestry resources), but this represents only a minor part of the total value of mangroves. Mangroves have often been considered as unproductive land and their destruction and degradation have been due to the preference for short-term exploitation for immediate economic benefit, rather than long-term sustainable exploitation (Saenger *et al.*, 1983).

By undervaluing mangrove ecosystems, development has too often favoured their rapid conversion and loss. Mangrove conversion usually leads to short-term economic gain at the expense of greater, but longer-term, ecological benefits and off-site values. For example shrimp farming in the late 1980s and early 1990s was perceived to have high profits and resulted in large-scale conversion of mangrove areas. Underestimation of the total economic value of mangroves and of the impacts of human activities is major factors contributing to the widespread loss and degradation of mangrove ecosystems (Gilbert and Janssen, 1998). Serious environmental, social and economic impacts are associated with the decline and degradation of mangroves. For example in Vietnam where mangrove loss due to defoliants used during the Vietnam war, logging and aquaculture have led to coastal erosion, salinity intrusion and decline in natural shrimp and mud crab (*Scylla* species) populations (Hong and San, 1993).

Human Security

More precisely, the very concept and definition of 'human security' became familiar with the UNDP's Human Development Report 1994, which suggested:

"Human security can be said to have two main aspects. It means, first, safety from such chronic threats as hunger, disease and repression. And second, it means protection from sudden and hurtful disruptions in the patterns of daily life" (UNDP, 1994, p. 23).

Incorporating the environmental dimension in this overall framework, Tariq Banuri argued:

"Security denotes conditions which make people feel secure against want, deprivation, and violence; or the absence of conditions that produce insecurity, namely the threat of deprivation or violence. This brings two additional element to the conventional connotation (political security), namely human security and environmental security (Banuri, 1996)."

Natural and human induced global environmental change belongs to the class of risk with high probability of occurrence and damage potential but in such a remote future that for the time being no one is willing to perceive the threat. Until recently, the effects of man's activities on climate variation was perceived as negligible and so climate was generally taken for granted and there was little thought that the climate could be a problem with severe impacts (Ojo, 1987) but today, because climate and environmental change affects the very basis of human existence the connection between environmental changes and human security has become very strong.

Although the probability of occurrence and the damage potential are well known and clear, there is always a time lag between trigger and consequence, which create a fallacious impression of security (German Advisory Council on Global Change, 1998). Most disasters (including flood, droughts, desertification, land degradation, subsidence, etc) are not random events without underlying causes; they are sudden manifestations of slow but continuous degradation processes (UNEP/GRID-Arendal, 2005). Climate and environmental change processes lead to changes in the biophysical life support system including land surface (vegetation), water resources, soil and atmosphere which constitute the elements that support the long term sustainability of life on earth. It also affects the social and economic structure and framework, and by implication survival, of people. These changes are driven by combination of both natural and anthropogenic causes.

Forests need human security as much as human security needs forests. Forest resources and services play an important role the achievement of human security. Whether it is through the provision of livelihood resources (*e.g.* food, medicine, cooking fuel, construction materials, etc.) or ecosystem services such as water purification, climate regulation and erosion control, forests are undeniably linked to the well-being and prosperity of people. When the availability of these resources and services is undermined, the resulting insecurity can range from hunger and susceptibility to disease, to loss of income and livelihoods, to social tensions and open conflict, and as seen in the examples above–to acute vulnerability to natural hazards. The calamity of the Indian Ocean tsunami showed a reason to reassess the role of forests in natural disaster prevention and mitigation.

Man and Mangrove

Humans inhabit mangroves in many places, usually belonging to traditional local communities that harvest fish and other natural resources but in recent years the coastal areas have become under intense pressure for development. Mangroves have been over exploited or converted to various other forms of land use, for example agriculture, aquaculture, salt ponds, terrestrial forestry, urban and industrial development and construction of dikes and roads (Macintosh, 2002). Mangroves can be affected by several different activities simultaneously or over time as land use patterns change. Population pressure, poverty and lack of livelihood options, changes in values and 'open access' nature all contribute to over-exploitation. As well as the mangroves losing areal coverage they have also declined in terms of biological diversity and economic value, due to excessive harvesting of the most valuable trees and a shift in forest structure towards younger trees and secondary growth as the larger trees were removed.

Causes of Mangrove Loss

The fundamental cause of mangrove forest loss is the increase in human population living near the coastal zone. Ong (1995) considers that burgeoning populations are possibly the biggest cause of mangrove destruction and degradation. An estimated 60 per cent of the global population lives within roughly 100 km of the shore (WRI). This means that 3.4 billion people rely heavily on marine habitats and resources for food, building materials, building sites and agricultural and recreational areas. They also use coastal areas as a dumping ground for sewage, garbage and toxic wastes. Moreover, many more people in the non-coastal population live in agricultural and urban communities concentrated along rivers and in the surrounding hills. Pollution and poor land use practices within these watershed areas affect downstream marine habitats because sediments and pollutants are ultimately washed into coastal waters.

Mangroves have been overexploited or converted to various other forms of land use, including agriculture, aquaculture, salt ponds, terrestrial forestry, urban and industrial development and for the construction of roads and embankments. Mangroves can be affected by several different activities simultaneously, or over time as land use patterns change. The threat is estimated to be from low to high and whether it is deemed to be increasing or decreasing. Over and above the chronic loss of mangrove area worldwide, mangrove habitats have also declines in terms of their biological diversity, forest structure and economic value.

In most costal zones of the world there has been marked decline in the mangrove area since 1960s. The mangrove destruction at places is over 50 per cent.

The main factor leading to the loss of mangrove biodiversity is habitat loss through conversion or degradation of the forest. Urbanization and off-site activities can lead to degradation through siltation and changes in water temperature and flow, other physical factors and salinity. Contaminants may be directly toxic to some marine organisms and their effects may be instantaneous or cumulative. Introduction of species can also cause habitat loss and loss of diversity. For example, the introduction of *Nypha fruticans* to West Africa where it is not found naturally has caused considerable problems through its continued spread (Wilcox, 1985). Global warming could be a significant threat to mangrove cover and biodiversity. Rising water (as a result of melting ice caps) could drown coastal mangrove. The presence of existing agricultural and urban development and dikes would in many cases prevent the establishment of new mangrove areas. These would alter the species compositions and perhaps trigger local and global extinction. (McCarthy *et al.*, 2001).

Historical knowledge of mangrove utilization can explain much about the current status, habitat condition and other biodiversity indicators of mangrove ecosystems, provide an insight into local cultures and attitudes to resource use, and what the impacts of previous management policies towards mangroves have been. Historical records indicate that the original extent of mangrove forests has declined considerably under pressure from human activity (Macintosh and Ashton 2003).

The east coast of India bordering the Bay of Bengal, supports rich tropical environment characterized by high biological diverse mangrove ecosystems. One of these, is the world largest mangrove forest, Sundarban in West Bengal. The other mangrove system is Bhitarkanika Mangrove ecosystem (second largest in Indian subcontinent) at Mahanadi delta in Orissa coast.

Both ecosystems are situated in the states where the maximum population living in and around mangrove forests is under below poverty line. The prime occupation being seasonal agriculture (mainly paddy). The population is highly dependent on

Threat	South and Southeast Asia	Africa	Central and South America
Natural disasters	Low-High *Increasing*	Medium *Increasing*	Low *Increasing*
Population pressure	High *Increasing*	High *Increasing*	Low-Medium *Increasing*
Over-exploitation by traditional users	High *Increasing*	High *Increasing*	Low *Stable-Decreasing*
Forestry	High *Stable*	Medium *Increasing*	Low *Stable*
Agriculture	High *Decreasing*	High *Increasing*	Low *Stable-Decreasing*
Aquaculture	High *Increasing*	Low *Increasing*	Medium-High *Increasing*
Salt Production	High *Decreasing*	High *Stable*	Low-Medium *Decreasing*
Mining	Low-Medium *Decreasing*	Medium *Increasing*	Low *Decreasing*
Urban and industrial development	High *Increasing*	Medium *Increasing*	Medium-High *Increasing*
Tourism	Low-Medium *Increasing*	Low *Increasing*	Low-Medium *Increasing*
Hydrological diversions e.g. dams	Medium-High *Increasing*	Medium-High *Increasing*	Low-High *Increasing*
Coastal pollution	Medium-High *Increasing*	Medium-High *Increasing*	Medium-High *Increasing*
Management shortcomings	Medium-High *Decreasing*	High *Stable*	Low-High *Stable*

Figure 21.1: A Summary Table of the Main Threats to Mangroves, by Region (Macintosh and Ashton 2003)

these estuarine forest resources (wood, fishery and others) for their livelihood. As a result there has been always over-exploitation of the already fragile ecosystem leading to the conflict between the man and the mangroves.

In context of the human security, the present paper emphasizes on the status of these two ecosystem *i.e.* Sundarban Mangrove ecosystem and Bhitarkanika Ecosystem

Case Study 1: Sundarban Mangrove Ecosystem, West Bengal

The Sundarban Forest is the largest estuarine mangrove forest in the world located in 24 Paraganas district of West Bengal (India) and Bangladesh. In India, it covers an area of 9630 km^2 out of which 4266 Km^2 is Mangrove forest (Naskar 2000). Due to it's beauty and richness of wildlife, it was declared a world natural heritage site by UNESCO in 1974 and a national park in 1984.

The Sundarban Mangrove Forests has been home for a wide biodiversity. It is natural habitat of many endangered species including the Royal Bengal tiger, Estuarine Crocodile (*Crocodilus porosus*), Gangetic Dolphin (*Platinista gangetica*), Snubfin dolphin (*Orcella brevirostris*), River Terrapin (*Batagur baska*), marine turtles like Olive Ridley (*Lepidochelys olivacea*), Green Sea Turtle (*Chelonia mydas*), Hawksbill Turtle (*Eritmochelys imbricata*), thus making it a natural biodiversity hot spot. The entire mangrove forest is called Sundarban owing to the dominance of the tree species *Heritiera fomes*, locally known as 'sundari' because of its elegance (Jain and Sastry, 1983). Sundarban flora is characterized by 26 True Mangrove species and 29 mangrove associates species (www.indiansundarban.org). Characteristic mangrove species include Garjan (*Rhizophora* sp.), Kankra (*Bruguiera sp.*), Goran (*Ceriops* sp)., and Baen (*Avicennia officinalis*). *Heritiera fomes* is scattered over areas of higher elevation, along with Keora (*Sonneratia* sp.), Gengwa (*Excoecaria* sp.) and *Xylocarpus* sp. The Golpata palm (*Nypha fruticans*) and Mangrove date palm or Hental (*Phoenix paludosa*) is relatively uncommon but occurs on wet mud-banks along the creeks. The Sundarban Mangrove Ecosystem is the region of transition between the fresh water of the Ganges distributaries system and the saline water of the Bay of Bengal. The river system of Hugli–Haldi, Muriganga, Saptamukhi, Thakuran, Matla, and Goasaba bring a huge sediment load with which is deposited at delta. The estuary annually receives 411 X 10^6 m^3 T (*i.e.*, 328 X 10^6 T sediment + 83 X 10^6 T solute load) from Hugli river alone. (Sarkar *et al.*, 2004) Therefore the different ecosystem processes are dominated by various factors such as river discharge, salinity, freshwater flushing, daily tidal water fluctuations, siltation, sedimentation and lateral erosion. However during past few years, the natural ecosystem has been disturbed due to anthropogenic activities.

The mangrove ecosystem has suffered from loss of habitat and contamination. Aquaculture, logging of mangrove forest for timber and fuel wood has caused the decrease in the forest cover area. They have suffered the repercussion of deforestation for the purpose of artificial pond construction for aquaculture of shrimp farming. A large number of populations are dependent on forest for fishing, honey collection, wood cutting, and shrimp fry collection. Fishing is second largest occupation after agriculture. This coastal environment suffers from environmental degradation due to rapid human settlement, tourist activity, deforestation, and increased agricultural and aquaculture practices. Moreover, the mighty Hugli river has supported multifarious industries such as paper, chemicals, textiles, pharmaceuticals, shellac, leather, plastic, tires, and cycle rims (UNEP, 1982) and thus carrying a huge pollution load to the mangrove ecosystem.

Resources and Utilization

The virgin forest of Sundarban attracted the attentions of adventurous settlers. Climatic hazards of swampy forestland, onslaught of carnivores and poisonous snakes, could not dampen the spirit of people from venturing this area, for settlement. The total population of Sundarban (India) has increased from 2,451,850 (1981), 5,715,030 (1991), to 6,909,015 (2001 census). Reclaiming the forestland for cultivating and exploitation of timber, fish, honey, etc., are found to be quite common in practice.

Position was further aggravated due to the physical and climatic hazards of the area such as occasional flood and intrusion of saline water into the agriculture fields which is one of the main problems to be faced. Single crop system was also found to be impractical and double crop system has to be introduced along with better embankment and the proper maintenance throughout the year. These embankments are also being used by the population for their (pedestal) movement from one place to another particularly during dry season. A committee was established to develop better communication, health, education etc. The purpose is to engage a large population to get employment through out the year and agriculture will be the supporting means of occupation along with other crops such as jute, tobacco, pulses etc. Plan was also drawn for the production of more milk, fish and poultry, salt manufacture, rice milling and development of cottage industries like handloom, carpentry etc., with a view to utilize local skills on top priority basis. This committee functioned till 1973 when for better development and co-ordination it was thought necessary to form a Board as per recommendation of the various committees. As such the Sundarban Development Board (S.D.B.) was formed in 1973 and entrusted with the responsibility of development and planning for this region.

The newly formed S.D.B. concentrated in formulating a 10-year Master Plan for the future development of the area. This Board also thought that the land should be utilized for high production and for the benefit of the small and marginal farmers; employment oriented projects have to be taken up. Allied activities like horticulture, animal husbandry, fishery, poultry, piggery and dairy etc., are also to be developed. Main trade Centres are the Kakdwip, Raidighi, Canning, Pathar Prathima and Hasnabad. In addition, there are principal towns viz., Canning, Jaynagar and Taki Namkhana and Pathar pratima.

Agriculture

Agriculture being the main occupation (88.53 per cent of the total population) of the local population, poverty was found to be the rule. It has been estimated that out of about 1460 sq miles, 360 sq miles are not suitable for cultivation and 125 sq miles are considered as wasteland. The cropped area is 975 sq miles (=737,967 acres or 15,223,300 hectares) of which 96 per cent of the area is covered under 'Arnan' paddy cultivation. The total production of paddy is estimated as 550–600 thousand tons annually. The cultivation is depending upon rainwater only as the ground water suitable for agriculture is available only at 800–1000 ft. at high land, 'Khesari' 'Jute', 'Tobacco' and 'Vegetables' are cropped along with 'Orchards'.

Although about 88 per cent of the inhabitants are dependent upon agriculture but 54 per cent do not have their own land and the marginal farmers are maximum in this area. Among landholders 70.31 per cent are having land within 1 hector, 13.401 per cent people are the owner of 1-2 hectares of land and only 5.60 per cent people possess land above 2 hectares (Mandal and Ghosh 1989).

Rise in the population has lead to an increase in the demand of food grain. Owing to the shortage of agricultural land as well as lack of monetary resources has forced the local population to clear the mangrove forest and make the land suitable for farming. Plaeological evidences have suggested the extension of Sundarban within the reach of Calcutta Municipalities. The history of restriction of Sundarbans is nothing but the history of conversion of mangrove forests for agricultural purposes. This reclamation process was initiated in 1770 and it continued till recent past. In the largest delta region of the world, existing between India and Bangladesh, 1,50,000 ha of mangroves were destroyed during the past 100 years, and these were mainly reclaimed for agriculture. The mangrove areas are deforested and reclaimed with rainwater by constructing embankments. Once the salt is leached to a sufficient level, the land is cultivated either with paddy or coconut. (Chaudhari and Chaudhary, 1994).

Salt-water menace was found to be the main hindrances for agriculture. So, special importance was always given on the formation and maintenance of embankments in this area. However, protection from salt water for agriculture crop through embankments raised several other problems afterwards as they inhibits the tidal flushing and the large volume of sediments carried by the innumerable rivers and creeks of this area instead of rising the level of delta mouth, settles on their own beds and gradually they became chocked up, causing the natural water flow to change their direction. Old river beds of Bidhyadhari and Matla of recent past exhibit very good testimony of this situation. This is one of the main reasons or ever-changing courses of rivers and creeks. Embankments and consequent rise of riverbed severally restricted the fresh water flow. As a result fresh water loving species namely *Heritiera fomes*, and *Nypha fruticans*, which were once predominately distributed in Sundarban, are being replaced by salt tolerant species viz. *Rhizophora* and *Avecennia*.

Fishery

A sizeable population in Sundarban directly or indirectly depends on fishery for their livelihood. Absence of proper infrastructure such as quick transport marketing facility, ice plant etc. remained a stumbling block for development of fishery as organized industry. Exploitation by middlemen and moneylenders made their condition deplorable one.

Sundarban is rightly considered as gold mine for fishery. From time immemorial local inhabitants as well as subsequent settlers contributed a lot for development of various indigenous means for exploitation of fish. Various rivers, river-lets and creeks of Sundarban having tidal saline water with flushing of sweet water provide a unique ecological niche for the propagation of fish. These vast resources for fishery have not been properly utilized for a long time. Even today the Gangetic delta is having 4,04,700 hectares of estuaries of which only 20,000 hectares are being utilized for fisheries. But the vast water resources of this area have a great potentiality for fisheries. The silted bed of Bidhyadhari and its surroundings became very important pocket for Bhery fishery especially after the technique of utilizing the Kolkata town sewage for fishery came into existence.

Recently the export market also opened a new chapter for development of Prawn industry in this area. The Bagda, *Penaeus monodon*; Fabricus and Galda, *Macrobracium roscnbergi* (De Man) are mainly being exported. Recent statistics shows that from Sundarban alone Prawn worth of Rs. 35 crores had been exported, out of total production Worth of Rs. 45 crores.

Collection of seeds of *Panaeus monodon* is one of the main sources of earning for the small and landless fisherman and women of this area. During Collection of economic prawn seeds, the rural people segregate the *Panaeus monodon* seeds and destroy other 90 to 95 per cent of fish and prawn seeds, which is leading towards the destruction of large number of estuarine species (Patronobis, 2004; Mahapatra *et al.*, 1999). Further, setting up of fisheries in rivers, canals, creeks and estuary, raising embankments along the major river systems against insurgence of saline water (Mandal and Ghosh, 1989; Bose, 2004).

Forest Product

In addition to the fishermen two more different categories of people are very important who live on forest wealth of Sundarban. They are the honey collector or 'Mouley' and the wood cutter or 'Bouley'.

The honey from *Aegiceras* and *Cynometra* are considered to be the best in quality.The major host plants of honeycombs are, *Excoecaria agalocha* (Genwa) contributing 39 per cent of the total honey production; *Avicenia* species (Baen) contributing 16 per cent; *Ceriops* (Goran) contributing 11 per cent; and *Rhizophora mucronota* (Garjan) contributing 10 per cent. Genwa branches are ideal for comb formation and these branches have maximum number of combs per unit area. It is interesting to note that Goran that occupies 90 per cent of the forest area, nests only 11 per cent of the combs. The reason could be the bushy structure of these *C. decandra* and unbranched and thick crown of *C. tagal*–suitable for formation of hives. Preferences for honeybees are trees along narrow creeks in forest blocks distant from human settlements.

Lucrative cost of timber and other forest resources remained a sufficient temptation for the poor mass with low per capita income to pilferage the forest. The cultivators were not the only ones trying, largely unsuccessfully, to clear the forest. The woodcutters also faced a challenge in cutting down trees that were enmeshed with one another in such a way that they could only be removed one piece at a time. Main forest products of Sundarban are the Timber-wood for paper pulp, Plywood, Firewood, Barks for tannin along with honey and wax.

Humans have long seen themselves within the man vs. nature paradigm, in which survival must be won through brutal ways if it is to be won at all. This perception is apparent in those records kept by foreign empires of the Sundarbans. However, the harsh realities of this region have been accepted and incorporated into the cultures of the local human inhabitants.

Not only did the control of the Sundarbans environment present a challenge for those intending to utilize its resources, but the existence of the wild animals was also seen to present danger.

As Hunter explained:

> *All woodcutters are very superstitious, and believe in the existence of numbers of forest spirits. None of them will go into the forest to cut wood unless accompanied by a fakir, who is supposed to receive power from the presiding deity-whom he propitiates with offerings-over the tigers and other animals....Before commencing work in any allotment, the fakir assembles all the woodcutters of his party, clears a space at the edge of the forest, and erects a number of small tentlike huts, in which he places images of forest-deities, to which offerings and sacrifices are made. When this has been done the allotment is considered free [of] tigers...."* (Greenough, 1987).

The timbers of *Heriteria fomes, Sonneratia, Xylocarpus, Avicennia* sp., *Excoecaria, Xylocarpus* and *Amoora* are sawable. They can be used for beam, rafter, post etc. The rhizoporaceae is also used as fuel wood. Xylocarpus is mainly used for the posts used in constructing houses. They are excellent for paneling. Good qualities of planks can be prepared form the timber of Avecinia, Sonneratia as well as Xylocarpus which can be used for making doors and other wooden furniture.The barks of Ceriops, Bruguiera and Rhizophora are used for extracting tannin. The Xylocarpus is used in the pencil factory, for toy making and as timber. The dwarf Palm or Golpatta (*Nypha fruticans*) is used as thatching materials.

The utility of mangroves in these deltaic islands lie with the protection of the coastal line as well as the reclamation of land. The exposed land after afforestation was also utilized for cultivation and settlement. Along with the procurement of wood and fish the people collect honey and wax from inside the forest. They play a conspicuous role in this specialized environment by obstructing the direction of current, depositing the silt and binding the soil. They are also useful for the gradual rise of land level which results in spreading the forest zone further towards the sea. This forest belt is also essential for maintaining optimum eco-climatic and socio-economic condition of the entire area. However the over-exposure and illegal and uncontrolled cutting of trees for monetary purpose has posed a serious threat to the already fragile ecosystem (Sen and Naskar, 2003)

Case Study 2: Bhitarkanika Mangrove

Bhitarkanika is a unique habitat with lush green mangrove forest on the sides of creeks and tidal mud flats located in Kendrapada district of Orissa. The mangrove forest in the deltaic region of the rivers of Brahmani, Baitarani-Dhamra deltaic is continuous and comprises of a single compact patch of estuarine forest in Orissa. This is the second- largest Mangrove ecosystem with diverse flora and fauna intricately linked with each other after the Sundarbans. A total of 64 species of plants were recorded from Bhitarkanika Mangrove Protected Area, which included 28 true mangroves, 4 mangrove associates and 32 others. The mean canopy cover of Bhitarkanika mangroves was found to be 33.25 per cent. The area was declared a wildlife sanctuary to protect the endangered Saltwater crocodile, (*Crocodilus porosus*), in 1975. This deltaic, estuarine-mangrove wetland system, harbours the highest diversity of Indian mangrove flora, the largest known rookery of the olive ridley sea-turtles in the world, the last of the three remaining population of salt-water crocodiles in India, the largest known population of king cobra, water monitor lizard, one of the largest heronry along the east coast of India and one of the highest concentration of migratory waterfowls–both ducks and waders. Along with the largest diversity of estuarine and mangrove obligate fisheries resource, Bhitarkanika also is rich in prawn fisheries. The mangrove and associated forests provide the subsistence requirement of timber, fuel wood, tannin, honey, and thatch roof for the local people and fodder for the local communities (Chadah and Kar,1999).

Initiatives of Protection

The Bhitarkanika area has had a strong protection policy since 1951. In 1975, the Government of India declared 670 km^2 of the area as a Wildlife Sanctuary under the Indian Wildlife Protection Act, 1972, with a core area of 145 km^2 that was upgraded to a National Park in 1988. In 1997, 1435 km^2 area was declared as Gahirmatha Marine Sanctuary, with a core area of 725.5 km^2. All these show that there is no market, information or intervention failure at the primary level in case of the Bhitarkanika area *i.e.* the inability of the government to implement the Wildlife Protection Act effectively and lack of inter-sectoral coordination. Between 1951-61 there was unprecedented growth of population in the area due to the resettlement of refugees from Bangladesh. Immigrants from Bangladesh settled in 1971 in the coastal areas, which were thickly forested with mangrove jungles, soon converting the land into paddy fields. Subsequently, prawn farming became popular in the eighties, since these areas are ideal for setting up prawn farms. Paradeep port was set up in 1965 and was responsible for the cutting down of the mangroves of the Mahanadi mouth area. However, the Bhitarkanika sanctuary (declared in 1975) area comprises of a much larger area of 672 square kilometers. There are as many as 410 villages located within the sanctuary having a population of approximately 0.25 million comprising of six Gram Panchayats. Between 1994-95 with scant regard to Wildlife and Forest Conservation Acts, the revenue department legalized a large number of illegal settlements within the Sanctuary area. As per law the creation of villages in the sanctuary limit was illegal and had to be taken into account during the finalization of rights in the Sanctuary leading to loss of Mangroves. Similarly despite the protected status of the Bhitarkanika area and the existence of a strong Maritime Act (1982) of the Government of Orissa and Orissa Marine Fishing Regulation Act (1982) and rules (1983) there are unabated development activities such as construction of port and defense structures and inshore fisheries using mechanized vessels in the area. This is the result of information failure on part of the fisheries, waterways, defense and other government departments. In the absence of valuation studies the forest department has been unable to articulate the importance of conserving this ecosystem in the face of developmental activities that promise higher turnover.

Traditional Uses of Mangrove Resources

The uses of the mangrove forests by the local people ranges from timber, poles and posts to firewood and fiber. Non-wood forest products include thatch, honey, wildlife, fish, fodder and medicine. Though the area as protected status and legally no extraction is allowed, the villagers living in mangrove areas use mangrove trees mainly for fuel wood, construction of houses and agricultural implements.

Fishery

The villages are basically agriculture based 60 per cent of the people are engaged in agricultural activities. A single crop is sown during the Kharif season due to lack of irrigation facilities and salinity of soil baring small tracts of land along riverbanks in some areas. Fishing is an important occupation in the area, it accounts for 5.56 per cent of the village people. Fishing is prohibited inside the park but not in the sanctuary area. Fishing leases are being granted to traditional fisherman to operate in the sanctuary area by the Orissa revenue department. Quite a significant amount of fishing activities were observed in the major rivers like Baitarni and Brahmani. The river Dhamra is in the vicinity of forest and hence no leases are granted but more cases of fishing offences are committed because of the presence of marine fishing trawlers that frequently violate the provisions.

Also, shrimp seedling collection, also called pin is a profitable business and is mainly carried out by the female members in the villages located on bank of river Dhamra. However even the mangrove area, estuaries and tidal creeks in mangrove areas in the dry seasons *i.e.* February to June are used by local fisherman particularly women folk, for mud crabs, prawn and other fishes. But majority of the populace do the fishing in the tidal waters on the side of dykes and roads along the rivers. However, fish productivity has declined dramatically in recent years due to the increased number of fishers, the loss of mangrove areas to shrimp farms and water pollution from these farms. In terms of the total value of catch, shrimps are one of the most important resources for these coastal fisheries, which are often associated with the mangrove. In narrow creeks, when the tides start receding the nets are spread to get the catch whereas in wider creeks like Khola, Dangamala and Gupti the fishermen use handnets and try to catch whatever they can. However, in Patasala river leading to Gahirmatha beach, bigger nets are put in place using buoys thus obstructing the route of migratory fishes, brackish water turtles and crocodiles. Ponds are sometimes prepared on land in the mangrove forest, where they cultivate fish and shrimps. Shrimp aquaculture production is heavily dependent on surrounding mangroves for provision of seed, food, clean water, etc.. When shrimp farms are made, they can often be built on too large an area, which means that not much mangrove is left in the surroundings.

Non Wood Forest Products

Permits were granted by the Forest Department for the collection of honey to ST community people, in two of the sampled villages, *i.e.* Dangmal and Satavaya. In Bhitarkanika, honeybees are abundant during the flowering season. The quality of honey found in the mangroves is different in different seasons. Every year during February-May, 3000-5000 kg of honey is produced in Bhitarkanika area. A local tribe called 'Daleis', who live in the area specialise as honey collectors. Only the males from the community go to the forest for honey collection. Since honey collection is essentially a seasonal work, the Daleis do not depend entirely on honey collection for their livelihood. They also work as labourers with the forest departments in agricultural activities, fishing, etc. One of the mangroves associates Myriostachya wighitiana locally called Nalia is used for weaving baskets, mats and other articles which arelong lasting. A number of people are still involved in the Nalia extraction and weaving business though the permits that were earlier given for their collection have long been stopped. The leaves of Phoenix paludosa stacked together in bundles of 13-20 served as shingles for thatching purpose.

Socio-cultural Values

The wetland has a good number of ancient monuments like palace of ex-zamindar, Shiva temple inside Bhitarkanika forest block, Jagannath temple at Righagarh and Keradagarh, Panchubarahi goddess temple at Satabhaya and others such small temples which are culturally significant to the inhabitants.

Pressures on the Mangrove Ecosystem

Rapid increase in population in the recent times has only increased the pressure on the mangrove resources in Bhitarkanika. In the late 1960s, the complex pressures resulting from population growth, urban expansion and coastal economic development brought about heavy exploitation and destruction of mangrove resources. The detrimental activities in this already fragile ecosystem included poorly executed illegal logging operations, road construction, and conversion of mangrove forests into shrimp farms, fishponds, and agricultural operations.

Dire poverty forces the people of Bhitarkanika village to frequent the forest in search of livelihood. Some of them risk-facing cyclones for fishing in the open sea while others enter the forest to collect honey, fuel wood and other minor forest produce. The vulnerable mangrove eco-system in Bhitarkanika is under severe stress due to such interference.

Deforestation and poaching is adversely affecting the Bhitarkanika Sanctuary, the world famous home of nesting Olive Ridley turtles, the salt-water crocodile and many migratory and resident birds. The Dangmal Forest Range has recorded the killing of about 125 animals (including 111 deer) in the last few months. According to a senior forest official, this number too may be far off the mark, adding that the Forest Department has been handicapped by covert support extended to illegal settlers and the prawn mafia.

No attempt has been made to evict the estimated 25,000 illegal settlers and prawn farms from the Sanctuary area despite a series of court orders and pressure from environmentalists worldwide.

Protection Against Natural Disaster–Storm Protection

The Indian Ocean is one of the six major cyclones–prone regions of the world. Such cyclones occur every year often with wind speeds of about 200 km/h accompanied by storm surges of about 4 m to 5 m high. They cause considerable damage to crops, different types of buildings/dwellings and structures, power transmission, communication systems resulting in heavy loss of life and property (Shanmugasundaram *et al.*, 1989).

The undisturbed and natural mangrove forests ecosystems act as seaward barrier and check the coastal erosion and minimize the tidal thrust or storm hit arising from the sea considerably (McNae, 1968). The degree of protection varies with the width of mangroves. Mangrove root systems retard water flow. Resistance to water flow serves to dissipate the energy of floodwaters, of particular service during cyclone (*e.g.* Gilbert *et al.*, 1998). A super cyclone in the month of October in 1999 hit the Orissa coast near Saharabedi, a village lying about 1.5 km away from the seacoast in Ersama Block of Jagatsinghpur district had a wind speed of around 260 km/h and a storm surge of about 10 m. This super storm traveled more than 250 km inland and within a period of 36 hrs ravaged more than 20 million hectares of land, devouring trees and vegetation, leaving behind a huge trail of destruction. This cyclone affected around 15 million people in 12 districts and caused the deaths of about 20,000 people and over 400 thousands cattle. The loss to property, crops and plantations, communication and transportation networks was colossal, the value of which is estimated to be over Rs 100 billions. The most severely affected districts were Balasore, Bhadrak, Cuttack, Ganjam, Jagatsinghapur, Jajpur and Kendrapara affecting around population a population of around 11 million people. Many reports have suggested that those villages that were in the shadow of mangroves suffered the least damage from the cyclone (UNDP, WII).

Assessment of Human Security in Sundarban and Bhitarkanika

During the last three decades, as much as 40 per cent of the mangrove area has been converted to agriculture or urban centres. As a result of continuous biotic pressure, the mangrove and other marine resources are experiencing habitat loss, changes in species composition, and shifts in dominance, loss in biodiversity and threat to survival. Some of the mangrove species are on the way to extinction. Examples are *Xylocarpus granatum, Bruguiera cylindrical, Sonneratia acida* and *Cynometra ramiflora*. Similarly, *Nypa fruticans, Heritiera minor* and *H. formes* have all but disappeared from the Sundarbans. The settlers have encroached about 5000 km^2 of Sundarban for their settlement and cultivation. At present within the settled areas, the forest becomes very meager. The settlers sometimes find difficult to procure firewood and fodder for the domestic animals and encroach protected forests. About 59 villages in Bhitarkanika area depend on mangroves for their livelihood. The other nearby mangroves from Chilka, Hetamundia, Kujang and from estuaries of Subernarekha and Jatadhar have already disappeared. The forests of Bagapatia and Sunei-Rupei were cleared for rehabilitation purposes. The conflicts between traditional and commercial fishing practices are also a major factor contributing to mangrove degradation.

Man needs mangrove and mangrove needs man. The extreme poverty of the two region viz. Sundarban and Bhitarkanika (24 Parganas-South and Kendrapada district respectively) has lead to over-exploitation of these valuable resources. Due to lack of alternative livelihood, the dependency on these fragile ecosystems has increased drastically. The continuous decrease of the forest cover will ultimately reduce these protective coastline belts, thus increasing threats from natural calamities like storm, cyclones and tsunamis. Due to the decline of the forest resources, there will be an increase in intersociety conflict, which ultimately will affect the personal security in human community.

Conclusion

The Sundarban and Bhitarkanika have undergone enormous change at the hands of its human inhabitants. The various challenges and threats of environmental change have highlighted the associations and influences of environmental changes on security in general and that of human security in particular. These linkages are diverse, and dynamics of their interactions across socio-political and ecological scales vary across time and space. Unprecedented biophysical and socio-cultural alterations are result of natural upheavals and human interventions. They have become potent and real concerns of security in the Indian coastal zones. Government sensibility is very much need in this regard for the sustainable management of these coastal zones.

There is an urgent need of reducing the dependency of local community by providing them with alternative livelihoods.

The emerging scenarios in these socio-ecological settings underscore the urgency and scope to understand and integrate the synergy of environment and security to build upon an adaptive, effective, equitable and sustainable management practice for these natural shields against disasters.

References

Aksornkoae, S., N. Paphavasit and G. Wattayakorn. 1993. "Mangroves of Thailand: Present status of conservation, use and management." In: ISME (1993). The Economic and Environmental Values of Mangrove Forests and their Present State of Conservation in the South-East Asia/Pacific Region. International Society for Mangrove Ecosystems, International Tropical Timber Association and Japan International Association for Mangroves. ITTO/ISME/JIAM Project PD71/89 Rev.1 (F). Pp.83-133.

Badola, R. and Hussain S.A. 2003. Valuation of the Bhitarkanika Mangrove ecosystem for ecological security and sustainable resource use. Study Report. Wildlife Institute of India (WII). Dehra Dun. 101 pp.

Banuri, T 1996. Human Security. In Nauman Naqvi (Ed.) Rethinking security, rethinking development, 163-167. Islamabad: Sustainable Development Policy Institute.

Bose, S. 2004. The Sundarbans biosphere: a study on uncertainties and impacts in active delta region. *Proc. 2 nd. APHW Conferenc*e. vol. I. Singapore. Proc. pp475-483.

Chadah, S. and C.S. Kar 1999. Bhitarkanika, Myth and Reality. Natraj Publisher, Dehra Dun. 388 pp.

Chaudhuri, A.B. and Choudhury, A. 1994. Mangroves of the Sundarbans. Vol. 1, India. IUCN, Bangkok, Thailand.

Dani, C.S. and S.K. Kar.1999. Bhitarkanika-A unique mangrove ecosystem. Nature and Wildlife Conservation Society of Orissa. 30-42 pp.

Gilbert, A.J. and R. Janssen.1998. Use of environmental functions to communicate the values of a mangrove ecosystem under different management regimes. Ecological economics. Vol. 25; 32-346 pp.

Greenough, Paul 1987 "Hunter's Drowned Land: Notes Toward the Intellectual History of the Sundarbans." Sundarbans Workshop, Nov. 20-21, 1987, Washington, D.C.: Unpublished paper.

Hong P.N. and San H.T. 1993. Mangrove of Vietnam. IUCN, 173pp.

Lugo, A.E. and Snedaker, S.C., 1974. The ecology of mangroves. Annual Review of Ecology and Systematics 5: 39-64.

Macintosh D.J., and Ashton E.C.. 2003 Draft Code of Conduct for the Sustainable Management of Mangrove Ecosystems.World Bank, ISME, Centre Aarhus.

Macintosh, D.J., Ashton, E.C. and Havanon, S., 2002. Mangrove rehabilitation and intertidal biodiversity: study of the Ranong mangrove ecosystem, Thailand. Estuarine and Coastal Shelf Science 55: 331-345.

Macintosh, D.J., Ashton, E.C and Havanon, S. 2002.Mangrove rehabilitation and intertidal biodiversity: a study in the Ranong mangrove ecosystem, Thailand. Estuarine, Coastal and Shelf Science, 55(3):331-346.

MacNae, W. 1968. A general account of the Flora and Fauna of Mangrove swampa and prospectives in the Indo-West Pacific region. Adv. Mar. Biol. Vol. 6; 73-270 pp.

Mahapatra, B.K., Chatterjee, P., saha, D. and Datta, N.C. 1999. Declining trend in the abundance of seeds of tiger shrimp, *Panaeus monodon* (fabricus) in the Sundarbans with suggestions for its restoration. In Guhabakshi, D.N., Sanyal, P. and Naskar, K.R. ed. Sundarbans Mangal. P 532-538

Managing Global Environmental Risks. Annual Report 1998.

Mandal, A.K. and Ghosh, R. K. 1989. Sundarban. A socio Bioecological Study Bookland Pvt Ltd. Calcutta, pp. 193.

Ojo, S.O. 1987. The Climatic Drama, Inaugural lecture Series, University of Lagos Press.

Ong, J-E. 1995. The ecology of mangrove conservation and management. Hydrobiologia, 295, 343-351.

Patronobish S..2004. *A stretch of mangrove forest in the Sundarbans*. Article in Web site on Sunderbans. (http://www.riversymposium.com/index.php?element=12.)

Primavera, J.H., 2000. Development and conservation of Philippine mangroves: institutional issues. Ecological Economics 35: 91-106.

Rajaram Satapathy. 1998. 'Orissa sanctuary falls prey to poachers'. Times of India. 13/11/98.

Rakshit, D.K. 1989. 'Some Aspects of Community Preparedness Plan against Cyclone.' Proceedings of the US-ASIA Conference on Engineering for Mitigating Natural Hazards/Damage, Bangkok.

Saenger, P, Hegerl EJ, Davie JDS, 1983. Global Status of Mangrove Ecosystems. In: Anon (ed) Working Group on Mangrove Ecosystems of the IUCN Commission on Ecology, IUCN, Gland, Switzerland, p.88.

Shanmugasundaram, J., S. Arunachalam, G. Annamalai and M. Arumugam 1989. "Cyclone Hazards to Structures and Methods of Mitigation".Proceedings of the Symposium on Preparedness, Mitigation and Management of Natural Disasters. Vol.New Delhi.

Spalding, M. D., Blasco, F. and Field, C., 1997. World Mangrove Atlas. The International Society for Mangrove Ecosystems. Okinawa, Japan, 178 pp.

UNDP: Human Development Report 1994: Human Security, UNDP, 1994.

UNEP–GRID-Arendal. 2005. Environment Times No 3, GRID-Arendal, Logum Park, values of a mangrove ecosystem under different management regimes. Ecological.

Wilcox, B. H. R., 1985. Angiosperm flora of the mangrove ecosystem of the Niger Delta. In: The Mangrove Ecosystems of the Niger Delta. Wilcox, B. H. R. and Powell, C. P. (Eds.) Proceedings of a workshop, University of Port Harcourt, Port Harcourt, Nigeria, pp.34-44.

Part IV

Coastal Ecosystem: Country Reports and Case Studies

Chapter 22

The Status of Biodiversity in the State of Kuwait

Mahmoud Al Khabbaz

National Biodiversity Committee, Environment Public Authority, State of Kuwait

E-mail: khabbaz@epa.org.kw; markhbz@hotmail.com

Introduction

The State of Kuwait is situated at the northwestern corner of the Arabian Gulf between latitudes 29° 22'N and longitudes 47° 58'E. The total land area of the country is estimated at about 17818 km^2 including the land areas of the offshore islands. To the north and the west, it shares a border of 240 km with the Republic of Iraq, and to the south and south-west it shares 250 km with the Kingdom of Saudi Arabia. On the east it has a coastline of 290 km on the Arabian Gulf facing The Islamic Republic of Iran. The Gulf is a shallow (maximum depth 31 m) sedimentary basin, with freshwater input mainly from the Tigris-Euphrates system, and subject to strong evaporation. Temperature and salinity in the Gulf range from high to very high during summer. Evaporation is greatest in southern sectors of the Gulf, where surface salinity is highest; this drives outward movement of dense water at depth through the Straits of Hormuz and inward movement of less saline surface waters from the Gulf of Oman.

Having realized the value of biological resources as an integral part of the natural heritage with great potential for yielding sustainable benefits, the state of Kuwait has signed the Convention on Biological Diversity at the United Nations Conference on Environment and Development (UNCED) in Rio de Janeiro, in June 1992.

Conservation of biodiversity is especially vital in arid regions, like Kuwait, where ecosystems are fragile and the renewable natural resources are scarce. Moreover, such ecosystems become devastated when subjected to crisis, such as the case of the Gulf War which resulted in total loss/damage and alteration in the biota of most ecosystems in Kuwait and Saudi Arabia. In such ecosystems the lost habitats coupled with losses in species diversity are difficult to restore and require long-term rehabilitation efforts. Changes and alterations in species biodiversity have to be carefully monitored, policies should be reviewed possible conservation measures should be prepared and strategies and action plans to conserve, manage, and rehabilitate resources should be formulated.

The desert and the marine ecosystems provide goods and services to the people (such as food, feed, fiber, aesthetic values and recreation opportunities) and therefore are of great commercial and social value. The harsh climatic conditions in Kuwait especially during the summer, cause the desert ecosystem to become totally denuded as most perennial plants become dormant. Yet, during rainy seasons (spring and winter) the desert becomes rich with green mantle and many plants and animals thrive. The Kuwait Marine Environment as part of the Arabian Gulf tropical lndo-Pacific region, is in contrast characterized by high species diversity and abundance.

This report consists of five major parts. The first two parts present general characteristics and biodiversity of both the terrestrial and the marine ecosystems of Kuwait, the third part highlights the major threats to biological ecosystems, and the fourth part presents institutional role in conservation and monitoring of biological diversity with brief description of policies and regulations issued by the government. The regional and international collaboration is discussed in the last part of the report.

The Terrestrial Ecosystem

Kuwait occupies the southern part of the Al-Dibdibba plain which is bounded to the north by the Mesopotamia flood plain of Iraq and to the west by the Al-Dahna sand area of Saudi Arabia. The surface of Kuwait is generally flat to gently undulating desert plain occasionally interrupted by low relief ridges, hills and escarpments. It slopes gradually northeastward with an average gradient of about 2 m/km. The landscape of Kuwait is controlled, to a great extent, by the occurrence of duricrusts.

In spite of its small area (about 17818 km^2) nine geomorphic provinces can be recognized in the desert of Kuwait: coastal flat, northern gravely flat, Al-Huwaimiliyah undulated plain, western calcretic plain, southern sandy flat, Jal Az-Zor escarpment, Wadi Al-Batin, and major ridges and depressions.

The vegetation comprises of perennial herbs and ephemerals that vary seasonally depending on winter rainfall. The vegetation of Kuwait is an important component of the desert ecosystem, it protects the soil from erosion, and provides important sources of feed for both livestock and wildlife species.

The desert ecosystem of Kuwait is deteriorating rapidly due to multiple interacting factors such as overgrazing, uprooting of woody shrubs, increasing activity for recreation, gravel quarrying, and other environmental and natural processes. Lack of law enforcement in protecting the terrestrial ecosystem is another cause for the noticeable destruction of the components of this ecosystem. The Iraqi occupation of Kuwait added devastating constraints and pressures on the desert ecosystem of Kuwait.

Desert Ecosystem Diversity

There are four distinct desert ecosystems identified in Kuwait: The sand dune ecosystem, the salt marsh and saline depression ecosystem, the desert plain ecosystem and the desert plateau ecosystem. Each ecosystem is characterized by the dominating plant community and the associated species. Five major communities are recognized *Haloxylon salicornicuin, Rhanterium epapposum, Cyperus conglomeratus, Zygophyllum qatarense,* and *Panicum turgidum.* The communities are associated with *Stipagrostis plumosa, Moltikiopsis ciliata, Plantago boissieri, Schimpera arabica, Arnebia decumbens, Astragalus spp.* and others (Halwagy and Halwagy 1974).

Some areas in the desert are designated as national parks/protected areas for the conservation of habitats. Each area has a different type of ecosystem/habitat. For instance, the National Park of Kuwait in the northwest of Kuwait Bay is about 250 Km^2 in area and includes the Jal Az-Zor escarpment, Urn Al-Rimam depression, the sand dune foot hills, the coastal habitats and the desert plateau habitat. Each ecosystem is characterized by different types of soils and plant communities. Other reserves include the Doha Bird Reserve and the Jahra Pool Reserve (Omar *et al.,* 1983).

Terrestrial Species Diversity

Flora

Kuwait's wild plant-life is one of its unique natural heritage. Plants are adapted to survive in the harsh conditions and extreme temperatures. The Vegetation comprises of perennial herbs and ephemerals that very seasonally depending on winter rainfall. Five major plant communities are recognized *Haloxylon salicornicum, Rhanterium epapposum, Cyperus congolmoratus, Zygophyllum qatarense,* and *Panicum turgidum.* The communities are associated with *Stipagrostis plumosa, Moltikiopsis ciliata, Plantago boissierii, Schimpera arabica, Arnebia decumbens, Astragalus spp.* and others.

Studies on plant species in Kuwait conducted in 1982 indicated that 285 species are indigenous to Kuwait. Many introduced species were excluded from the list as they were not found in the denuded lands. The list indicated that annual species are most dominating (179 species) followed by shrubs (44 species) and trees (3 species) (Omar 1982). More recent studies indicated an increase in the total number of recorded plant species totaling 386. The list includes native and exotic species that are spreading in denuded lands and cultivated areas. The data on the native plants of Kuwait needs more compilation and assessment especially at the autecological level.

Research on the productivity potential of the major plant communities in Kuwait are documented in several technical reports of Kuwait Institute for Scientific Research (KISR). However more research is needed to evaluate the status and condition of the plant communities especially due to the fluctuation in the amount of annual precipitation, and the intensive use of resources.

Fauna

Kuwait is home to numerous species of insects, animals and birds. Among the diverse insects the most attractive group is that of butterflies. Several beautiful varieties are found here and the best time to see them is spring. There are 40 recorded reptiles in Kuwait, comprising of a variety of snakes, lizards, geckos etc. Besides, there are around 28 listed varieties of mammals a number of them are endangered species. Some commonly found animals include hedgehog, wolf, wild cat, gazelle etc. Nearly

350 species of birds have been recorded here. Incidentally, most of them are migratory birds, only 18 are resident birds that live and breed locally.

On the endangered list of Kuwait are: 8 mammals, Fennec Fox (*Fennecus zerda*), Common Red Fox (*Vulpes vulpes*), Honey Badger (*Mellivora capensis*), Indian Grey Mongoose (*Herpestes edwardsi*) Wild Cat (*Felis silvestris iraki*), Caracal Lynx (*Caracal caracal*), Saudi Gazelle (*Gazella dorcas saudiya*), Sand Gazelle Saudiya (*Gazella subgutturosa marica*); 5 birds: Hubara Bustard (*Chamydotis undulata*), Crab Plower (*Dramas ardeola*), Cream-coloured Courser (*Cursorius cursor*), Brown-necked Raven (*Corvus ruficollis*), Pin-tailed Sandgrouse (*Peterocles alchata*); and several reptiles (*e.g.* Spiny-tailed Lizard *Ultramastix microlepsis*) (Al-Sdirawi 1989).

Further research is needed to monitor wildlife and to record their status and distribution. Measures are also needed to conserve their habitats and to re-enforce laws to prevent hunting and misuse of wildlife biodiversity.

Insect Fauna

The Insect Fauna of Kuwait started with the collections of H.P. Dickson, V. Dickson and G.E. Bryan (Al-Houty, 1989). Uvarov (1930) prepared a list of insects which he presented to the British Museum (Natural History). He drew a long list of species of the Coleoptera, Hemiptera and Neuroptera. In 1970, Brown listed the butterflies of Kuwait and Goodridge (1977) mentioned insect pests of a local tree. In 1989 Al-Houty published a list of insects which were collected from Kuwait during an extensive 8 year study. More than 570 species are reported from several locations in the State of Kuwait.

The Marine Ecosystem

The situation of the country in the northwestern corner of the Arabian Gulf adjacent to Shalt Al-Arab provides the marine environment with distinct features. The flow of Shalt Al-Arab influences the physical properties of seawater, topography and habitat characteristics of the northern area. It also creates environmental gradient between the northern area, including Kuwait Bay, and the southern area (which extends beyond the Bay to the southern borders).

Kuwait's coastline totals about 660 Km including the islands. The most characteristic feature of the marine ecosystem is Kuwait Bay which is a large engulfment shallow basin extending from open sea to the west. It is the main nursery ground for shrimps and many species of fin fishes. The northern shore of the Kuwait Bay is characterized by an extensive area of mudflats, which extend from Khor Ai-Subiyah to Sulaibikhat Bay. The mudflats also extend in the southern coast of Kuwait Bay and around Warba and Bubiyan Islands. The water turbidity in Kuwait Bay is comparatively higher than that of the southern region. The transparency ranges from about 133-147 cm in the north to 255-461 cm in the south. Whilst the annual salinity in the Bay is 39 ppt, it varies in the south to 42 ppt.

At the southern tidal flat area a complex system of creeks and lagoons are located in Al-Khairan which is near the southern border of Kuwait. Al-Khairan includes the following creeks: Khor Al-Nhaim, Khor-Iskandar, Khor Al-Muffatah and Khor Al-

Ami. The main creeks Khor Al-Muffatah and Khor Al-Ami stretch few km inland. These two *Khors* represent a hypersaline environment with salinity reaching more than 75 ppt at the interior part of the tidal flat, and 40 ppt at the inlet. The substrate at the inlet is mainly sandy that gradually changes into muddy sand and then muddy.

Marine Ecosystem Diversity

In spite of the comparatively small area of the marine environment it comprises of diverse habitats including the following areas:

Intertidal Area

The major types of shore grades that are represented along the coastal line include: mudflats, sandy shores, sandy rock flats and rocky shores at Doha, Ras Ajuza in Kuwait Bay, Beda'a, and Mina Saud in the southern area. The most productive and diverse tidal flat is the mudflat which supports a diverse marine life. It is an important transit area for wading and migratory birds and might be an important fish spawning and nursery ground. The highest part of the mudflats is occupied by salt marshes of five major plant communities that are characterized by dominant species such as: *Halocnemum strobilaceum, Nitraria retusa, Tamarix aucherana, Seidlitzia rosmarinus* and *Zygophyllum qatarense.* Landward of the salt marshes extends a wide area of sabkha flats (it forms flat plains, with crusts of sodium chloride and gypsum, with important and quot; algal mats and quot; a few centimeters thick, beneath which is a black reducing layer) and below the halophytic zone algal mats, composed of blue-green algae associated with diatoms, are usually found.

The Intertidal area is inhabited by diverse benthic communities. Macrofaunal community is considered impoverished; area of high number of organisms dominated by one or few species. The extensive study of the entire Kuwaiti Intertidal zones (except the islands) conducted by Al-Bakri, *et al.* (1985) showed that 235 taxa of macrofauna inhabit the Intertidal zones of Kuwait. These taxa belong to 12 families, 48 orders, 23 classes and 11 phyla. In terms of abundance according to number of taxa the phyla are Mollusca, Annelida, Anthropoda, Chordata, Echinodermata, Porifera, Coelenterata, Echiurida, Bryozoa, Sipunculida and Platyhehninthes. Johnes (1986) recorded 304 macrofaunal species belonging to 12 phyla as well as 45 species of marcoflora from rocky and sandy shores of Kuwait.

Mudflats

Mudflats form an important part of the Intertidal region of Kuwait Bay and support an extremely varied animal and plant life and are one of the richest habitats of the Kuwait marine environment. Of prime importance are the three species of mudskippers, these gobiid fishes are *Periophthalmus koelreuteri, Boleophthalmus bodarti* and *Scartelaos viridis* (Clayton, 1986).

Meiofauna is monitored periodically by the Environmental Protection Authority (EPA). In general the meiofaunal community is dominated in terms of number of individuals by 3 major taxonomic groups; Foraminifera, Gastropoda, and Nematoda. About 200 species are recognized belonging to 12 phyla. However, the most diverse taxonomic groups are Polychaeta (22 species), Ciliata (20), Rotifera (17), Nematoda (17), Foraminifera (17), Gastropoda (15), and Bivalvia (13).

Coral Reefs

Kuwait's corals are located in the most northern areas of the Arabian Gulf. There are three well developed coral islands with their associated reefs: Kubbar, Qaru and Umm Al-Maradim. The other areas *e.g.*, Qittat Urayfjan, Taylor Rock, Mudayrah, Qaru NW, Umm Al-Maradim NW, Deres Rock, Getty, Twin Reef and Single Reef, are platform Reefs. The coral reef community is of low diversity, in comparison with coral reefs of the Red Sea or those around Jena and Karan islands in Saudi Arabia. About 29 species of corals from 31 genera are recorded by Downing (1985). The communities are either dominated by the genus *Porites* or the genus *Acropora*. About 199 coral reef associated fishes have been identified. They comprise more than 50 per cent of the fish species recorded in Kuwait.

Species Diversity in the Subtidal Habitats

Species diversity of planktons, shrimps and fishes are of particular interest to researchers and to concerned governmental agencies. Selected groups of these populations are monitored systematically and information about their abundance, seasonal variation and distribution is available.

Plankton

Plankton species are recognized as indicators of environmental stresses and are therefore monitored monthly by the Environment Protection Department of the Ministry of Health. Baseline data on their abundance, seasonal variation and distribution are available in EPD, KISR and Kuwait University (KU). Phytoplankton populations are dominated by diatoms which constitute about 90 per cent of the population. About 117 species have been identified, 69 diatoms, 36 dianoflagellate and 12 blue-green algae. The highest recorded phytoplankton annual density was about 20,236 cell/l, and in August–September bloom it was about 119,616 cell/l (EPA Annual Report 2005).

Zooplankton

In EPA studies the highest recorded annual density for Zooplankton was about $50264/m^3$ and highest recorded value in summer bloom was $86015/m^3$. About 82 species were recorded of which 40 species were *copepods* in addition to 12 other-crustacean and 30 non-crustacean species. In an extensive study carried out in Kuwaiti waters Michel *et al.* (1986) reported more than 95 species of Zooplankton in 15 phyla. The copepods represented by over 40 species were the most diverse and numerous group, constituting 83.1 per cent of the Zooplankton. Next in abundance were Larvacea (7.1 per cent), Cladocera (4.7 per cent), Mollusca (2.9 per cent), Chaetognatha (1.2 per cent), Decapoda (0.5 per cent), and Coelenterata, Ctenophora and Ostracoda (0.1 per cent each). The numerically dominant Copepods species were *Paracalanus Crassirostris, Oithona spp., Temora turbinata, Euterpina acutiferons* and *Microsetella rosea.;* among the Non-Crustaceans larvaceans *Oikopleura spp.;* and the *Cladoceran penilia avirostris* from the Other- Crustaceans (EPA Annual Report 2004).

Fishes and Shrimps

Kuronuma and Abe (1986) recorded 465 fish species in 101 families and 24 orders in the Arabian Gulf. Only 7 fish species, *Pseudochromis persicus, Cheilodipterus bipunctatus, Petrus belayeur, Upenus oligospilus, Blennius persicus, Istigobius day*, and *Callinymus persicus* are endemic to the Gulf. They constitute only 1.5 per cent of the known fish species in the region. Fish statistics is conducted by PAAFP and KISR. The data include the shrimps *Penaeus semisulcatus, Metapenaeus affinis* and *Penaeus japonicus;* as well as 22 commercially important fish species, *Hilsa ilisha, Lisa macrolepis, L. subviridis, Pampus argenteus, Epinephelus tauvina, E. jayakri, Otolithes argenteus, Formio niger, Latiganus coccineus, Scomberomorus guttatus, S. comnerson, Caranx malabariens, Plactorhynchus cinctus, Lithrinus ralloptenus, Johnius maculatus, Rachycentron canadus, Acanthopagrus latus, Saurida tumbil, Nemipterus japonicus, Arcyrops filamentosus, Pomadasys argenteus, Eleutheronema tetradactylum.*

Major Threats to Biological Diversity in Kuwait

Many factors are greatly effecting biological diversity in Kuwait. Among these are the following:

1. The relatively small area of the State of Kuwait both in land and sea has resulted in great demand for suitable sites for development of urban, industrial and touristic areas. The pressure is particularly great on the coastal areas since development is taking place alongside the shorelines. Valuable shoreline communities have already been eliminated by landfill and the coastal ecosystems,especially the muddy Intertidal zones are altered by construction of marinas and artificial islands. Likewise vast areas of the natural desert ecosystem are lost due to urban expansion; roads construction and oil production, transport and storage activities.
2. Inappropriate fishing practices over many years *i.e.* use of bottom trawls and fishing in breading areas has greatly threatened biodiversity in the sea. Much damage is also caused by sports fishermen, particularly damage caused by anchoring in coral reef areas.
3. Uncontrolled wildlife hunting, uprooting of woody shrubs and trees, seasonal camping, and over-gazing have hastened the process of land degradation and altered biological diversity.
4. Pollution, particularly oil pollution in the marine environment and recently in the desert environment, pause a major threat to biodiversity in Kuwait. Several species of corals have died due to oil pollution. Other pollutants specially chemical and biological pollutants are also threats to biodiversity especially in shallow-coastal area communities.
5. Lack of in-depth and continued studies to survey the Species diversity in different ecosystems have lead to lack of appreciation for biodiversity in Kuwait. This is particularly true for diversity of marine benthos. Such lack of information is due in the first place to lack of specialists in the taxonomy of minor phyla. Even large taxa such as polychaetes, amphipods and sponges are not well identified.

6. Lack of law enforcement and legislation that calls for preservation and protection of biodiversity have lead to improper and unwise utilization of natural resources and adversely effected the biota both at sea and land.
7. Import of large numbers of exotic plant species as part of the country's greening program, also, marine fish species as for fish culture pauses a threat to indigenous biodiversity since these exotic species are often carriers of pests, bacteria and viruses.
8. Lack of clear policy and strategy for the conservation and management of resources lead to non-sustainable use of resources and irreversible losses of biota.
9. The 1990 Iraqi invasion to Kuwait and the subsequent war activities resulted in major environmental crisis that affected the Arabian Gulf regional environment its air, land, water, and biota. Among which the injury caused to the region's natural resources from the torching of Kuwait's oil fields, the fallout of thousands of tons of oil and soot, and the deliberate dumping of an estimated 11 to 12 million barrels of oil into the Gulf from several tankers and loading facilities off the coast of Kuwait. These massive oil spills destroyed whole ecosystems that had developed and prospered over many millennia along the coast.

Institutional Arrangements and Legal Aspects

Many governmental and non-governmental organizations in Kuwait have mandates to use natural resources, therefore, involve in activities related to biodiversity. However, only four governmental institutions:

Environment Public Authority (EPA); Public Authority for Agriculture and Fish Resources (PAAFR); Kuwait University (KU) and Kuwait Institute for Scientific Research (KISR) are responsible by Amiri Decrees for management, conservation and research on biodiversity in the State of Kuwait.

The non-governmental societies such as the Kuwait Environment Protection Society, the Agriculture Engineering Society and the Kuwait Centre for Volunteer Works are also active in pursuing biological diversity conservation and management.

The National Biological Diversity Strategy of Kuwait

Kuwait has signed the International Convention on Biological Diversity (CBD) which was launched at the Earth Summit in the 1992, and it has been ratified the convention since 2002. Since 1996 Kuwait has embarked on the preparation of a National Biodiversity Strategy, central activity called for in the Convention's articles. The National Strategy, initiated and funded by United Nations Development Program (UNDP), is being coordinated through the offices of the Environment Protection Authority (EPA), the Wildlife Committee in Kuwait, with technical assistance from the International Union for Conservation of Nature (IUCN). The overall objective of the Strategy is to conserve Kuwait's biological diversity, ecological systems and processes.

The strategy formulates an integrated approach to biodiversity issues in all sector and at all levels of governance in Kuwait. The intention is to develop a concise, non-technical, multi-sectoral planning document which can be readily understood at all levels. The strategy presents the primary goals and objectives of biodiversity conservation, set out the justification issues. The strategy also includes an action plan, with appropriate measurable targets to redress the situation. The strategy contains a series of guidelines for policy and decision makers to try to ensure that biodiversity considerations are fully integrated into the national planning and development process.

The strategy is based on a thorough analysis of the best information available; this is particularity true with regard to an understanding of the status and trends of Kuwait's biological resources, the socio-economic dynamics which contribute to biodiversity loss, and the prevailing policy and legislative environment.

The State of Kuwait has ratified three Wildlife Convention in the year 2002:

1. The International Convention on Biological Diversity (CBD),
2. The Convention on International Trade in Endangered Species of Wild Fauna and Flora (CITES), and
3. The Regional Convention for Wildlife Conservation in the Arab Gulf State Countries.

The Activities of Governmental Organizations Including PAAFR, EPA, KU and KISR

Environment Public Authority (EPA)

EPA was established since 1996 by law (former, Environmental Protection Council, EPC, 1980-1996) to ensure environmentally sound land use and to establish protected areas to preserve wildlife flora and fauna. For the biodiversity conservation issues, two National Committees were formed in EPA to follow up the outcoming decisions from the Conference of Parties for the CBD and CITES, and The Regional Convention for Wildlife Conservation in the Arab Gulf State Countries, and to assure the commitment to the action plan of The National Biological Diversity Strategy of Kuwait, furthermore, to coordinate activities and follow up on issues related to Biodiversity. Members of the National Biodiversity Committee include representatives of EPA, Ku, KISR, PAAFR, Kuwait Municipality, Ministry of Foreign Affairs, Kuwait Environment Protection Society, and the Kuwait Centre for Volunteer Works.

To fulfill its responsibilities, EPA has processed two Law Decrees:

1.. Decree Law No. 210/2001 empowered EPA environmental regulations; it contains a chapter on Biodiversity.

2.. Decree Law No. 93/2003 regarding the regulations to the governmental agencies and the wildlife traders to assure the commitment to the Convention on International Trade in Endangered Species of Wild Fauna and Flora (CITES).

The Public Authority for Agriculture and Fish Resources (PAAFR)

PAAFR was established in 1983 in accordance with Law No. 94. The Authority assumed the management and operation of agriculture facilities which were previously attached to the Ministry of Public Works.

PAAFR supervises all production activities on state level, landscaping activities in coordination with Kuwait Municipality (KM) and grazing lands in collaboration with concerned bodies in the State. PAAFR was also given the responsibility to establish national desert parks and to undertake their supervision. It was also mandated to control desertification by using all methods for its prevention in coordination with concerned bodies and in accordance with plans made by Kuwait Municipality.

The Law No. 71, 1988 empowered PAAFR to ban grazing in certain areas, issue grazing permits and confiscate livestock in case of violation.

Amiri Decree Law No. 112, 1976 regarding Agricultural Quarantine. This Law prohibits the import of plants or any agriculture products infected by pests and gave PAAFR the responsibility in regulating and planning their entry into the State.

To fulfill its responsibilities, PAAFR issued a number of ministerial decisions including the following:

1. Decision No. 242/1989 regarding licensing of grazing permits.
2. Decision No. 244/1989 regarding penalties for violations of the Decree No. 41/1988 regarding range management.
3. Decision No. 22/1981 regarding prohibition of fishing vessels using certain techniques in designated area.
4. Decision No. 9/1983 regarding fishing fin fishes of certain length.
5. Decision No. 10/1983 regarding preventing fishing at designated areas in the territorial waters.
6. Decision No. 11/1983 regarding prohibition of fishing within 3 miles from the main land.
7. Decision No. 9/1985 regarding regulating the agriculture quarantine.
8. Decision No. 33/1992 regarding mesh size of nets used in catching shrimp and Fin fishes.

Academic and Research Institutions

Kuwait University (KU) and Kuwait Institute for Scientific Research (KISR) are conducting research dealing with biodiversity on species and ecosystem level and participate in decision making regarding the issue. Moreover, they assist the concerned agencies by conducting studies, on management and conservation of protected areas.

Regional and International Co-operation

As a member of the GCC and the United National Organizations, Kuwait is collaborating with other Nations in Biodiversity issues and Conservation of wildlife in terrestrial and marine ecosystems through the following specific organizations and programs:

Regional Organizations

1. Regional Organization for the Protection of the Marine Environment (ROPME).
2. Environment Committee of GCC Countries. The Committee is now in the process of finalizing the consolidated agreement for the protection of wildlife in the region.
3. The Council of Environmental Affairs Ministers–Arab League.

International Organizations

1. United Nations organizations including United Nations Environment Programme (UNEP); United Nations Educational, Scientific and Cultural Organization (UNESCO), World Health Organization (WHO), United Nations Development Programme (UNDP) and United Nations Food and Agriculture Organization (FAO).
2. International Union for Conservation of Nature and Natural Resources (IUCN).

References

Al-Bakri., Foda, M., Behbehani, M., Khalaf, F., Shublaq, W., El-Sayed, M.I., Al-Sheikh, Z., Kittanah, W., Khuraibit, A., and Al-Kadi, A. 1985. Environmental Assessment of the Intertidal Zone of Kuwait. Final Report, Project EES-35 Environment Protection Council and Kuwait Institute for Scientific Research, Volume 1.

Al-Houty, W., 1989. Insect Fauna of Kuwait, Kuwait University 196 pp.

Al-Sdirawi, F.A. 1989. Wildlife Resources of Kuwait. Historic Trends and Conservation Potentials. PhD. Dissertation. UMI. University Microfilms International. USA.

Biodiversity in The State of Kuwait, Country Report. Prepared by The Wildlife Committee Environment Protection Council Kuwait, 1994.

Brown, J.N.B 1970. Preliminary list of Butterflies of Kuwait. Ahmadi Nat. History Fld. Stud. Newsl. No. 2:7.

Clayton, D. A., 1986. Ecology of mudflats with particular reference to those of the northern Arabian Gulf in Marine Environment and Pollution, Riad Halwagy, David Clayton and Manaf Behbehani eds. Kuwait University, pp. 83-96.

Downing, N. 1985. A Study of the coral reef fisheries. Kuwait Institute for Scientific Research. KISR MB42, Kuwait.

EPA Annual Reports, Kuwait, 2004 and 2005.

Goodridge, E.J. 1977. Insects. Ahmadi Nat. History. Fld. Stud. Grp. No. 16.

Halwagy, P., and Halwagy, M., 1974. Ecological Studies on the desert of Kuwait. II. The vegetation. J. Univ. Kuwait. (Sci.)1:87-94.

Jones, D. A., 1986. A Field Guide to the Seashores of Kuwait and the Arabian Gulf. Kuwait University.

Jones, D. A., 1986. Ecology of the rocky and sandy shores of Kuwait in Marine Environment and Pollution. Riad Halwagy, David Clayton and Manaf Behbehani eds. Kuwait University, pp. 69-81.

Kuronuma K, and Abe, Y., 1986. Fishes of the Arabian Gulf. Kuwait Institute for Scientific Research, Kuwait.

Michel, H.B., Behbehani, M., Herring, D., Arar, M., Al-Shoushani M. and Brakoniecki, 1. 1986. Zooplankton diversity, distribution and abundance in Kuwaiti waters. Kuwait Bulletin of Marine Science (8): 37-105 pp.

Omar, S.A., 1982. Baseline information on the native plants of Kuwait. Kuwait Institute for Scientific Research. Technical Report. KISR1790, Kuwait.

Omar, S.A., and Al-Shuaibi F., 1986. Criteria for development and management of Kuwait's first national park/nature reserve. Vol. III. Summary and Recommendations. Kuwait Institute for Scientific Research. Final report No. KISR2164. Kuwait.

Uvarov, B.P., 1930. Insects of Kuwait. In Dickson, H.R.P. The Arab of the Desert. George Allen and Unwin Ltd, London.

Chapter 23

Status of Coastal Ecosystem in Mauritius

M. Nallee
Ministry of Agro-Industry and Fisheries (Fisheries Division), Mauritius
E-mail: mnallee@mail.gov.mu

Introduction

Mauritius is the central member of the Mascarene group of islands and is located at 20°S and 58°E. The Republic of Mauritius has an Exclusive Economic Zone (EEZ) of 1.9 million km^2 (Figure 23.1) and the coastline of Mauritius is 322 km long. It is almost fully encircled by coral reefs, 150 km long and enclosing a lagoon area of 243 km^2. Breaks occur in the reef on the southern and western coasts where there is no continental shelf, and at river mouths where the high infusion of fresh water does not favour the growth of corals. The climate is between tropical and subtropical with a cyclonic season, which starts in November and ends on 15th May. The wind blows mostly from the southeast and the water temperature of the ocean varies between 22 and 28°C. The tide variation is weak–about 0.7 m during the spring tide.

Mauritius is surrounded by 49 offshore islets, and two groups of outer islands Agalega and St. Brandon. Agalega comprises two sandy cay islands covering 21 km^2 surrounded by 100 km^2 of coral reef, with a lagoon of 25-100 m wide. St Brandon Shoals comprises 55 low-lying islets and sand cays covering 3 km^2 within a 100 km long coral reef system, and a shallow water area extending over several thousand km.

Mauritius has an extremely rich coastal zone consisting of near-shore wetlands and mangroves, lagoon coral, fringing coral reef and all their associated marine life. All these components are interrelated–wetlands provide a natural buffer, controlling surface water run-off to the lagoon by neutralising pollutants, nutrients and sediments which damage the lagoon ecosystem; mangroves provide a habitat for juvenile fish

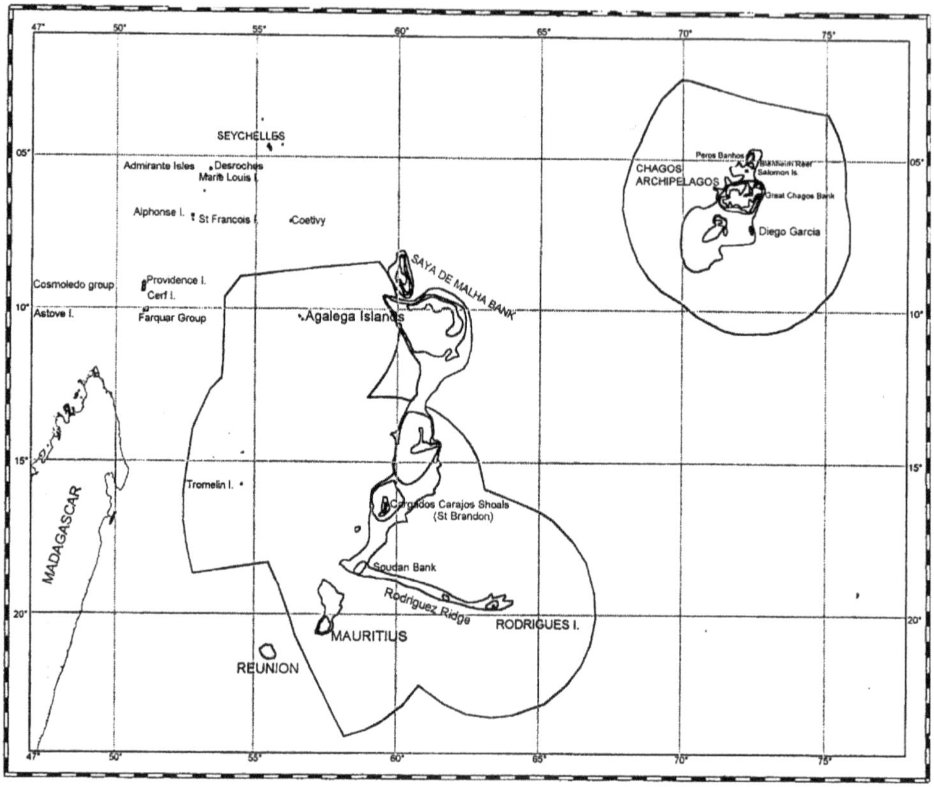

Figure 23.1: Exclusive Economic Zone of Republic of Mauritius

and invertebrates; the fringing coral reef protects the coastline of Mauritius from the sea and is pivotal to the ecology of the ocean.

Foreign tourists also exploit the marine and coastal ecosystem for recreational purposes, with tourism being one of the most vibrant industries in the Mauritian economy. Tourists enjoy the beaches, water sports, sport fishing, dolphin watching, snorkelling and diving. For example, in 2004, 109,000 people visited the Blue Bay Marine Park. The sub-tropical location of the island, its natural beauty, turquoise lagoon and warm climate together with a hospitable population constitute the necessary ingredients for a successful tourism development. The long-term success of the industry is intimately linked to the quality of the lagoon, its biodiversity resources and associated ecosystem services including the natural environment, although at present there is little 'eco-tourism'. There are also plans to develop the outer islets of Agalega and St. Brandon, Rodrigues and Outer Islets for tourism. The beach and the lagoon are also popular leisure sites for Mauritians.

As a result of intense pressure being exerted on the coastal zone from sea-based and land-based activities, the marine environment has been affected in certain areas. In the long run the whole lagoon may be degraded putting at risk its economic potential if these activities are not controlled. In order to maximise the economic and social benefits from the lagoon, a holistic approach in the management of the intense and often conflicting pressures on the lagoon is essential.

The Coral Reef Ecosystem

The coral reef is one of the most biologically productive and diverse of all ecosystems, providing food and shelter for most of the fish species caught in the shallow tropical coastal waters. It also acts as a barrier against aggression of the sea and contributes to the formation of sandy beaches and lagoons. However, the coral reef is slow growing and fragile ecosystem prone to natural and anthropogenic disturbances.

Fringing Reefs occur in shallow water near to land, extending to depths of 15-45 metres. The coral reef encircling Mauritius and protecting it from the sea is a fringing reef. The seaward reef front is where virtually all scuba diving takes place in Mauritius.

Patch Reefs are found in relatively shallow water around Mauritius where the underlying seabed has at some time been close enough to the surface for corals to grow. Many patch reefs in Mauritius have been seriously impacted by pollution and physical damage. Blue Bay and Balaclava, the sites of the two marine National Parks, are good examples of patch reefs.

Atolls start as fringing reefs around volcanic islands, forming atolls as the island gradually submerges. The offshore islands of the St. Brandon archipelago include 22 atolls.

Reef flats are formed as the fringing reef pushes steadily seaward leaving behind limestone areas that are eroded almost flat by the sea, and are essentially an inter-tidal zone. Reef flats do occur around the island of Mauritius but are more significant around Rodrigues.

Barrier reefs are developed typically on the edge of a continental shelf. They may have an atoll-like geological history. Barrier reefs are found far from the main shoreline and are separated from the shore by deep water. They occur in the south and in the south west of the island.

A total 159 species of Scleractinian corals (hard corals) have so far been recorded. Out of 340 species of fish which have been identified, 42 are of economic importance within the inshore area, with a different composition and relative abundance in the near-shore waters of each island. The effect of over-fishing of *Lethrinids* is apparent on the fringing reefs of Mauritius with a population explosion of sea urchins *Diadema* spp. and *Echinometra* spp. in evidence.

Five species of Penaeid shrimps (*Penaeus monodon, P. latisulcatus, P. canaliculatus, P. indicus* and *Metapenaeus monoceros*) as well as two species found in deeper water have been identified. Other invertebrates include octopus, mussels, oysters (the endemic *Crassostrea cuculata*), barnacles and clams.

Several species of sea stars and echinoderms are reported in the Mauritian waters. The coral eater *Acanthaster plancii* (crown of thorns (COT)) is occasionally sighted on the fringing reefs.

Mauritius has a rich algal flora. Over 160 genera of marine algae have so far been identified from the coastal waters. The marine floristic records of Mauritius date back to 1875 and the Mauritius Herbarium have a collection of more than three hundred marine algae.

Sea cows, once common in the lagoons of Mauritius are extinct, and marine turtles are rarely seen due to lack of beaches and intense hunting pressure. However, they still use St. Brandon Shoals and Agalega (and Rodrigues) shores for breeding. Seventeen marine mammal species have been recorded in Mauritian waters–mostly during their migration to and from warm tropical waters from Antarctica for calving. Some dolphins are resident in Mauritian waters and encountered more frequently than whales although the breeding and nursery grounds of the dolphins have not yet been located.

Of the outer islets, the marine ecosystem around St. Brandon is still virtually ecologically intact and very few coral reefs anywhere in the world still harbour such large populations of the bigger reef fish, corals and smaller sea creatures. The vegetation formations are also largely intact and consist of pantropical coastal species. St. Brandon is one of the most important tropical seabird sites with significant breeding populations of eight species. Seabird numbers appear to be in decline mainly because of killing and disturbance by fishermen and the effects of introduced rats. Nesting sea turtles are still numerous but numbers are also in decline.

Impacts on Reefs

It is a fact that coral reefs throughout the world are degrading at an alarming rate. It has been estimated that as much as 10 per cent of the global reefs has been degraded beyond recovery and that another 30 per cent will collapse within the next 10-20 years. Coral reefs are under constant man-induced and natural disturbances. Anthropogenic activities may cause extensive degradation from nutrification, over fishing and sedimentation. Tourism is the most recent threat to reefs causing major problems such as anchor damage, extensive hotel and beach development along the coast and dredging of bathing areas and ski lanes. Some other disturbances such as rise in surface sea temperature; coral bleaching, coral diseases, red tides and outbreaks of crown-of-thorns starfish may be linked to human activities. Natural physical disturbances may be caused by climatic factors such as storms, cyclones and floods.

Measures to Protect Coral Reefs

To safeguard the marine environment and its associated ecosystems, a number of measures have been taken, namely:

1. The removal and sale of corals and shells have been banned.
2. A closed season for net fishing is observed.
3. All types of underwater fishing have been declared illegal.
4. A very tight control on the import, storage and utilization of dynamite to prevent its use in fishing is exercised.
5. The mining of corals for lime production has been prohibited.

Wetlands

As defined under the Fisheries and Marine Resources Act of 1998, a wetland is (*a*) an area of marsh whether (*i*) natural or artificial; or (*ii*) permanently or temporarily with water which is static or flowing, brackish, or salty; and (*b*) includes areas of marine water. There are 44 coastal wetlands in Mauritius. Wetlands act as an interface between terrestrial and marine systems and play an important environmental and economic function. They reduce erosion, act as a nursery for fish stocks, protect the coast and filter runoff before it enters the sea.

Measures have been initiated to preserve the wetland at Rivulet Terre Rouge Estuary Bird Sanctuary (RTREBS) which was declared a Ramsar Site in 2001, the first for Mauritius. The wetland, on the edge of Port Louis, covers 26 ha and comprises mud flats and muddy sands. The estuary is abundant in marine organisms, and forms a refuge and important feeding ground for migratory birds. It is at constant risk from development.

Part of another wetland, Mare Sarcelles (20 ha) at Bras D'Eau, has been declared a Reserve under the Wildlife and National Parks Act of 1993.

Mangroves and Sea Grass Ecosystems

Two species of mangrove, *Rhizopora mucronata* and *Bruguiera gymnorhiza*, grow around Mauritius. Over the years the extent of mangrove cover around the island has significantly decreased (20 km^2 in 1987 and 14 km^2 in 1994) through cutting for firewood, for construction purposes and for providing boat passage.

Mangrove habitats in Mauritius are severely threatened by conversion to residential, commercial, industrial and agricultural real estate by land filling. Major threats to sea grass usually go unnoticed because unlike mangroves and coral reefs, they are not visually obvious to most observers. Damage results from excessive siltation, turbidity, water pollution, bottom trawls, marine excavation and filling operations in the sea.

Recently this trend has been turned round with large-scale replanting programmes.

Causes of Biodiversity Loss

Unsustainable use of resources: Lagoon fishing currently exceeds sustainable levels of exploitation; the size of the fish has decreased over the years and the total catch is decreasing with increase in effort. This has resulted in a decline in fish catch per fisher day from 5.1 kg in 1988 to 3.6 kg in 2004

Corals are also being damaged by fishermen who scrape them with anchors, traps and boat poles. The turtle fishery has been prohibited by law under the Fisheries and Marine Resources Act 1998. Corals and shells are also protected by law and their removal is not allowed except with a permit issued by the Ministry of Fisheries. However, species from the region are still found on the market. Some sea weeds are used as bait in the basket trap fishery.

Development: Land-based human activities, including construction, pollution from sewage and agriculture, and unsustainable use of coastal resources, are having considerable impacts on the coastal zone, especially in terms of lagoon water quality. Much of this is a result of pressure for tourism and residential development, particularly in the north of the island. At present, the development of infrastructure associated with tourism is not well coordinated. Non-compliance with established guidelines also occurs as many land developers consider them too stringent. At the same time there is pressure for more land development. A review of the existing planning norms focusing on development density, height of complexes and compulsory architectural designs by qualified professionals is considered a priority.

Environmentally sensitive areas are being degraded or lost mainly due to construction for housing or tourist accommodation. Areas of particular concern are wetlands, including marshes and mangroves. Most of the wetlands at Grand Baie, Pereybere, Tombeau Bay, and Flic en Flac have been reclaimed for hotel and residential development. The Ministry of Environment estimates that 20 per cent of wetland had been filled in the northern tourist zone of Mauritius and 50 per cent in Flic on Flac. Half of remaining wetlands (marshes and mangroves) are under pressure mainly in north, south-west and Belle Mare tourist zones. There are also plans to develop tourism on St. Brandon.

Principal effects of inappropriate construction include reduction of biological habitat and of the amenity value of natural areas, increased pollution and sedimentation of the lagoon when important wetland functions such as reducing nutrient loads and retaining sediment are lost, flooding of developments built on former wetlands that lack adequate drainage.

Beach erosion: Whilst some beach erosion is natural, it is exacerbated by hard structures on the beach, such as sea walls and recreation jetties, which restrict the natural, *accrete–erode* movement of sand, and cause erosion by reflecting wave energy during storms. Many of these structures do not conform to existing planning guidelines. For example, the appropriate set-back for hard structures has varied between 15 m and 30 m of the high water mark. This phenomenon is particularly acute in the coastal areas of Flic en Flac and the Northern Tourist Zone. A survey of public beaches carried out by the Ministry of Environment in 1997 identified seven critical sites of beach erosion (covering 5,300 m) and four non-critical sites (covering 1,220 m). A further study was commissioned in 2002 on coastal erosion.

Sand mining: One of the important effects of sand mining has been the severe biological impact through loss of habitat and changing currents. Most of the inland deposits have been over-exploited and although it is now banned, the lagoon has also been a major source for sand mining. This has had negative impacts on the marine environment, such as disturbance of the ecosystems, destruction of habitats, destruction of feeding and breeding grounds for fish, movement of fish, smothering of corals by silt deposits, and weakening of the reef structure. Sand mining has also been linked to loss of beaches at Poste Lafayette and Riambel. In 1995, twenty-five co-operative societies involved in sand mining removed 800,000 tonnes from the lagoon, compared to 300,000 tonnes in 1990 and 76,000 tonnes in 1987.

Pollution: In many coastal locations and some established settlements such as Grand Baie, rapid development of housing and commerce has outstripped the rate at which local authorities have been able to provide environmental services and community facilities, especially sewage systems. The effects include a reduction in scenic attractiveness and hence amenity value of the coastline, restricted public access to the beach, and pollution of coastal waters with sewage and solid wastes. For example discharge of effluent is released through four short sea outfalls directly into the sea in the Port Louis region, or via surface water and sewage seepage from shore-side residences.

Industrial effluents discharges as well as agricultural run off by freshwater systems to the sea are also cause for concern, causing siltation of the lagoon and degradation of water quality, that can lead to eutrophication and algal blooms ('red tides'), depletion of oxygen content leading to mortality of marine life (very frequent in the Port Louis region), irreversible damage to the reef ecosystems, and the introduction of potentially invasive species through ballast water discharge from ships. For example at some locations on the eastern coast, high nutrient levels, due to the heavy use of inorganic fertilisers on coastal strip onion plantations, have caused eutrophication of the lagoonal waters causing the growth of the nuisance algae (*Ulva* sp.) that can cover inshore corals and be washed ashore causing odour and aesthetic problems.

However, results for 2001-2003 from long-term monitoring of the quality of coastal and lagoon waters around Mauritius under the Coastal Environment Research Project at fourteen sites around Mauritius suggests that parameters were generally within the required limits at most of the sites.

It is likely that pollution is the major cause of biodiversity loss for freshwater systems. Fertilisers and pesticides from agricultural run-off that enters the rivers may affect water quality. Discharge from textile industries is another point source of pollution.

Invasive alien species: IAS threatens biodiversity in all the aquatic zones. Many introduced species have been recorded in freshwater systems, including the western Mosquito fish (*Gambusia affinis*) renowned for its damaging effect on the freshwater fauna, the Nile perch (*Tilapia* sp), and the Golden Apple Snail (*Pomacea bridgesi*) which appears to have naturalised recently having been first recorded in the wild in 1999. It is a known agricultural pest elsewhere and given its sheer large size and prolific reproductive abilities, it is likely to cause significant ecological impact in freshwater bodies here. The Water Hyacinth (*Eichornia crassipes*), and many other water weeds are also present, and river banks tend to be highly invaded by alien species (Page and d'Argent 1997).

Ballast water discharge from ships is known to be a potential source of invasive marine species, and has been raised as a problem for Mauritius, through a workshop in 2003 organised by the Ministry of Land Transport and Shipping.

Shell extraction: While the extraction of shells and corals from the lagoon is illegal, the importation of shells and corals into Mauritius is permitted (32,570 units were imported in 2004 from Madagascar and Philippines). As a consequence, it is likely that there is still a high incidence of illegal collection of corals from the lagoon in Mauritius and that these are sold together with imported ones. There is some local artisanal trade, although much of the jewellery is also imported. Many sensitized foreign visitors have recently reacted to the sale of corals and shells in Mauritius, resulting in strong protests being levelled to the authorities.

Uncontrolled operations of pleasure crafts: The uncontrolled operations of an increasing fleet of pleasure craft in the lagoons are a source of growing anxiety. As well as damages caused by irresponsible anchoring and illegal coral extraction (undertaken or encouraged by boat skippers), the safety and security of users is also of concern to all tourism and environment stakeholders.

Improper beach management: Proper beach management is severely lacking in Mauritius. The use of prime beach areas for parking, cooking and an increasing number of commercial activities, has deprived these zones of their recreational/ leisure values. The cleaning up of beaches is necessary as is a move towards professional beach management.

Climate change: Climate change is a global concern and as an island state accelerated sea level rise is likely to have short- and long-term consequences on Mauritius. Coral bleaching was observed in late February 1998, which coincided with the abnormally high temperatures and heavy rainfall. Coral bleaching was studied within the Balaclava and Blue Bay Marine Parks and the results showed bleaching had affected 39 per cent and 31 per cent of the live corals in the two parks respectively.

Coral bleaching: In 2003, the reefs in Mauritius suffered sporadic bleaching. Surveys were carried out at four sites, namely Ile aux Benitiers, Belle Mare, Poudre d'Or and Albion and the percentage of completely bleached corals were 85, 27, 17 and 16 per cent respectively. Branching and tabular corals were mostly affected in the back reef while massive corals were affected in the fore reef. Bleaching of corals was observed by late February. By June, 95 per cent of the bleached corals had recovered, 2 per cent were recovering and 3 per cent had died.

Coral bleaching was also observed in 2004 and surveys at Belle Mare site showed that above 60 per cent of corals were completely bleached in the month of March 2004 but by July 2004, the corals had recovered to a certain extent. Those that had died had filamentous algae were growing on them. Branching and tabular corals were the most affected.

Impact of Management on Biodiversity

Identification and Monitoring

Identification

The Mauritius Oceanography Institute (MOI) is developing a taxonomy-based biogeography online database of marine organisms of Mauritian maritime zone. It will store literature based on taxonomic and geographical data for species reported from Mauritian Waters. The end-product, which will be made available online and on CD-ROM, will be a standard reference for research and training in the field of marine species diversity in the vast Mauritian maritime territory. There is also an on-line check list of marine fish of Mauritius as well as two field guides to fish and corals produced by the Albion Fisheries Research Centre.

Ecosystem Level

The Albion Fisheries Research Centre of the Ministry of Fisheries has recently prepared thematic maps of the coastal area around Mauritius and Rodrigues. These maps depict the different coastal habitat up to the reef and will be used for monitoring changes in the coastline and habitats.

A preliminary survey has been carried out to identify all the coastal wetlands of Mauritius. The National Park Conservation Services is in the process of recruiting a consultancy team to carry out a comprehensive survey of all wetlands for the Republic of Mauritius. This will include Rodrigues and Agalega.

In August 2005 the Mauritius Oceanography Institute published a map of the marine habitats on the south-east coast of Mauritius.

Monitoring

There is on-going monitoring programme of the coral ecosystem at 12 selected sites around Mauritius (Albion, Pointe aux Sables, Trou aux Biches, Anse la Raie, Trou d'Eau Douce, Bambous Virieux, Bel Ombre, Poudre D'Or, Belle Mare, Le Goulet, Baie du Tombeau and Ile aux Benitiers). Since the early nineties observations on coral spawning, growth and recruitment of coral larvae have been carried out regularly between October and December. Coral bleaching was observed in 1998 and a study is ongoing.

A long-term monitoring programme has been established for the Blue-Bay and Balaclava Marine Parks, with monitoring being carried out regularly at the different permanent stations. Data on coral, benthos and fish populations are collected, compiled and analyzed. Physical, chemical and bacteriological properties of water are also monitored.

In order to detect variations over time and space in seawater quality, monitoring of the physical-chemical parameters is being carried out in the above monitoring sites, as well as other selected sites (Balaclava, Tombeau Bay/Terre Rouge Bird Sanctuary, Port Louis Harbour, Albion/Belle Eau estuary at the fore reef, back reef, shore reef, and sea-grass stations), on a quarterly basis.

In-situ Conservation Management

Coastal and Marine

About 21 artificial reefs, especially on the west coast have been created by sinking unused ships. These sites are visited regularly by divers, and the success of the projects is monitored, albeit infrequently, by Fisheries Division.

There have been outbreaks of COTS at Trou d'Eau Douce and Ile aux Benitiers in 1998, 2003 and 2004. Control is carried out using the 'dry acid' injection method.

Protected Areas

Marine Protected Areas

Under section 7 (1) and (2) of the Fisheries and Marine Resources Act 1998, a Marine Protected Area may be designated as a Marine Park, a Fishing Reserve or a Marine Reserve.

Two marine parks at Blue Bay (353 ha) and Balaclava (485 ha) were proclaimed in 1997 as National Parks under section 11 (1) of the Wildlife and National Parks Act 1993. Both were declared Marine Protected Areas (MPAs) and designated as Marine Parks in June 2000 under the Fisheries and Marine Resources Act 1998. The Balaclava Marine Park is as yet unmanaged, while the Blue Bay Marine Park has been zoned and management is operational, although there is still no management plan.

Fishing Reserves at Port Louis (331 ha), Grand Port (1,828 ha), Black River (797 ha), Poudre d'Or (2, 542 ha), Poste La Fayette (280 ha) and Trou d'Eau Douce (574 ha) were also proclaimed as Marine Protected Areas in June 2000. The main objective of a Fishing Reserve is to protect and conserve the marine habitats, which are ideal nursery grounds for juvenile fish.

Realizing the importance of mangroves, a propagation programme was initiated in 1995 with the aim of reconstituting mangrove stands in areas where they have been destroyed or where natural regeneration was slow. The public has been sensitised to the importance of mangroves and is being urged to protect and propagate them. A mangrove propagation programme has also been established by the Ministry of Fisheries and mangrove propagules are planted at different coastal regions. To date 17 sites in the coastal region around Mauritius have been propagated and a total area of about 12.8 ha has been covered by mangrove plantations. Mangroves are protected under the Fisheries and Marine Act (1998).

Wetlands

The functions of wetlands in the Grand Bay area has been found to be vital as the northern part harbours the best beaches and lagoons which are potential assets for the tourism sector. In view of this the Ministry of Environment commissioned a study in 2002 to advise and recommend remedial measures for the protection of the wetlands in the Grand Bay area. An Implementation Committee has been set up at the Ministry of Environment and NDU and is following up on the recommendations. The study has earmarked all the important wetlands which badly need protection and actions are being initiated to declare these sites as reserves so as to provide them a legal protection. NPCS, in collaboration with the Fisheries division, has submitted documents to Ramsar Bureau to proclaim a second site, Blue Bay Marine Park, as an international Ramsar site.

The NPCS, with the collaboration and support of the RAMSAR Bureau, organised a two day workshop on 'Wetlands Conservation and Management' in April 2001 to discuss the required strategies towards wetlands conservation and management.

NPCS has also initiated several actions to comply with the Ramsar Strategic and Action Plan 2003-2005. The following has been undertaken.

1. A definition for wetlands as applied in the local context has been worked out from the international definition provided by Ramsar.

2. A National Ramsar Committee was established in 2003 to advise on all aspects related to requests for development on wetlands. The committee comprises members from all institutions involved in wetland matters.

3. The National Ramsar Committee has prepared a wetland bill as the legal framework on this area is too weak to control development on wetlands

Alien Species

There is no active management against aquatic alien species although stakeholders from the aquatic sectors are represented on the National Invasive Alien Species Committee. A task force on the Monitoring of Ballast Water, consisting of representatives from various institutions, has been set up by the Ministry of Shipping. There is an urgent need to undertake a baseline port survey to identify naturally occurring and exotic marine species. Results of the survey will also help analyse possible means of introduction and the implications for port management.

Sustainable Use of Components of Biological Diversity

Artisanal fishery: Since 1996 management measures have been taken relieve the fishing pressure in the lagoon. The effectiveness of these measures is being monitored by AFRC. These include:

1. A buy-back policy for large nets and gill-nets, and compensation of fishermen who relinquish their net licences;
2. A closed season of five months (beginning of October to end of the following February) when fishing with large nets and gill nets is prohibited. This coincides with the peak spawning period. During this period, net-fishermen receive an allowance of 135 rupees per day);
3. A control on the number of nets in the fishery;
4. A limit on the net length and regulation of the mesh size. It is set at a size just beyond first maturity so as to enable the species to spawn at least once;
5. A minimum size for harvest of oysters; crustaceans in the berried state are prohibited for capture;
6. Banning of spear, dynamite and cast-net fishing;
7. Placing of 21 Fish Aggregating Devices (FADs) 1.67-12.3 nautical miles outside the lagoon;
8. Training of fishers in modern off-lagoon fishing with a stipend of 250 rupees per day;
9. Soft loan facilities for cooperatives to purchase equipment for off-lagoon fishing.

Efforts are being made to restock the lagoon with experimental seed productions for the Mangrove Crab (*Scylla serrata*), Giant Tiger Prawn (*Penaeus monodon*), and the Gueule Pavée (*Rhabdosargus sarba*) while culture of Berri Rouge (*Oreochromis* sp.) and Crayfish (*Cherax quadricarinatus*) are on-going. In 2003 a total of 292,000 fry of Gueule Pavée were produced and released in the lagoon at four sites, including Ferney, Tamarin, and Albion. Some 45,650 juveniles of the Tiger Prawn were released at Grand Sable, Petit Sable and Bambous Virieux, more than double that in 2002, although the quantity is still not big enough.

Outer Islets

The marine environment of the outer Islets of St. Brandon and Agalega are relatively undisturbed. Except for fishing no other marine-based activity has been carried out so far. The islands were once important nesting sites for the Green Turtle (*Chelonia mydas*) and the Hawksbill Turtle (*Eretmochelys imbricata*). These species were threatened by over-exploitation and since 1983 they have been declared protected species and their capture has been prohibited.

A project to place permanent mooring sites in the ocean to prevent anchor damage to coral reefs is carried out by the NGOs Reef Mauritius and MMCS.

Sand Mining

The Government took a policy decision to ban sand mining in 2001. Following this ban, the four sites, Grand Gaube, Poudre D'Or, Mahebourg and GRSE, were monitored to observe the condition of the seabed and recolonisation of the areas by marine organisms. Observations showed that the marine ecosystem is recovering slowly. Rock sand is now being used in the construction industry. Those engaged in this sand mining were compensated and many have been given training in FAD fishery.

Incentive Measures

There is a bad weather allowance, a controversial allowance to registered fishers to compensate for days when the weather is too bad to fish (135 rupees per day). This system has become a perverse incentive by encouraging people to become registered fishers, and thus increasing the pressure on the lagoon fish stocks. It is also expensive-costing the Government MRU 32 million in 2004.

Research and Training

Coral Reef Monitoring

This project ran from 2003-2005 funded by the IOC to collect regional data on coral reefs. The programme included training of more than 25 people (from government and NGOs) in monitoring techniques, data processing and logistical support (including equipment and training in diving). The data is used in the Global Coral Reef Monitoring Programme.

Study of the Lagoon-beach Interactions at Specific Sites Around Mauritius for a Better Control of Coastal Erosion

The main aim of this project is to characterise processes leading to coastal erosion at specific places around Mauritius through a geomorphological study of the beach in conjunction with a hydrodynamic and sediment dynamic study of the lagoon.

Public Education and Awareness

NPCS has constructed a visitor's centre at RTREBS site to promote visitors and carry out education campaigns for school children. In this context, NPCS celebrates the World Wetland Day every year in February with a series of activities to sensitise the school children and the public at large on the importance of wetlands. There is strong involvement of NGOs in the region including "Mouvement pour le Progres de Roche Bois" and the fisher community which forms a major part of the population in the vicinity. NPCS has produced a 23 minute documentary on wetlands with the collaboration of Mauritius College of the Air. NPCS also encourages NGOs to embark on projects related to wetlands. Recently an NGO was awarded a small grant from Ramsar to carry out a project related to public awareness and training of wetland leaders.

AFRC produces regular publications, including two field guides and there is an annual open day at AFRC to sensitise public on the fisheries sector. There are two dedicated education officers who, in addition to the other staff, give talks to visitors at AFRC and Blue Bay Marine Park (which also has a visitor's centre), at schools and hotels. There is also a sensitisation campaign for fishermen.

Some NGOs are involved in awareness-raising activities *e.g.* CEDREFI carried out at 2 year project in the fishing village of GRSE to sensitise villagers of the importance of a sustainable way of living; EcoSud carries out awareness raising in schools in the Mahebourg area; Reef Mauritius and MMCS produce posters, games and leaflets, and carry out sensitisation in key areas (*e.g.* dolphin watching at Tamarin).

Impact Assessment and Minimizing Adverse Impacts

Pollution

1.2-1.4 billion MRU have been invested since 2000 in a sewage masterplan. This includes wastewater project, including tertiary level treatment, a long sea outfall at "Montagne Jacquot" (where the tender has already been floated) and the cessation of use of the short outfalls at Pointe aux Sables and Bain des Dames.

Erosion

The Ministry of Environment and NDU commissioned a study in 2002 on coastal erosion around Mauritius. The final report was submitted in 2003 by Baird and Associates. A committee has been set up to implement the recommendations made by the consultants. The committee comprises the Ministries and institutions involved in coastal zone management and meets every 3 to 4 months. So far, the committee has increased the setback distance from the high water mark from 15 m to 30 m.

Public Beaches

A Beach Authority has been set up (2002) to provide for a legal framework for the management and control of the Public Beach in Mauritius and Rodrigues.

Development

The Ministry of Environment and NDU has commissioned a draft technical report for the preparation of a strategic EIA for the identification of potential sites for marinas, ski lanes and bathing areas. A report was prepared by local and international consultants and submitted in 2003.

The Ministry of Fisheries has a detailed website which lists all publications. The Ministry also produces annual reports and research reports, and information on fisheries is used in the Environment Indicator System being developed by the Ministry of Environment.

Legislation, Policy and Institutional Framework

Legislation

Legislation is fragmented and dispersed across a number of different acts and regulations, with powers to enforce existing and enact new legislation, being conferred on several different ministries resulting in an apparent lack of co-ordination and co-operation between them. The existing legislations are also inconsistent in several respects. Whenever possible such pieces of legislation need to be strengthened and harmonised in order to provide adequate protection to the biotic resources. The main acts relevant to marine, coastal and freshwater biodiversity protection include:

Environment Protection Act, 2002

The Environment Protection Act (2002) specifies that an EIA is required for land clearing and development in environmentally sensitive areas such as water catchment areas, waterlogged areas, mountain slopes and islets, and wetlands

The Act provides for the creation of an Integrated Coastal Zone Management Unit within the Ministry of Environment to coordinate the various actors. This specifies the creation of a Committee which in respect of the coastal and maritime zone, is tasked with developing management plans, monitoring coastal waters and generally conducting studies and making recommendations on matters such as beach erosion, and conservation in the islets and the outer islands. The ICZM Committee was set up in 2003 and is comprised of all Ministries, institutions involved in the management of the coastal zone and 6 NGO's designated by the Minister.

The Act vests powers in the Minister to make regulations for the prevention of pollution in the coastal and maritime zones, being mandated to make such regulations as he thinks fit to control various types of pollution and *"such other matters relating to the preservation and conservation"* of the coastal and maritime zone. Although this appears to be a powerful mandate it has not yet been used to any substantial degree, and decisions for individual items are largely taken independently by the agencies concerned.

Regulations have been put in place for levels of inorganic and organic parameters in inland surface water for effluent discharge [The Environment Protection (Standards for effluent discharge) Regulations 2003, The Environmental Protection (Effluent Limitations for the Sugar Industry) Regulations, 1997], for drinking water

[Environmental Protection (Drinking Water Standards) Regulations, 1996], and for water quality (Guidelines for Inland Surface Water Quality, 1998).

Mauritius Oceanography Institute Act, 1999

The Mauritius Oceanography Institute (MOI) was established in January 2000 by the proclamation of the MOI Act (Act No. 24 of 1999). The need to rationalise and co-ordinate research and development activities related to Oceanography was the motive for the setting up of the MOI.

A major responsibility of the MOI is to monitor the marine environment around Mauritius, Rodrigues and the Outer Islands, and advise the Government on appropriate policies and strategies for the intelligent management of the living and non-living resources under its jurisdiction.

Wildlife and National Parks Act 1993, The National Parks and Reserve Regulations of 1996, and The Wildlife Regulations of 1998

This is the principal legislation for the protection of the flora and fauna with the Wildlife Regulations of 1998 giving effect to the CITES Convention in Mauritian law.

The act specifies that camarons (*Macrobrachium lar*) less than 8.5 cm, and berried female camarons cannot be caught or sold, and that camarons and shrimps cannot be fished for with explosives, poisons or using artificial light. It also specifies that unless the watercourse is on private land, a permit is required for fishing.

Forest and Reserves Act, 1983

River Reserves are protected under the Fourth Schedule of the Forests and Reserves Act (1984). This protection restricts changing the vegetation cover of the river banks, thus helping to reduce soil erosion and water pollution.

The Fisheries and Marine Resources Act, 1998 and MPA Regulations, 2001

The Act provides for the management, conservation and protection of fisheries and marine resources, and the protection of the marine ecosystems. In addition to providing for general enforcement and compliance measures, it also deals with protection of the aquatic ecosystem against pollution, exploitation of mangroves, construction of any structures etc. The Act also includes proclamation of marine protected areas in Mauritian waters. The Regulations include specifying associated land area for the MPA, setting up of an MPA Fund and conservation measures.

National Coast Guard Act, 1988

The National Coast Guard has the responsibility for enforcement of any law relating to the protection of maritime zones. In addition, they have the powers to prevent any activity which is likely to constitute a threat to maritime zones including pollution.

Maritime Zones Act, 1997, Updated 2005

This provides for the Prime Minister to make regulations for the preservation and protection of the marine environment and the prevention and control of marine pollution.

Continental Shelf Act, 1970

Section 5 gives power to the Minister responsible to prescribe necessary measures in the safety zone for the protection of living resources of the sea and the natural resources of the continental shelf from harmful effects

Pas Geometriques Act, 1895

This Act provides for 81.21 m as an area of reserved lands (as public domain) along the sea coast extending from high water mark towards the coast.

Planning and Development Act, 2004

The objectives of the Act include, amongst others to provide in relation to land development for the promotion and co-ordination of the orderly and economic use and development of land; the proper management, development and conservation of natural and man-made resources for the purposes of promoting the social and economic welfare of the community and a better environment; use of land for public purposes; and ecologically sustainable development.

Central Water Authority Act, 1981

This act, lays down the responsibilities of the CWA. These include the study and formulation of policy in relation to the control and use of water resources including the protection of wildlife, afforestation and control of soil erosion, disposal of industrial waster, abatement and prevention of pollution of water. The CWA also has the power to stock rivers and water courses with fish, undertake measures for the prevention of malaria and other diseases and discontinue supply of water of any consumer that discharges polluted water.

River, Reserves and Canals Act, 1863

This act specifies actions, buildings and livestock activities that are prohibited within a given distance from the water body as they can cause water pollution.

Public Health Act, 1925

This act empowers the authority to remove any nuisance that pollutes a water body.

International Legislation

Convention of Biological Diversity;

Convention on the Prevention of Pollution from Ships (MARPOL);

United Nations Convention of the Law of the Sea;

Civil Liability for Oil Pollution Damage (CLC) 69 Protocol 1976;

International Fund for Compensation for Oil Pollution Damage (FUND) 1971 and protocol of 1976;

Convention on Migratory Species;

African Eurasian Waterbird Agreement.

Institutional Framework

The following governmental institutions are involved in coastal, marine and freshwater issues:

Institution	*Responsibility*
Department of Environment	EIA, environmental quality
Ministry of Agro-Industry and Fisheries	Fisheries and lagoon eco-system
Department of Local Government	Shore development, beaches
Ministry of Housing and Lands	Physical planning policy
Ministry of Land Transport, Shipping and Ports	Marine pollution from vessels
Mauritius Ports Authority	Port functions, shipping
Ministry of Tourism and Leisure	Coastal hotels and recreation
Prime Minister's Office	Meteorological services,
	National Coast guard/police
Beach Authority	Shore development and beach management
Central Water Authority, Ministry of Public Utilities	Freshwater distribution
Water Resources Unit, Ministry of Public Utilities;	Assessment of freshwater resources
Waste Water Authority; Ministry of Public Utilities	Waste water management
Outer Islands Development Corporation	Outer islands management

Chapter 24

Status of Coastal Ecosystem in Myanmar

Thuzar Win
Remote Sensing Centre, Ministry of Science and Technology, Yangon, Myanmar
E-mail: thuzar@surferboard.com

Introduction

Union of Myanmar is located on the mainland Southeast Asia, and the largest country among the SE Asia countries. Myanmar covers a total land area of 676,577 Km^2. The longest distance from north to south as the crow flies is 2,050 Km, and from east to west is 925 Km.

Boundaries and the Neighboring Countries

Myanmar shares common land boundaries; on the northeast with China, on the east with Lao, on the southeast with Thailand, on the west with Bangladesh, and on the northwest with India. The south is demarcated by the Andaman Sea and Bay of Bengal is on the southwest of the country.

In addition, Myanmar shares common maritime boundaries with Thailand at the southern extremity in the Andaman Sea, and with Bangladesh on the west in the northeastern part of the Bay of Bengal.

The Physiography

The landscapes of Myanmar are extraordinarily diverse. From east to west, Myanmar can be subdivided into four major geomorphic belts which are trending north-south; *viz.*,

1. Eastern Highlands,
2. Central Lowlands,
3. Naga-Chin-Rakhine Ranges, and
4. Rakhine Coastal Belt.

Generally, elevated plains and high mountain ranges alternate with low, hilly and open lowlands.

The *Eastern Highlands* occupies the eastern half of the country and boasts hilly mountainous regions including the eastern Kachin Ranges, Shan Plateau and Tanintharyi Ranges. The eastern Kachin Ranges demarcate the Myanmar-China frontier boundary. The Shan Plateau is an elevated plain with hilly landscapes. To the south, the Tanintharyi Ranges and its western coastal area extend through the southern Thailand into Malaysian Peninsular. The Thanlwin (Selween) River originated from the China cross through the eastern part of the Shan Plateau and joins into the Andaman Sea beyond the Tanintharyi coast.

The *Central Lowlands* intervening the Eastern Highlands and western mountainous ranges comprises the basins of Ayeyarwady (Irrawaddy), Chindwin and Sittaung Rivers, which are flowing southward and join into the Andaman Sea. The Chindwin River joins the Ayeyarwady River at the centre of the Lowlands and the end of the Ayeyarwady River is represented by a delta known as Ayeyarwady Delta (about 200 km in length as well as in width). The mouth of the Sittaung River is a tide-dominated estuarine opens to the Gulf of Mottama (Martabin). As the name imply, the Central Lowlands embrace the open and broad hilly central plain of the country and the northern extremity of the Lowlands is fringed by the mountainous ranges while the southern end is broadly opened toward the Andaman Sea. The ranges at the northern fringe of the Lowlands are the southern continuation of the Himalayan Ranges.

The *Naga-Chin-Rakhine Ranges* is the western mountainous ranges of Myanmar and the ranges are found as continuation of the Himalayan Ranges. From the Myanmar-China-India frontier area, these ranges extend southwestward through the Myanmar-India frontier area and then curve to the south through the western part of Myanmar, finally into the Andaman Sea in which the ranges emerge again as a group of Andaman Islands.

The *Rakhine Coastal Belt* is the westernmost and narrowest landscape which extends southward from the Myanmar-Bangladesh frontier area to the Negaris or Maw din point.

Maritime Zones

In accordance with the Law of the Sea, Myanmar has already claimed the maritime zones, such as, Territorial Sea, Contiguous Zone and Extended Economic Zone (EEZ) in 1977. The maritime boundary between Myanmar and India in the Andaman Sea has already agreed. Myanmar continental shelf embraces an area not less then 280,000 Km^2 and the area-coverage of the EEZ is not less than 486,000 Km^2, respectively.

Myanmar Coastal Zones

Myanmar has a long coastline, which stretches, from the Myanmar-Bangladesh frontier area, at the western boundary of Myanmar, to the Myanmar-Thailand frontier area, at the southern end of Myanmar and total length of the coastline is approximately 2,832 Km. Off the whole Myanmar coast are about a thousand of islands and there are over 800 reported islands in the Meek Archipelago. From northwest to southeast, Myanmar coastal zones can be recognized as:

1. Rakhine coastal zone (about 940 km),
2. Ayeyarwady Delta coastal zone (about 460 km),
3. Mottama (Martabin) Gulf coastal zone (about 232 km), and
4. Tanintharyi coastal zone (about 1200 km).

The *Rakhine coastal zone* demarcates the northeastern margin of the Bay of Bengal and includes the mainland Rakhine Coastal Belt, Rakhine coastline and Rakhine continental shelf. The coastline stretches 740 km from the Naïf River to Negrais point. Off the coast are islands and rocks including two large islands, Yanbye Island and Man-aung Island. The Rakhine continental shelf covers about 36,000 Km^2 (Aung Htay Oo, 2000).

The *Ayeyarwady Delta coastal zone* includes Ayeyarwady Delta coastline and Ayeyarwady deltaic offshore shelf. The Ayeyarwady Delta coastline is between the Negrais point and the mouth of Yangon River. On the west, the delta is fringe by the Rakhine Ranges and on the east by the Sittaung River basin. The delta is a flat alluvial plain with a network of distributaries of the Ayeyarwaddy River and total area-coverage of the delta plain is approximately 35,138 Km^2. The annual sediment discharge of the Ayeyarwady River has been estimated as 250 million ton. The delta is advance seaward at the rate of 5 km per a hundred years (ESCAP, 1995).

The *Mottama Gulf coastal zone* is the shortest portion of the Myanmar coastline and includes Mottama coastline, mouth of Sittaung River or Mottama Gulf and its shallow-marine terraces. The southeastern end is represented by the mouth of Thanlwin River.

The *Tanintharyi coastal zone* is the longest zone of Myanmar and demarcated by the mouth of Thanlwin River in the north and Kawthaung River in the south. The length of the coastline is about 1200 km. The zone includes Tanintharyi coastline, Myeik Archipelago and Myeik offshore shelf. The shelf is bounded in the west and north by the Andaman spreading system. The coastal plain is narrow and gradually rises towards the east to become the Tanintharyi Ranges with highest elevation of 2,073 m. The Myeik Archipelago contains about 800 islands embracing an area of approximately 34,310 Km^2. Coral reefs surround the outer islands and mangroves cover much of the inner islands. Majority of the islands are uninhabited apart from the temporary fishermen camps. The scenery of the islands is outstanding beauty with densely forested hills and occasional sheltered coves provide good anchorage.

Economic capital, Yangon has a port with major facilities and the adjacent port, Thilawar has modern facilities especially for the containers. To meet the increased requirements of the expending economics and trade, several deep-sea ports are being developed along the Rakhine coast and Tanintharyi coast.

The People

There are about 54 million people living in Myanmar; the country boasts a great cultural diversity and special region-specific qualities, charming towns and attractive landscapes.

There are a large number of regional written languages and "Myanmar" is the official language. There are eight major ethnic minorities: Kachin, Kayah, Kayin, Chin, Bamar, Mon, Rakhine, Shan and many other ethnic groups traditionally resident in Myanmar.

Myanmar is administratively divided into 7 States and 7 Divisions, viz., Kachin State, Kayah State, Kayin State, Chin State, Mon State, Rakhine State, Shan State, Ayeyarwady Division, Bago Division, Magway Division, Mandalay Division, Sagaing Division, Tanintharyi Division, and Yangon Division.

Education, Science, Technology and Research

Myanmar is a country, which highly values education and vocational training, research and sciences. Our Ministry, Ministry of Science and Technology, together with other Ministries decide and implement goals and measures in conjunction with the Higher Education and Institutional administrative and academic boards. Our Ministry was established in 1996 composed of six departments. Under the human resource development sector, our Ministry is providing well-trained and skilled engineers, computer scientists, technicians and researchers in the various fields, such as, Engineering Technologies, Information Technologies, Computer Sciences, Biotechnology and other higher technologies. At present, our ministry already established 4 Technical Universities, 2 Universities of Computer Sciences, 1 University of Aeronautical and Space Technology, 26 Government Technical Colleges, 24 Government Computer Colleges, 11 Government Technical Institutes.

After completing primary school, about one fifth of children attend Government Technical Institutes, and Colleges. This imparts a basic engineering and technical education to its pupils. Successful completion of colleges enables admittance to the Universities.

Our Department

My department is Remote Sensing and Data Processing Centre, located in Yangon. In Myanmar, remote sensing techniques have been extensively employed in various Ministries and Departments. Using aerial survey and satellites data, our department is doing research on the natural resource management and exploration. The remote sensing data and interpretation are processed at our research centre by the aid of GIS.

Ocean Observation and Hydrographic Survey in Myanmar

Adjacent Seas

Bay of Bengal, which is the northeastern part of Indian Ocean, is adjacent to the Rakhine Coastal Belt and provides monsoon rains to Myanmar during May to October annually. Among the cyclones generated in the northeastern part of the Indian Ocean, few directed to the Rakhine Coastal Belt but no longer strengthen when they passed over the mountainous Rakhine Ranges.

The Rakhine offshore shelf is relatively narrow and up to the 60 km in width. The common depth along the shelf edge is about 80 m and the step continental slope down to the abyssal plain about 800 m in depth. Canyons and submarine slumps are encountered during bathymetric survey. The narrow coastal zone (Rakhine Coastal Belt) is bounded in the east by the mountain ranges of the Naga-Chin-Rakhine Ranges. The finer muddy sediments deposited on the Rakhine shelf are derived from the Rakhine Ranges.

Hydrographic conditions of the Rakhine coast are strongly influenced by the Monsoon. During the rainy season, the surface waters are extensively mixed with fresh water originating from the rivers runoffs. The southward flow of diluted seawater reduces the salinity to 18 ppt in the near-shore areas whereas in the dry season a northward flow of high salinity water 34 ppt was reported. Indications of local upwelling were recorded in the southwestern offshore area of Man-aung Island. Organic production is also higher during this period. Due to the rock types building up the Naga-Chin-Rakhine Ranges, the sediments deposits by the rivers flowing westward from the Rakhine Ranges are finer and the runoff becomes turbid in Monsoon period. Therefore most of the beaches of the coastline are muddy except well known recreation centre Ngapali, Chaungtha and Ngwe-saung Beaches. Later, the sediments deposited at the shelf edges are redistributed onto the abyssal plain underlain by the world's biggest submarine fan, Bengal Fan.

The Rakhine offshore shelf is one of the hydrocarbon prospect areas in Myanmar. Therefore, hydrographic survey for safe navigation, geological and geophysical surveys for hydrocarbon exploration and exploitation are being carried out. In addition, the underlying continental margin is geologically (or tectonically) active and its southern continuation in Sumatra area has triggered 2004 Tsunami and several deadly earthquakes. Therefore, marine geological research works are being carried out along this marginal zone along which ocean-generated natural hazards can be occurred.

Andaman Sea is an intervening shallow marine between Myanmar coastal zones and Andaman Islands. The Andaman Sea embraces the Ayeyarwady deltaic offshore shelf, Gulf of Mottama and Myeik offshore shelf.

The western margin of the Ayeyarwady deltaic offshore shelf is bounded by the group of Cocos Islands, which are, submerged extension of the Naga-Chin-Rakhine Ranges. The southern and eastern margin of the offshore shelf is demarcated by the Andaman spreading rift system and its northeast-trending canyon.

The Ayeyarwady delta is found to grow southward into the Andaman Sea and deltaic sediments are being deposited on the inner offshore shelf. On the other hand, the marine current system prevailed on the deltaic offshore shelf re-distributes the sediment to the Gulf of Mottama. Hydrographic conditions of the Ayeyarwaddy Delta coast are strongly influenced by the Monsoon originated from both Andaman Sea and Bay of Bengal. During the rainy season, the surface waters are extensively mixed with fresh water originating from runoffs from the distributaries of

Ayeyarwaddy River. The northeastward flow of diluted long-shore currents reduce the salinity along the near-shore areas and distribute the sediment to the Gulf of Mottama. Therefore, Letkokgone Beach, a beach east of Yangon near the mouth of Sittaung River is a muddy beach always turbid with the transported finer sediments.

Hydrographic conditions of the Gulf of Mottama coastal zone are strongly influenced by the Monsoon generated from the Andaman Sea. During the rainy season, the surface waters are extensively mixed with fresh and turbid water originating from runoffs from the Sittaung and Thanlwin Rivers. The tide system in the Gulf of Mottama redistributes the sediment deposited by Sittaung and Thanlwin Rivers together with those sediments transported by long-shore currents from the Ayeyarwady Delta coast.

The sediment deposited on the northern part of the Myeik offshore shelf is provided by the Thanlwin River while the sediments accumulated on the southern part of the shelf and Myeik Archipelago are deposited by the several small rivers flowing from the Tanintharyi Ranges. Hydrographic conditions of the Tanintharyi coast are influenced by the Monsoon but presence of archipelago reduces the velocity. During the rainy season, the surface waters are extensively mixed with fresh water originating from runoffs from the small rivers. The adjacent mountains in the east of the coastal region provide relatively clean sediments onto the Myeik offshore shelf and hence beautiful fine sandy beaches are occurred along the coast, such as "Setsè Beach".

Myanmar Coastal and Offshore Resources

Rakhine offshore shelf, Ayeyarwady deltaic offshore shelf and Gulf of Mottama have high potential of hydrocarbon prospect. In addition, due to the underlying rock types and mineralization along the Tanintharyi Ranges and coastal zone, secondary placer deposits and sedimentary deposits are occurred along the coast. Moreover, high nutrients fluxed by the major rivers, marine organisms are niche in these waters, especially in the Ayeyarwady deltaic offshore shelf, Gulf of Mottama and Myeik offshore shelf of the Andaman Sea. Therefore, for the fishery development, environmental conservation and exploitation of geological resources, ocean observation and hydrographic survey are being carried out in this region.

Hydrographic Surveying

Hydrographic Surveys are most commonly conducted by Myanmar Naval Hydrographic Department and Ministry of Transport. Coordination of the organization joined with the goal of improving hydrography and safe navigation is conducted by the International Hydrographic Organization. Recently, Shallow Water Mapping which is collaborated with the Hawaii University was carried out in the northern part of the Andaman Sea and two reconnaissance surveys for hydrographic and bathymetric surveys have been done.

Nautical Charts

Based on the British Admiralty charts, recent survey data were added and new charts are being emended at the Cartographic Section of the Naval Hydrographic Department.

Oceanographic Observation and Research

Universities, Institutes and governmental organizations have various sorts of marine science researches and observation not only for the research and development but also for the environmental conservation and monitoring of the ocean-generated natural hazards is being carried out.

Government Interventions for Ocean Observation

Myanmar offshore areas have highly potential of natural resources, such as hydrocarbons, sedimentary ore deposits and aquatic life. Located off the Ayeyarwady delta in the Gulf of Mottama and Andaman are numerous gas-fields, such as, Yedagon Gas-field and Yadana Gas-field. Exploration and exploitation of geological resources and production of aqua-products are under the guidance and management of respective governmental organizations. Some are mainly run by the government enterprises and some are joint-ventured with private enterprises.

Introduction of the Project

Fishing, coastal forestry, mining, tourism, industry and hydrocarbon exploration are some of the variety of economic activities utilizing the land and marine resources of the coastal zone and offshore shelf. Unlike other ASEAN countries, Myanmar is in many respects in the initial stages of development. Now a day, it has marked improvement of infrastructure in the coastal and offshore areas which will attract increased private sector investment both by local and foreign investors. Therefore, the country has the opportunity to set up environmental management plans for sustainable developments and hence ocean observation and oceanographic survey become fundamental roles.

In order to effectively manage and protect the environment, to conserve the natural resources and to establish the sustainable development in coastal and offshore areas, integrated and multi-disciplinary ocean observation and oceanographic survey is being planed to carry out with the co-operation of various governmental agencies, departments and ministries.

Background and Initiative Researches

Two geological, geophysical and bathymetric surveys collaborating with France scientists were carried out in the Myanmar-Andaman Sea and Rakhine offshore area in May 2000 and recently in May 2006. After processing, the first survey data are now in a phase of publication in International Scientific Journals.

Shallow Water Mapping in Myanmar-Andaman Sea is being carried out with collaboration among the Myanmar governmental agencies, departments, institutions and Hawaii University in 2005.

Another joint oceanographic research was initiated by a workshop on the India-Myanmar Joint Oceanographic Studies Program in Yangon in 1998, and the second workshop was in 1999, in Goa. The proceeding joint oceanographic survey was carried out in January, 2002. During this cruise, southern part of the Rakhine offshore area and Andaman Sea were survey and both sediments and water samples were taken. The scientific reports contributed from the survey were already presented in Yangon, in March, 2006. In addition, collaborating with other neighboring countries and United Nation agencies, early warning system for the ocean-generated natural hazards is being planed to set-up.

Development Targets

At present, oceanographic surveys and researches are being carried out in accordance with independent aims and objectives by the different agencies, departments and institution. However, the contributions and results implement an integrated coastal zone and ocean management and development programme under the inter-ministerial co-operation.

Major Features/Achievements

Research and development in relation with these oceanographic and geological survey implements the sound management on coastal and marine environment, opportunity for the training of young scientists from the universities, experience on application of advance technology, research contribution on the ocean processes and management of offshore resources.

Promotion of public awareness on proper management and conservation of coastal zone and marine environment as well as monitoring of ocean-generated natural hazards through mass media as well as grass root level information dissemination.

Future Programs and Directions

Different agencies, which are responsible for the particular resources usually, manage and operate independently of each other but the mechanism for inter-ministerial co-ordination on natural resource management issues need to be established. Activities in the coastal and offshore areas are cross-sectoral, imposing the need for inter-departmental and inter-ministerial co-ordination and planning and hence in accordance with the "Implementation of the Sustainable Development Strategy for the Seas of East Asia (SDS-SEA)" a multidisciplinary programme for marine environmental management including coastal management, ocean-generated natural hazards, environmental protection and conservation for the fishery, reefs and aquatic vegetation, exploration and exploitation of the geological resources and safe navigation, oceanographic surveying and ocean observation is being planed to be carried out as national level.

Chapter 25

Syrian Coast and Environmental Challenges

Ahmad Kara-Ali
High Institute of Marine Research, Latakia, Syria
E-mail: ahmad-karaali@hotmail.com

ABSTRACT

The shoreline is poor with gulfs and bays, in river estuarine and in other area suitable for biological species nursing and reproduction. The water of the Syrian coast is characterized by salinity (nearly 40 per cent), In addition, the intensive agricultural activities along the Syrian coastal area result in excess and uncontrolled use of fertilizers and pesticides, which pollute the ground and seawaters. The most affected marine community is that living in shallow coastline area due to human activities and to the direct impact of the anthropogenic pollution where in addition to large number of industrial sites; had been identified close to the sea, wastewater discharge that contributes to increasing BOD, COD, TSS, besides other more complex and harmful pollutants. The severity of problem causing Eutrophication, affecting sea uses, affecting bathing water quality in the adjacent swimming sites due to water currents and causing death to marine organisms.

Keywords: *Syrian coast, Marine pollution, Biological species.*

Introduction

The Syrian coastal line is about183 km in length, and the coastal area forms only 2 per cent of total country area but it is a home for 11 per cent of the total population. The coastal area produces about 11 per cent of national production (38 per cent of cement production and 50 per cent of national oil refining; for example).

The shore is mostly rocky, the sea is generally deep and the continental shelf is narrow, the shoreline is poor with gulfs and estuaries and in other areas suitable for biological species nursing and reproduction, the water of Syrian coast is characterized by high salinity (nearly 40 ppt).

The intensive agriculture along the Syrian coastal area result in excess and uncontrolled use of fertilizers and pesticides which pollute the seawater.

The most affected marine community is that living in shallow water coastline areas due to human activities and to the direct impact of the anthropogenic pollution (Ibrahim and Kassab 2002), the coastal region of Syria is divided into two administrative regions; Lattakia and Tartous.

This division is made due to differences between two regions regard of the type of industry operated Table 25.1.

Table 25.1: Type of Industry Operated in the Two Administrative Regions Lattakia and Tartous

Industry	*Lattakia*	*Tartous*	*Presence of Treatment Facility*	*Pollutant Discharge into the Sea*
Oil Refinery	0	1	Facility is not in use	Petroleum hydrocarbon = 500 tons/year Suspended solids = 150 ton/year
Oil loading terminal	0	2	None	No quantitative data
Commercial port	1	3	None	No quantitative data
Phosphate terminal	0	1	None	No quantitative data
Electrical power production	0	1	Physical treatment	$30m^3/h$ waste water $30m^3$/week sludge
Electrical engines factory	1	0	None	$75m^3$ industrial waste/day
Aluminum factory	1	0	None	No quantitative data
House machines planets	4	0	None	No quantitative data
Wood factory	1	0	None	No quantitative data
Paint and plastic	44	19	None	No quantitative data
Olive oil extractors	95	213	Collection and drying	No quantitative data
Food and soft drink processing	65	26	None	No quantitative data
Tobacco factory	1	0	None	No quantitative data
Luboil	1	0	None	No quantitative data
Leather	5	0	None	No quantitative data
Cement plants	0	1	Facility is not in use	65 per cent of national production
Textile	2	0	Waste water treatment	50 m^3/day waste water
Paper and pulp	4	0	None	No quantitative data

As it can be seen in Table 25.1 fairly large number of industrial sites especially in Lattakia (battery factor, wood processing, food processing, textiles, engines factory etc.) had been identified close to the sea; Small-scale industrial units including olive-oil-extractors are dominated along the coastal area. Thermal power generation station, an oil refinery, a cement factory, organic chemicals and oil and phosphate terminals are also identified on the coastal region of Tartous.

The vast majority of the above mentioned factories discharge wastewater which contributes to increasing BOD,COD, TSS; besides of other more complex and harmful pollutants.

Contaminants and Sources

Sewage

Waste waters of coastal cities and municipalities into the sea via few discharging points.

It is estimated that a total of 141.6 m^3 and 81.6 m^3 of domestic waste water are discharge into the sea daily in Lattakia and Tartous regions respectively (WHO/ MAP 2002). This results in a heavy loads reached into the sea which have an adverse effect on the quality of the water (UNEP/MAP1994). Estimation carried out on the waters of the two administrative regions, Lattakia and Tartous, revealed a heavy loads of BOD (7367 and 5582 tons/year, respectively) and of COD (14739 and 12200 Tons/year respectively);(WHO/MAP 2001).The nature and severity of the problem can be identified Table 25.2.

Table 25.2: Nature and Severity of the Problem on Coastal Line and Coastal Sea

Source	*Nature and Severity on Coastal Line*	*Nature and Severity on Coastal Sea*
1. Major cities	1. Local effect on the natural appearance of coastal line, through drifting back the sewage to the shore by the sea waves	1. Causing Eutrophication (high impact in discharge points during summer times)
2. Surrounding municipalities	2. Affecting coastal uses	2. Affecting sea uses
3. Coastal rivers collecting sewage from mountain municipalities	3. Affecting local biodiversity on the coast	3. Destroying biodiversity in the coastal area
	4. Affecting the basing of the rivers used as discharging sites (Fahoum 1997)	4. Affecting bathing water quality in the adjacent swimming sites due the water currents

Urban Solid Waste

Solid of the coastal cities are dumping into landfills located close to the sea, it is estimated that is deposited in the coastal area and affected the biological activity of the sea Table 25.3 and the life of marine vertebrates such as marine turtle.

In addition the southern surface–water-current drifts contribute the pollution by solid waste.

Table 25.3: Nature and Severity of the Problem on Coastal Line and Coastal Sea

Source	*Nature and Severity on Coastal Line*	*Nature and Severity on Coastal Sea*
1. Coastal municipalities	1. Destroying on the natural appearance of coastal line	1. Causing death to marine organisms
2. Tourist leaving solid waste on beaches	2. Affecting coastal line uses for farming, tourism.	2. Affecting sea uses
	3. Affecting local biodiversity on the coast	3. Decreasing the biological process of the sea bottom for long term
		4. Affecting bathing water quality in the adjacent dumping sites due to water currents

Estimated of Pollutants from Land Based

Many researches study of organic and inorganic pollutants in the coastal area Table 25.4.

Table 25.4: Estimated of Pollutants from Coastal Sea

Parameters	*Spatial Coverage*	*Levels*	*References*
Heavy metals	Sediment of Lattakia coast	Se: 0.394–0.0137 µgm/g Pb: 13.59–128.17 µgm/g Zn: 0.24–5.96 µgm/g Cu: 0.13–26.80 µgm/g Cr: 14.00–583.94 µgm/g Cd: 0.042–0.258 µgm/g	Aghiourli 2004
	Sediment and Biota along the Syrian coast	The highest Pb concentration was in sediment of Tartous Port 358.5 ppm followed by sewage discharging point of Jableh 118.47 ppm	Ibrahim 2001
Total petroleum hydrocarbon	Sediment of Alkabir Alshimaly and its Estuary	0.08-2441 µgm/g	Fahoum 1997

Heavy Metals Reached into the Sea from Industrial Sewage

(estimated from the hot spots of the regions WHO/MAP2001) Table 25.5.

Table 25.5: Quantity of Heavy Metals from Industrial Sewage in to the Sea

Lattakia	*Tartous*	*Heavy Metals (Kg/year)*
131	214.6	Hg
20	7.7	Cd
52	24.1	Pb
132	96.9	Cr
642	503.7	Cu
475	674	Zn

Hazardous Wastes

The severity and impact of hazardous wastes can be regard as in Table 25.6.

Table 25.6: Nature and Severity of the Problem on Coastal Line and Coastal Sea

Source	*Nature and Severity on Coastal Line*	*Nature and Severity on Coastal Sea*
1. Agriculture discharge	1. Affecting ground water quality	1. Changing water quality and marine ecosystem.
2. Industrial discharge	2. Affecting biodiversity in the area	2. Destroying the vegetation cover and animal community, especially in and around river estuaries

Physical Alteration and Habitat Destruction

There is a lack in the quantitative field data on the level and extent of physical alteration and habitat destruction (Tables 25.7–25.9). Thus, information were presented in this section on the basis of the present status of the coastal area.

Table 25.7: Shoreline Construction/Alteration

Problem	*Causes*	*Significance*
1. Destroying the nature and . appearance of the shore	1. Garbage (plastics etc.) and tars	1. Effect on tourism
	2. Tourist constructions and the consequent pollution	2. Destroying the resting and breading ground of marine mammals and turtles.
	3. Industrial development and planet buildings (*e.g.* cement factory in Tartous).	3. Loosing the traditional shores.
	4. shore filling by dumping waste material .	
2. Habitat degradation and destruction	1. Petroleum hydrocarbon pollution	1. Effect on biodiversity
	2. Tar ball pollution	2. Effect on shore useges

Table 25.8: Mineral and Sediment Extraction Alteration

Problem	*Causes*	*Significance*
1. Altration of sediment . chemistry	1. High input of heavy metals (especially Lead; Ibrahim 2001) and other chemicals	1. Bioaccumulation of pollutants through food chain .
	2. High input of petroleum hydro-carbons due to oil spills and discharge (Nasser and Houijeh, 2002)	2. Effect on marine biodiversity.
2. Alteration in sediment structure	1. Coastal erosion and dumping soil and building material into the shore	1. Effect on biodiversity

Table 25.9: Marine Waters and Coastal Watershed Alteration

Problem	*Causes*	*Significance*
1. Alteration in marine water chemistry	1. Input various polltants from land based sources	1. Bioaccumulation of pollutants through food chain.
	2. High input of petroleum hydrocarbons due to oil spills and discharge (Nasser and Houijeh, 2002)	2. Water Eutrophication in some hot spots and consequent elevated levels of BOD.
		3. Death of marine organisims.
		4. Effect on marine biological diversity and equilibrium.
2. Alteration in marine water biology	1. Microbiological pollution of bathing water caused by sewage discharges	1. Effect on marine life
	2. Eutrophication	2. Effect on tourism
3. Alteration in physical and chemical characteristics of coastal watershed	1. Dumping various pollutants (esp. side products of olive oil extructions)	1. Effect on biodiversity
		2. Effect on tourism

Biological Alteration

Many species of biota are known to be endangered due to habitat degradation and/or human activity they are listed in Table 25.10 besides the area concerned.

Table 25.10: Endangered Species of Biota in Syrian Coast

Sl.No	*Administrative Region*	*Area*	*Species*	*References*
1.	Lattakia	Om Attiur	Crustacean:	Al-Nomeh *et al.*, 1997
			Ocypode cursor	
			Reptiles:	
			Caretta caretta	
2.			Mollusk:	
			Dendropoma Petraem	
			Connus mediterraneus	
3.		Ibn Hani	Echenodermata:	National country
			Sea Urchin	study 1998
4.		Joun Jableh–	Higher Plants:	
5.		Latakia	*Posidonia oceanica*	
			Zostera marina	National country
			Cymodocea nodosa	study 1998
			Reptiles:	
			Caretta caretta	

Contd...

Table 25.10–Contd...

Sl.No	Administrative Region	Area	Species	References
6.		The whole coast	Fish:	National country
			Caranx crysos	study 1998
			Umbrina cirrosa	
			Dicentrachus labrax	
			Dicentrachu puncta	
			Epinephelus aeneus	
7.	Tartous	South of Tartous city	Reptils:	
			Caretta caretta	
8.		Arwad coast	Sponges:	
			Hippospongia communis	
			Spongia officinalis	
			Mammals:	
			Monachus monachus	
		Al-Nawras beach	Fish:	National country
			Caranx crysos	study 1998
			Umbrina Cirrosa	
			Dicentrachus labrax	
			Dicentrachu puncta	
			Epinephelus aeneus	
	Tartous	South of Tartous city	Reptiles:	
			Caretta caretta	
		Arwad coast	Higher Plants:	
			Posidonia oceanica	
			Zostera marina	
			Cymodocea nodosa	
		Banias coast	Crustacean:	Al- Nomeh
			Ocypoda	*et al.*, 1997

Syrian had initiated several basic baseline projects covering research studies on pollution monitoring along the Syrian coast (Med Pol phase II 1981-1996).

In this framework, the following programs were carried out:

1. The coastal water was being on monitoring since 1986 (UNEP/MAP 1994). Fifty nine stations (along the coast of Syria) the program was designed for microbiological pollution monitoring, in addition to samples taken for monitoring heavy metals in biota and petroleum hydrocarbons and nutrients in sea water.

2. A comprehensive field study on the national coastal water using Russian reaserch Vessel VITYAZ-24 was operated during February and March 1992 to measure pollution, in this national project, hydro-chemical parameters (mainly nutrients) in addition to organic (phenols and aliphatics hydrocarbons) and inorganic pollutants (Pb, Cd, Zn, Cu, Ni, and Fe) were assessed.
3. A national project had also been conducted to assess the effect of pollution on marine community (UNEP/MAP 1996a) where the project was carried out on the coastal water of Lattakia and planktonic community was chosen to measure the effect.
4. Hot spots project has been executed along the Syrian coast (WHO/MAP 2001) and 4 areas had studied and ranked according to the level of pollution.
5. In addition to the above mentioned programs, Several research projects had been initiated to assess the pollution levels marine ecosystem, representative of such studies carried out during the last 10 years are presented in Table 25.11.

Currently, four research projects are already going on at the high institute of marine research in Latakia and dealing with trend monitoring of organic and inorganic pollution along the Syrian coast.

In addition, in the year 2003, and in the contest the Med Pol. Phase III, Syria has completed the draft project of the national marine monitoring program to deal with the Land-based sources of pollution, monitor hot spots areas, compliance monitoring for recreational waters and trend monitoring of coastal waters.

Table 25.11: Representatives of the Studies Carried Out in Syria on Marine Pollution During the Latest Year

Sl.No.	*Parameters*	*Matrix and Place of Study*	*Reference*
1.	Heavy metals	Biota (fish, mollusks and echinodermata) of Latakia	Al-Noumeh *et al.*, 1997
2.	Inorganic nitrogen, orthophosphates	Seawater of Latakia, Banias,Tartous Phosphate terminal of Tartous port Sediment of Lattkia, Tartous	Omran 1995 Al-dash 1994
3.	Total petroleum hydrocarbons	Seawater of Latakia, Banias, Tartous Seawater of Latakia, Banias, Tartous	Hourieh 1995 Nasser and Houiejeh 2003
4.	Phenols	Seawater of Latakia, Banias,Tartous	Hourieh 1995
5.	Aliphatics hydro-carbons and PAH	Sediment of Alkabeer Alshimaly river and its eastuary	Fahoum1997
6.	Organic hydrocarbon	Seawater, sediment and biota of Banias coast	Ammar 2002
7.	Chlorinated pesticides and PCBs	Sediment of Alkabeer Alshimaly and Sediment of Latakia Coast	Kara-Ali 2000

Biomarkers

The information about biomarkers in the coastal area of Syria is very limited there are few researches which going on at the High Institute of Marine Research Table 25.12.

Table 25.12: Biomarkers in Some Fish from Lattakia City

Lattakia Area	*Species*	*Levels*	*Reference*
Wadi Kandil	Fish: Chelon labrosus (Mullet)		Ikhtiar 2005
	Male:	Cytochrom P-450: 0.00769-0.02473n mol/ml	
		Activity Resourfin: 1.425 nmol/min/ml	
		EROD Activity:5.365 nmol/min/ml	
		Activity protein P-450: 0.133-0.6012 nmol/min/ml	
	Female:	Cytochrom P-450: 0.00275-0.022n mol/ml	
		Activity Resourfin: 1.096-3.479 nmol/min/ml	
		EROD Activity:9.351-163.51 nmol/min/ml	
		Activity protein P-450: 0.0777-1.168 nmol/min/ml	
Commercial port	Fish: Chelon labrosus (Mullet)		
	Male:	Cytochrom P-450: 0.0341-0.02473n mol/ml	
		Activity Resourfin: 0.11-9.683 nmol/min/ml	
		EROD Activity:4.344-542.410 nmol/min/ml	
		Activity protein P-450: 0.036-3.278 nmol/min/ml	
	Female:	Cytochrom P-450: 0.00495-0.148 n mol/ml	
		Activity Resourfin: 1.37-7.123 nmol/min/ml	
		EROD Activity:1.509-87.114 nmol/min/ml	
		Activity protein P-450: 0.0915-3.112 nmol/min/ml	

The system of legislative power in Syria includes, first of all, the parliament (committee of environment) whose task is to manipulate individual environmental problems. Various institutions and ministries are also involved in pollution control.

Several legislations were imposed in Syria in order to protect and conserve the marine environment Table 25.13.

Table 25.13: Current Legislations Related to Marine Pollution Control

Sl.No.	*Legislation*	*Purpose*
1	Ministerial Decree No. 460 (1964)	Management of fishing activities in marine water
2	Legislation No. 30 (1964)	Regulates the basic issues regarding protection of aquatic organisms and ecosystem through ensuring prevention of harmful human activities
3	Act No. 10 (1972)	Imposes various measures to protect marine environment from oil pollution
4	Ministerial Decree No. 43 (1986)	Management and safe operation of fish farms
5	Presidential Decree No. 11 (1991)	Claims industrial plant owners to carry out a survey on the environmental impact Assessment in order to meet the relevant environmental conditions in order to obtain operating license.

Conclusion

There is national action plan for 2010 including the following steps:

1. Separate industrial wastewater from domestic water or storm water
2. Building treatment plants for waste water of coastal cities
3. Rehabilitation of treatment of oil refinery and thermal power generation station
4. Building treatment plants for all factories in coastal area.
5. Reduce of air pollution by:
   (*a*) Change of generation power of oil refinery and thermal power generation station from fuel to gas.
   (*b*) Usage filters for cement Factory.
6. Rehabilitation of dumps of urban solid waste in coastal cities.

References

Aghiourli, A.A.M. 2004. Contribution to study and the assessment of changes in the chemical pollutants levels in Lattakia coast.M.Sc. thesis, Tishreen University.

Al-Nomeh, M., Othman, I. and Al-Assafin, I. 1997. Benthos of Om-Attiur. Syrian Atomic Energy Commission Technical report HTZS-TNBA 138.

Ammar, I.A. 2002. Study on zoon benthos in Banias coast and effect of petroleum hydrocarbons on them. PhD. Thesis, Tishreen University.

Fahoum, I.Y. 1997. Determination of some organic compounds from industry in sediments of Al-Kabir Al-Shimali river and their effect on marine environment. M.Sc. thesis, Tishreen University.

Ghaleh, M. 2002. New migration record in the Syrian coast, *Fitularia petimba*. Tishreen University.

Hourie, J.A. 1995. Investigation of petroleum hydrocarbons and phenols in Syrian coastal waters M.Sc. thesis, Tishreen University.

Ibrahim, A.A. 2001. Status of pollution along the Syrian coast. Research Poject. Tishreen University.

Ikhtiar, S. 2005. Biomarker and bioaccumulation of some chemical pollutant in Syrian marine ecosystem. PhD. thesis, Tishreen University.

Kara-Ali, A. 2000. Analytical study organic pollutants and pesticides in coastal region of Syria. PhD. thesis, Tishreen University.

Nasser, M. and Houijeh, I. 2002. Evaluation of the state of petroleum hydrocarbons in some marine organism and sediments. Tishreen University.

National country study of biological diversity in the Syria 2000. State Ministry of Environment/UNEP.

Chapter 26

Effects of Coastal Hazards on Lagoonal Ecosystems in the Centre of Vietnam and Measures for Mitigation

Nguyen Huu Cu

Institute of Marine Environment and Resources (IMER), Hai Phong City, Vietnam

E-mail: cunh@imer.ac.vn

Introduction

The Central coast of Vietnam is characterised by numerous and diverse coastal bodies of water, among which are 12 lagoons situated in a range of 11^0N lat. (Ninh Thuan province)–16^0N lat. (Thua Thien Hue province) with a density of one lagoon in every 57 km in length of the Central coastline (685 km). These lagoons are all productive ecosystems, from which dozens of thousand tons of aquatic living (biotic) products a year are exploited. Besides this, a large amount of the other biotic factors is got from the brackish aquaculture. Lagoonal ecosystems are composed of sub-ecosystems such as lagoonal deltas, mangroves, sea-grass beds, tidal flats and channels, etc., due to which they have a high biodiversity with fresh, brackish (even euryhaline) and marine species.

The lagoonal ecosystems are, however, sensitive to coastal hazards generated from both natural processes (coastal hydrodynamics, tropical climatic disturbances, meteoro-hydrologic events) and human activities through local socio-economic operations. Coastal hazards, which occur frequently to the Central coast as well as coastal lagoons in form of shore erosion, unstability of lagoonal inlets, sand movement from barriers and dunes, infilling the lagoonal basin, flooding, desalting, salt intrusion, etc., greatly influenced the lagoonal ecosystems, leading to the decrease in aquatic living resources and environment quality as well as the quality of community life surrounding the lagoons. The chapter is aimed mainly to discuss measures to be taken for the mitigation of effects of coastal hazards on lagoonal ecosystems in the Centre of Vietnam.

Overview of Coastal Lagoons in the Centre of Vietnam

With a total of 436.9 sq km in area of the water surface, 12 coastal lagoons in the Centre of Vietnam (Figure 26.1) are diverse in size (Table 26.1, Figure 26.2) very small, small, medium and large, in which the Tam Giang-Cau Hai lagoon is the largest and even comparable to large ones of the world; in shape-elongate, equilateral or semi-equilateral; in typology-nearly closed, partly closed and closed (Table 26.2). Though formed in the same way of island connecting spits of sand, morpho-dynamics of lagoonal inlets is seasonally changeable, either unstable or relatively stable. The inlets laying between rocky formations of the lagoons formed in relation to the equilibration of bays or embayment or laying between one rocky side and the other side of a spit with a seasonally inconsiderable long-shore drift of sand are relatively stable (Table 26.3). Contrary, the other inlets situated between the frontal spits of the lagoons closed to the development of deltas equilibrating large indents with a seasonally intensified long-shore drift of sand are unstable, characteristic of the Tam Giang-Cau Hai and Truong Giang lagoons.

Table 26.1: The Area and Dimensions of Coastal Lagoons in the Centre of Vietnam (Nguyen Huu Cu, 1995, 1999)

Sl.No.	*Lagoon*	*Area (sq. km)*	*Basin*				*Inlet*		
			Length (km)	*Width (km)*	*Mean Depth (m)*	*Max. Depth (m)*	*Length (m)*	*Width (m)*	*Depth (m)*
1.	Tam Giang-Cau Hai	216	68	2-10	1.6	4.2	TA. 600 TH. 100	350 50	2-11 1
2.	Lang Co	16	6	3.5	1.2	2.0	1000	150	1-8
3.	Truong Giang	36.9	10	5	1.1	2.0	AH. 500 TaH. 400	400 200	4 1
4.	An Khe	3.5	3	1.1	1.3	2.0	3000	150	1
5.	Nuoc Man	2.8	3	1.0	1.0	1.6	300	120	1.5
6.	Tra O	14.4	6	2.5	1.6	2.2	5000	150	1-4
7.	Nuoc Ngot	15.6	8.5	2.5	0.9	1.4	2000	125	1.6
8.	Thi Nai	50	15.6	3.9	1.2	2.5	1200	900	7
9.	Cu Mong	30.2	17.6	2.2	1.6	3.5	500	350	5
10.	O Loan	18	9.3	1.9	1.2	2.0	6300	50	1.5
11.	Thuy Trieu	25.5	17.5	0.3-3	1.5	4.0	1000	1000	4
12.	Nai	8	6	3.5	2.8	3.2	2500	500	4-6

Note: Inlets-TA: Thuan An; TH: Tu Hien; AH: An Hoa, TaH: Tam Hai.

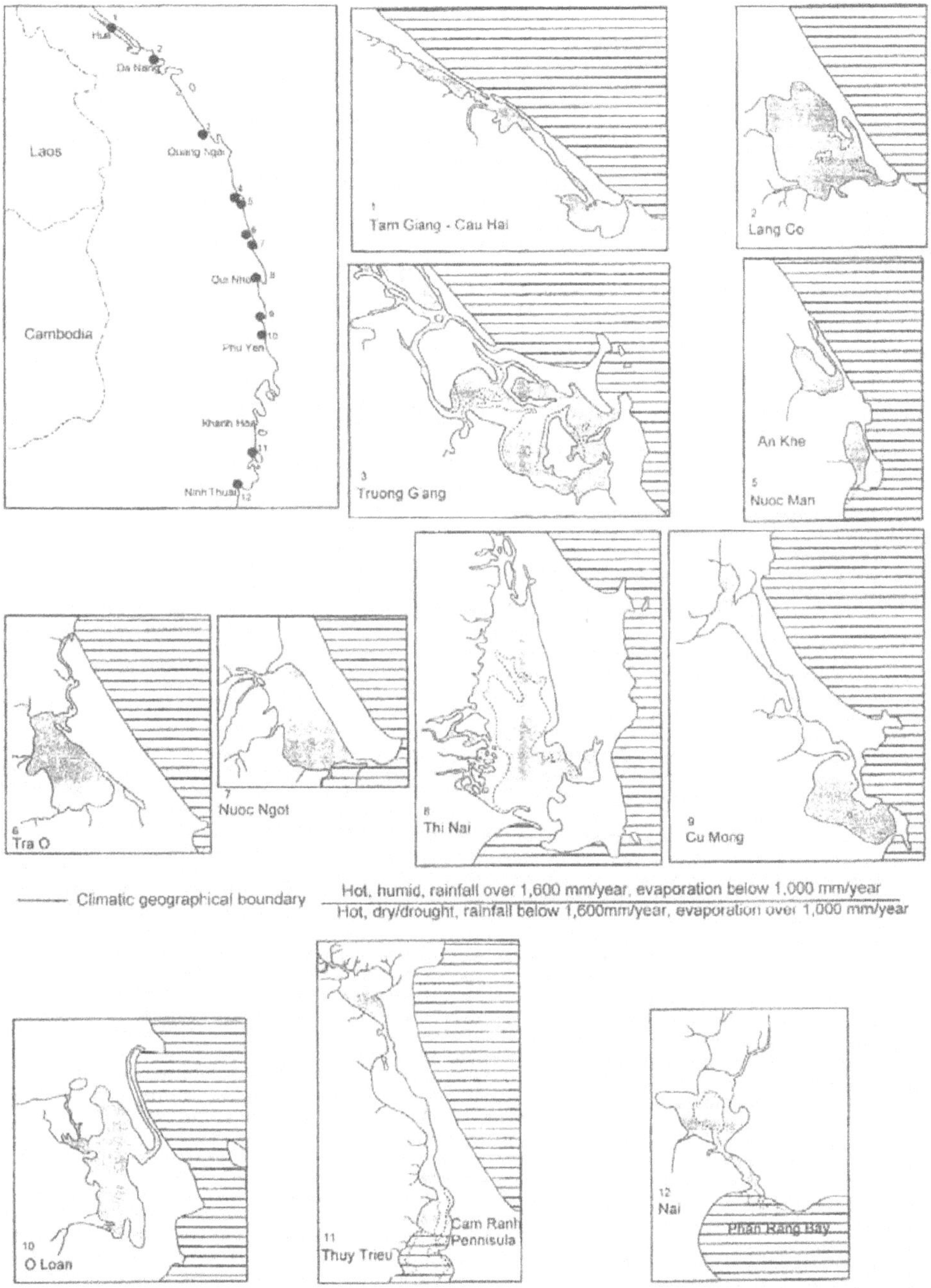

Figure 26.1: Maps of 12 Lagoon of Central Vietnam (Nguyen Huu Cu, 1999)

	10 Km²	20 Km²	50 Km²
Very small	**Small**	**Medium**	**Large**
An Khe, Nuoc Man, Nai	Lang Co, Nuoc Ngot, Tra O, O Loan	Truong Giang, Cu Mong, Thuy trieu	Tam Giang- Cau Hai, Thi Nai

Figure 26.2: Degree of Coastal Lagoons in Size

Table 26.2: Typology of Coastal Lagoons in the Centre of Vietnam (Nguyen Huu Cu, 1995, 1999) in Comparison with Others of the World (Nichols and Allen, 1981)

Lagoons Types of the World	*Lagoonal Types of Vietnam*		
	Nearly Closed	*Partly Closed*	*Closed*
Estuarine			
Open	Tam Giang-Cau Hai, Truong Giang, Thi Nai, Cu Mong, Thuy Trieu and Nai	Lang Co, Nuoc Man, Nuoc Ngot, O Loan	
Partly closed			
Closed			Anh Khe, Tra O

Table 26.3: Changeable State of Lagoonal Inlets

Sl.No.	*Lagoon*	*State of Inlets*	
		Relatively stable	*Unstable*
1.	Tam Giang-Cau Hai		Thuan An Tu Hien
2.	Lang Co	Lang Co	
3.	Truong Giang	An Hoa	Tam Hai
4.	An Khe		Phuoc Dien
5.	Nuoc Man	Sa Huynh	
6.	Tra O		Chau Truc (Hara)
7.	Nuoc Ngot	Degi	
8.	Thi Nai	Quy Nhon	
9.	Cu Mong	Tu Nham	
10.	O Loan		Phu Son
11.	Thuy Trieu	Cam Phuc	
12.	Nai		Ninh Chu

The nature of lagoonal water is basically characterized by the seasonal and spatial changes in the salinity that it is much higher in dry season than in rainy season, decreasing towards the river mouths and the innermost, and by the phenomenon of stratified water. In rainy season, the salinity in some lagoons is in a range of 0.5–15$^{0}/_{00}$, meanwhile, the salinity in some others in dry season can be up to 30$^{0}/_{00}$.

The lagoonal environment quality is decreased at a low level, manifested in a high concentration of oil and nitrate in water, some heavy metals in sediments, and a high density of coliform in water. The decrease in the environment quality is related to high inhabited areas surrounding the lagoons, activities of port and waterway, aquatic product processing, industry, tourism, etc.

Coastal Hazards Affecting the Lagoonal Ecosystems

Among the coastal hazards often occurring to the coastal zone of Vietnam are 4 typical hazards affecting strongly the lagoonal ecosystems, which are as follows:

1. Frontal shore erosion
2. Sand movement (flying in dry season and flowing in rainy season)
3. Unstability of inlets (seasonal changes in its location and morpho-dynamics)
4. Inundation by flooding.

The Tam Giang–Cau Hai is typical of its size (a large one of the world), natural resource potentials for the development of local economy, and also of being affected heavily by the whole coastal hazards mentioned about.

These effects and measures to be taken for mitigating can be summarized in the Table 26.4.

Hazard Management and Rehabilitation

Problems of Management and Rehabilitation

Among the numerous issues of hazard management and rehabilitation of coastal lagoon ecosystem in the Centre of Vietnam are 3 notable problems that require great attempts from researchers as well as managers, and that should be solved in the framework of ICZM actions. The problems are the following:

1. Stabilising the lagoonal structure and typology. A coastal lagoon is composed of 4 structural units of basin, frontal barrier of sand, inlet (one or more) and sheltered shore (rocky and/or clastic), each of which takes a role in the existance and typology as well as ecological succession and health.
2. Stabilising the environment quality and rehabilitating the environment function of coastal lagoons.

   The environmental nature of coastal lagoon is often affected by seasonal changes of both regional and global climate, and human activities, leading to the decrease in the environment quality at different levels among the lagoons.

3. Stabilising the development of natural resources and rehabilitating the ecological function of coastal lagoons.

Table 26.4: Effect of Coastal Hazards on the Lagoonal Ecosystem and Measures for Mitigation

Sl.No.	*Coastal Hazards*	*Effects*	*Measures of Mitigation*
1.	Frontal shore erosion	Destruction of barrier beaches and inhabitation.	Shore protection by revetment, artificial nourishment of beaches with groins, development of green belts as a set-back line.
		Intensification of longshore drift of sand, leading to the deformation of the entrance to the lagoon, narrowing and even closing the inlet. In case of being closed, the nature of environment ecosystem and aquatic living resources can be basically changed in form of water desalination, decrease in the number of marine species and the biomass of valuable species, including deterring boatmen from connecting with the sea, etc.	Stabilising the inlet by constructions and dredging the entrance.
2.	Sand movement by flying and flowing	Infilling populated and farmed areas with sand. shallowing the lagoonal basin and channel with wash-over fans, covering sea-grass beds with sand, invalidating facilities of transportation and irrigation.	Development of vegetated areas of sand, restriction of sand movement by building up frontal barriers and rain fall reservoirs on sand areas.
3.	Unstabitility of inlet	As mentioned above, the environment quality, ecological heath and living resource values are closely dependent on the situation of inlets. The unstability of inlets is in form of steadity moving by predominant longshore drift of sand, being narrowed or closed by suddenly intensified longshore drift of sand in dry season, being enlarged again and even opening new inlets by thixotropy in rainy season.	The most important measure is how to stabilise the inlets by dredging periodically the entrance to the lagoon, hardening the sides of the entrance, deterring the progradation of tidal deltas both inside (flood tidal delta) and outside (ebb tidal delta)
4.	Inundation by flooding	The inundation of lagoons and their surrounding low lands has caused water pollution, losses of cultivated aquatic products as well as agro-products, water desalination and migration of marine species, and changes in the species composition of the lagoonal fauna, phytoplankton and food chains.	Intensifying the water circulation in the lagoons and the water exchange with the sea by way of reducing fixed fishing weirs, stabilising the inlets.

Orientation of the Hazard Management and Coastal Lagoon Ecosystem Rehabilitation

In an endeavour to solve the problems, the following are environmental management and ecological rehabilitation actions that have been oriented in compatibility with the nature of coastal lagoons and their seasonal changes in resources and environment quality, with the need for developing coastal socio-economy as planned until the year 2010 and 2020, and with the feasibility.

1. Adjustment of natural hazards is shown in the Table 26.5, with 4 prior actions. Stabilising the surface of sand barriers by farming, planted vegetation cover as a green belt, enlarging inhabited areas, etc is to deter the sand movement in form of flying sand in dry season and flowing sand in rainy season. The shore protection from erosion by building groins, revetments, green belts, etc. is to reduce the sediment (sand) load of long-shore drifts that serves as one of the main reasons for closing and deforming lagoonal inlets. The regulation of water source in the whole catchment by the restoration of highland forests, reforestation, equity of land and water use is to restrict flooding, soil erosion, inundation, salt intrusion in dry season and desalination of lagoon water in rainy season, and to reduce the variation of species composition between the seasons.
2. Management of socio-economic activities.

Table 26.5: Prior Actions for Hazard Management and Ecological Rehabilitation

Sl.No.	Lagoon	Stabilising the Sand Barrier Surface	Shore Protection from Erosion	Inlet Protection from Closing, Deformation, Replacement	Regulation of Water Source in the Whole Catchment Area
1.	T.Giang-C.Hai	+	++	+++	++
2.	Lang Co	++	+	+	+++
3.	Truong Giang	++	+	+++	+
4.	An Khe	++	+	+++	+
5.	Nuoc Man	++	+	+	+++
6.	Tra O	++	+	+++	++
7.	Nuoc Ngot	+++	+	+	++
8.	Thi Nai	++	+	+	+++
9.	Cu Mong	+	+	+	+++
10.	O Loan	++	+	+++	++
11.	Thuy Trieu	+++	+	+	++
12.	Nai	+	+	++	+++

Note: Prior level _ +++: High; ++: Medium and +: Low.

The following is a framework for 6 prior actions managing socio-economic activities, which can be shown in the Table 26.6.

1. Management of all kinds of wastes (both point and non-point source) from inhabited areas and production bases in situ, surrounding the lagoons and in the whole catchment area is to maintain the quality of lagoonal water and sediments and to conserve the lagoonal biodiversity with its sub-ecosystems, species composition and aquatic living resources, especially economic and endemic (*Cyprinus centralis*) species.
2. Management of fishery operations is to conserve aquatic stock for both lagoons and adjacent coastal marine regions by way of fishing restrictively and seasonally, forbidding destructive fishing, aquaculture in a rationale scale, limitation to the extensive aquaculture with impoundment, conservation of lagoonal wetlands and their biodiversity.

Table 26.6: Orientation for the Management of Socio-economic Activities in/and Related to Coastal Lagoons

Sl.No.	*Lagoon*	*Management Actions*					
		Wastes	*Fishery*	*Agriculture*	*Port and Waterway*	*Tourism*	*Exploitation of the Catchment Area*
1.	TG-CH	++	+++	++	++	+	++
2.	Lang Co	++	++	+	+	++	+++
3.	T. Giang	++	++	++	+++	+	++
4.	An Khe	+++	++	+	+	+	++
5.	Nuoc Man	+++	++	+	++	++	++
6.	Tra O	++	+++	++	+	+	++
7.	Nuoc Ngot	++	+++	++	+	++	+
8.	Thi Nai	++	++	+	+++	++	++
9.	Cu Mong	++	++	+	++	+	+++
10.	O Loan	++	+++	+	+	++	++
11.	Thuy Trieu	++	+++	+	++	++	++
12.	Nai	++	++	+++	++	++	+

Note: Prior level _ +++: High; ++: Medium and +: Low.

3. Management of agriculture activities surrounding the lagoons is to reduce the residue of pesticides in lagoonal water and sediments, to balance the use of fresh water and lands among the needs for sectors, and to conserve the value of lagoonal wetlands.
4. There are numerous harbours in the lagoons and even ports (with ships of 400-5,000 DWT and up to 20,000 DWT until the year 2010) in some large lagoons (Tam Giang–Cau Hai, Truong Giang, Thi Nai and Thuy Trieu).

This operation is a source of releasing oil into the water, so the oil concentration near the ports and harbours is often high and higher than the permitted limit for aquaculture (0.05 mg/l). This operation should be controlled strictly with monitoring sites.

Conclusion

1. Coastal lagoons in the Centre of Vietnam are common, diverse in size and typology, rich in natural resources for the development of socio-economy, but sensitive to coastal hazards, among which are 4 typical ones-frontal shore erosion, sand movement, unstability of inlets, and inundations by flooding. The environmental and ecological sensitivity is manifested in its seasonal changes and man-made injuries.
2. Among the issues of hazard management and rehabilitation of lagoonal ecosystems are 3 notable problems of stabilising the lagoonal structure and typology, environment quality and natural resource development, rehabilitating ecological and environmental functions of coastal lagoons that should be solved on the basis of the adjustment of natural hazards and management of all human activities with their own details.

References

Nguyen Huu Cu, 1999. Generalisation of environmental and resource studies of coastal lagoons in the Centre of Vietnam. Mar. Resour. Env., tom. VI, p. 126-142, Sci. Techn. Pub. House, Hanoi.

Nguyen Huu Cu, 2000. A strategic orientation for controlling floods in the Centre of Vietnam. Mar. Resour. Env., tom. VII, p. 289-302, Sci. Techn. Pub. House, Ha noi.

Nguyen Huu Cu, 2001. Overview of coastal hazards in Vietnam. 6th Asean Sci. Techn. Week Conf. Sept. 17-19th, 2001, Negara, Brunei Darussalam.

Nguyen Huu Cu *et al.*, 2002. Impacts of human activities on the geological environment of the Tam Giang-Cau Hai lagoon. Mar. Resour. Env., tom. IX, p. 103-120, Sci. Techn. Pub. House, Ha noi.

Nguyen Huu Cu *et al.*, 2003. Research on coastal lagoons of Central Vietnam as a guide to management. Present knowledge and perspectives. Techn. Rep. N⁰ 86, CNR, Italia.

Nguyen Huu Cu, 2005. ICZM of Vietnam. Proc. Conf. VAST, tom III, p. 245-256, VAST press, Hanoi.

Nguyen Huu Cu *et al.*, 2006. Problems of environmental geology of the Vietnam coastal areas and adjustable measures. Jour. Mar. Sci. Techn. 1 (T.6), p. 25-41. Sci. Techn. Pub. House, Ha noi.

www.ingramcontent.com/pod-product-compliance
Ingram Content Group UK Ltd.
Pitfield, Milton Keynes, MK11 3LW, UK
UKHW050620290826
14794UKWH00003B/105